PROBABILITY AND RANDOM VARIABLES

Mathematics and its Applications
Series Editor: G. M. BELL, Professor of Mathematics, King's College (KQC), University of London

Statistics and Operational Research
Editor: B. W. CONOLLY, Professor of Operational Research, Queen Mary College, University of London

Mathematics and its applications are now awe-inspiring in their scope, variety and depth. Not only is there rapid growth in pure mathematics and its applications to the traditional fields of the physical sciences, engineering and statistics, but new fields of application are emerging in biology, ecology and social organisation. The user of mathematics must assimilate subtle new techniques and also learn to handle the great power of the computer efficiently and economically.

The need of clear, concise and authoritative texts is thus greater than ever and our series will endeavour to supply this need. It aims to be comprehensive and yet flexible. Works surveying recent research will introduce new areas and up-to-date mathematical methods. Undergraduate texts on established topics will stimulate student interest by including applications relevant at the present day. The series will also include selected volumes of lecture notes which will enable certain important topics to be presented earlier than would otherwise be possible.

In all these ways it is hoped to render a valuable service to those who learn, teach, develop and use mathematics.
For full series list see end of book.

PROBABILITY AND RANDOM VARIABLES

G. P. BEAUMONT, B.Sc., M.A., M.Sc.
Senior Lecturer in Statistics
Department of Statistics and Computing Science
Royal Holloway and Bedford New College, University of London

ELLIS HORWOOD LIMITED
Publishers · Chichester

Halsted Press: a division of
JOHN WILEY & SONS
New York · Chichester · Brisbane · Toronto

First published in 1986 by
ELLIS HORWOOD LIMITED
Market Cross House, Cooper Street, Chichester, West Sussex, PO19 1EB,
England

The publisher's colophon is reproduced from James Gillison's drawing of the ancient Market Cross, Chichester.

Distributors:

Australia and New Zealand:
Jacaranda-Wiley Ltd., Jacaranda Press,
JOHN WILEY & SONS INC.
GPO Box 859, Brisbane, Queensland 4001, Australia

Canada:
JOHN WILEY & SONS CANADA LIMITED
22 Worcester Road, Rexdale, Ontario, Canada

Europe and Africa:
JOHN WILEY & SONS LIMITED
Baffins Lane, Chichester, West Sussex, England

North and South America and the rest of the world:
Halsted Press: a division of
JOHN WILEY & SONS
605 Third Avenue, New York, NY 10158, USA

© 1986 G.P. Beaumont/Ellis Horwood Limited

British Library Cataloguing in Publication Data
Beaumont, G. P.
Probability and random variables. —
(Ellis Horwood series in mathematics and its applications)
1. Probabilities   2. Mathematical statistics
I. Title
519.2   QA273

Library of Congress Card No. 86–4728

ISBN 0–85312–768–9 (Ellis Horwood Limited — Library Edn.)
ISBN 0–7458–0051–3 (Ellis Horwood Limited — Student Edn.)
ISBN 0–470–20307–2 (Halsted Press)

Printed in Great Britain by Unwin Bros. of Woking

Contents

Preface

This book is intended for first-year students in Universities, Polytechnics and Colleges of Education. Although the treatment is mathematical, it is not intended to be severely rigorous. It requires Advanced Level Mathematics, together with easy double integration and some idea of convergence. A section which may be omitted on a first reading is marked with an asterisk.

Courses in Probability are notoriously difficult to teach. The same students who confidently undertake courses in mathematics often hesitate and stumble when tackling problems in this area. The difficulties experienced appear to spring from the diminished role played by formal manipulation and the increased need to identify the logical implications of the information provided. This text lays stress on the study of illustrative examples and the completion of typical problems. In an attempt to lighten the reader's burden in the latter respect, brief solutions and comments have been provided for most of the problems. For the A.L. questions, however, as a result of a restriction imposed by some Boards, only an answer and a brief hint has been supplied. *All solutions and answers appearing are of course the sole responsibility of the author, and are not to be ascribed to any of the examiners concerned.*

The presentation of the material, which has developed from part of an earlier work by the author, would be conveyed by the description 'Probability with Statistics in Mind'. Statistical techniques are now used in most fields of scientific endeavour. The availability of computers, with their library programs, now permits an uninhibited application of statistical analysis. This buoyant state of affairs is not without its risks. Extensive data snooping can lead to claims perilously akin to endorsing the winner of a race after the result has been declared! Every statistical test applied to data in search of significant features yields results which cannot be interpreted without some probabilistic assessment of their value. In this spirit, we point out ways in which the mathematical results might be applied.

Acknowledgements

I am indebted to the following persons and sources for permission to publish.

To the University College of Wales, Aberystwyth, the Universities of Hull, London and Surrey and the Queen's University of Belfast for questions from past examination papers.

To the Oxford & Cambridge Schools Examination Board for questions from past examination questions.

To E. Parzen and John Wiley & Sons, New York and London, for the gist of a remark on the vexed question of 'Friday the thirteenth' in *Modern Probablity Theory* and its applications (1960).

To T.W. Feller and John Wiley & Sons Inc., New York, for extracts from Table 3 of Chapter 4, Table 2 of Chapter 6 in *An Introduction to Probability Theory and its Applications*, Vol. 1 (1957).

To Professor B. Conolly for the use of freshly calculated tables of the normal distribution function and percentage points of the normal distribution.

I am grateful to Mrs B. Rutherford for once again agreeing to shoulder the burden of the typing.

I wish to thank again Professor K. Bowen for reading most of the material and for his spirited attempts to put some order in the solutions provided. He was quick to detect occasions when all aid short of actual help was proffered!

1

Introduction

We shall be using the word 'chance', and must rely on the reader already having grasped its meaning. For it is notorious that every determined effort to define a basic idea eventually involves using terms which are practically equivalent to the matter under discussion. We come to realize the role played by chance through our experience that not all events are either impossible or certain. There has been some reluctance fully to admit anything of the sort, as witness the attempt made by man to ascribe full power over the future to the gods. More to present taste is the continual shrinkage in the areas rule by chance brought about by the ceaseless expansion of knowledge. The confident prediction of the time of a solar eclipse is typical of the kind of scientific advance which leads man to hope that, if only he knew enought, all would be foreseeable. In this view, chance appears as a phenomenon associated with ignorance and inability to control a situation.

Consider the sex of a baby not yet born. This is determined by a complicated biological process which to date is beyond our control. Thus, it is not possible to say in advance whether a conception will produce a boy or a girl, sometimes hopes are fulfilled, sometimes not. The possibility of a mathematics of chance arises from an interesting observation, namely, that in every year a large population will produce *approximately* the same proportion of boys. Thus, a kind of stability arises from the seeming chaos at the individual level.

Interest in the mathematics of chance was vastly stimulated by the activities of card and dice players, whose games partly owed their attraction to the chance elements involved. Indeed, some features of the materials employed in games, especially those relating to symmetry, appeared to afford an explanation of the records of results. In games of chance, much attention must be devoted to questions of 'fairness'. Since an advantage sometimes accrues to the first mover or player, this honour has itself to be decided by a preliminary procedure, such as tossing a coin.

When a coin is tossed to decide which of two players is to start a game, the procedure is generally assumed to be fair to both sides. The meaning of 'fair' is that both players have the same chance of starting first. What if the *coin* is not fair, in the sense that it is more likely to give heads than tails? Intuition suggests that if one of the parties is unaware of this and calls at random, then the procedure remains fair. On the other hand, the coin may be unbiased but the tossing procedure be controllable so as to obtain any desired result. This also can be met by delaying the call until the coin is in the air. So far the emphasis has been on the precautions necessary to ensure that both sides really have the same chance − where chance is supposed to be a word that is understood in some obvious sense. Probability is a measure of chance, and we shall propose general rules for calculating the probability of combinations of simple events. Two distinct but related views seem plausible concerning the meaning of 'the coin is fair'. In the first view, the emphasis is on a single toss and states that after exhaustive tests there is no mechanical reason why the coin should come down heads rather than tails − hence the chances are even. In the second view, the behaviour of the coin in a long series of tosses is examined. It would then be held that the chances of heads or tails are even if the proportion of heads appears to tend towards one half as the number of tosses increases. It is not, of course, possible to carry out an infinite sequence of tosses and the proportion of heads does not tend to a half in quite the same sense as a mathematical sequence tends to a limit. A set of tosses which begins H, T, H, H, T, T, shows proportions of heads 1/1, 1/2, 2/3, 3/4, 3/5, 1/2, and after the fourth toss we are further from 1/2 than after the third toss. That the proportion of heads should really tend to 1/2 was not always felt to be so 'obvious' and has been the subject of experiment. We have talked about two views, but there is a strong temptation to declare that the second phenomenon is deducible from the fact that there are two sides to a coin and if the coin is not biased, these are *equally likely* to turn up and *hence* the limiting proportion of heads in a long series of toss must be 1/2. Unfortunately, the phrase 'equally likely' has been incorporated and it might be held that the only real test of whether the two sides *are* equally likely is to observe the proportion of heads in a long series of tosses!

The tossing of a coin is a simple example of a large class of games of chance with certain common features. Each game is decided on the results or outcomes of one or more trials, where a trial might be rolling a die, tossing a coin, or drawing a card from a pack. If the outcomes are distinguishable, we say they are mutually exclusive, and if they are the only possible results they are also said to be exhaustive. There may be more than one way of listing the outcomes. If we draw a card from the pack, the outcomes red, black are mutuaally exclusive and exhaustive, but so are the outcomes Spades, Hearts, Diamonds, and Clubs. These outcomes can be still further decomposed and there are advantages in using a set of outcomes which are indecomposable, when each outcome may be called a simple event. For drawing a card from a pack, we can list 52 mutually exclusive and exhaustive outcomes, one for each different card in the pack. The trials are also said to be independent if the result of one trial does not depend on the outcome of any previous trial, or any combination of previous trials.

Suppose in a series of n independent trials the outcomes $O_1, O_2, \ldots, O_k$ are mutually exclusive and exhaustive and that O_i has appeared f_i times. Then f_i/n, the relative frequency of O_i, satisfies

$$0 \leqslant \frac{f_i}{n} \leqslant 1.$$

It is a matter of observation that as n increases, f_i/n appears to settle down to a limiting value p_i where

$$0 \leqslant p_i \leqslant 1.$$

This apparently provides a suitable basis for assessing numerically the chances of the outcomes, for to each outcome O_i we can associate the number p_i, called the **probability** of O_i. Since

$$\sum_{i=1}^{k} f_i/n = 1$$

then

$$\sum_{i=1}^{k} p_i = 1.$$

Furthermore, for each pair of outcomes O_i, O_j,

$$\frac{f_i + f_j}{n} = \frac{f_i}{n} + \frac{f_j}{n}$$

which tends towards $p_i + p_j$. Hence the probability of the compound outcome O_i or O_j is $p_i + p_j$. The qualification 'apparently' is necessary for:

(a)   the independence is difficult to guarantee, for instance, apparatus is subject to continual wear; and

(b)   since the series of trials can never be infinitely long, the probabilities p_i are virtually unknown and we must be content with estimates.

How then shall the actual probabilities be estimated? One method is to use a previous record of trials and use the relative frequencies of the outcomes. It is also possible to exploit certain geometrical or mechanical features. Thus, if a die is to be rolled, each side, viewed as an outcome, may be assigned a probability of 1/6. However the probabilities are assigned, three properties must be observed.

(i)   Each probability should be between zero and one inclusive.
(ii)   The probability of a compound outcome should be the sum of the probabilities of its constituents, if these be mutually exclusive.
(iii)   The probability of a certain outcome should be one.

There may be many ways of assigning probabilities which are consistent with these requirements.

For the statistician the interest in probability arises from the frequently observed fact that the phenomenon of long-run stability of the relative frequency of outcomes appears in fields embracing all the physical and social sciences. This was established by arduous study of experimental data and records. Thus an essential part of life insurance as a business is to be able to estimate the life expected for a new applicant of a determined age, sex, profession, and health record. This would be impossible to assess without finding the proportions of survivors for particular lengths of time in previous generations of people of the same general category.

In the larger context, what has been called a trial in a game of chance is termed an experiment. This term will cover a very wide range of situations from the simple weighing of an object, where the outcome is a weight, to the drawing of a sample of persons from a population, the outcome perhaps being the percentage which voted in the last election. All such experiments are to be thought of as repeatable in the sense of repeated trials, and this view will be maintained in a hypothetical sense even where the particular experiment might be held in a certain sense to be unrepeatable. Thus, an experiment which seeks to estimate the effect of school milk on the weight of children over a stated age range just cannot be repeated on the same children. In such a case we must also think of the wider population of children not used in the experiment.

2

Probability

2.1 AXIOMATIC APPROACH

We wish to discuss elementary ideas in probability from an axiomatic basis. The previous discussions show us how to frame our axioms so that any deductions made from them bear a satisfactory relation to the real world. The position is similar to that in geometry or mechanics. In Euclidean geometry there are undefined elements known as points, lines, and planes, together with a list of axioms satisfied by these elements. From these axioms, it is possible to deduce theorems about figures composed of the basic elements. The axioms and definitions have clearly been 'drawn from life' in the sense that they assert properties of the elements which appear to be self-evident. The questions as to whether the axioms are consistent with each other and whether they are sufficient to describe the properties of Euclidean space as we find them are not readily answered. As late as the nineteenth century attempts were made to *prove* the uniqueness of the parallel through a point to a given line. It was finally realized that for Euclidean geometry this property had to be included as an axiom. Denial of uniqueness gives another geometry.

2.2 SAMPLE SPACE

In order to avoid continually referring to particular games or experiments, it is useful to employ an abstract representation for a trial and its outcomes. Each distinguishable and indecomposable outcome, or simple event, is regarded as a point in a sample space, S. Thus, for the experiment of drawing a card from a pack the sample space contains 52 points. *Every collection of simple events or set of points of S is called an* **event**. The word event now has a double interpretation. In the everyday sense, it means any statement about the result of an experiment, such as 'the card drawn was a Diamond'. It also means that set of points in the sample space, each of which corresponds to a simple event, in which the card is a Diamond. A simple event is also an event. How many distinct events are there in a sample space containing k points? In making up a set, we may say of every point in turn that either it is included or it is excluded, that is, there are two ways of disposing of each point. Hence, there are in all 2^k ways of disposing of all the points. This procedure, however, includes the case when all the points of S are rejected. The set then contains no points — it is the empty set. This set, denoted by $\varnothing$, will be included, for completeness, as an event. No trial of the experiment can produce an event which corresponds to the empty set since such a set contains no points corresponding to any of the outcomes. We may also select all the points of S, hence the whole sample space is an event in S. If we allow the agreement about the empty set, there are 2^k distinct events in S. In the case when two coins are tossed one after another, $k = 4$, corresponding to HH, HT, TH, HH, and there are $2^4 = 16$ events that can be distinguished in the sample space. There are two ways of looking at this collection of events. Suppose we think of a label for an event, say 'at least one of the coins showed heads' then we can pick out the points of S which belong to the set of outcomes which imply that statement. These are HH, HT, and TH. Alternatively, we may select some points and then search for a meaningful label. Thus, if we take the two points TT, HT, then this may be labelled 'the second coin tossed resulted in tails'.

We have declared that the points in a sample space represent the distinguishable and indecomposable outcomes. This definition seems harmless enough but in fact needs qualifying by the phrase 'as far as we can see'. Consider two cards drawn from a pack. If these are drawn one at a time without replacement, then we can distinguish $52 \times 51 = 2652$ outcomes and these can be recorded as ordered pairs such as (Ace of diamonds, 7 of hearts). This particular simple event is one of the 51×13 sample points which make up the event 'the second card was a heart'. However, if the two cards were drawn together then their order cannot be discerned and only $26 \times 51 = 1326$ different pairs can be distinguished. The corresponding points can be labelled with sets such as {Ace of diamonds, 7 of hearts}. A related point arises if we draw two balls at once from three red and two white balls which are otherwise similar. We can observe three cases, namely 0, 1 or 2 white balls. But in any application which required the preponderence of reds to be reflected, it would be an advantage to have a (conceptual) number of the balls from 1 to 5 and then ten pairs of balls could be distinguished.

2.3 COMBINATION OF EVENTS

We can perform operations on the sets which are called events to produce sets which are also events.

The **intersection** *of two sets* E_1, E_2 *is the set of points of S which belong to both* E_1 *and* E_2 *and is an event* (*written* $E_1 \cap E_2$) Thus the intersection of the sets $\{HH, TH, HT\}$ and $\{HT, TT\}$ is the set containing the single point HT. This event may be called 'heads on the first coin and tails on the second coin'. It may happen that the two sets have no points in common, that is, their intersection is the empty set. We now see why, for completeness, we decided to count the empty set as an event. By repetition, the intersection of any finite number of events is an event.

Another operation is to form the **union**, written $E_1 \cup E_2$ of two sets E_1, E_2. *This is defined as the set which contains all the points of S which are in either* E_1 *or* E_2 (*or both*). Thus, the union of the events $\{HH, TH, HT\}$ and $\{HT, TT\}$, in the present example, is the event $\{HH, TH, HT, TT\}$, which contains every point in the sample space and may reasonably be called 'the certain event'.

Finally, for any event E we obtain the **complement**, written E', by selecting all those points of S which are *not* in E. This event might be called 'not E'. Thus, the complement of $\{HH, TH, HT\}$ is the event $\{TT\}$ and corresponds to the event 'no head'.

Apart from its generality, the representation of events as sets shows more clearly how the operations of 'and', 'or', 'not' may be combined.

2.4 VENN DIAGRAMS

The set E_1 is said to be contained in E_2, written $E_1 \subset E_2$, if every point in E_1 is also a point in E_2. If $E_1 \subset E_2$ and E_1, E_2 are events, then if event E_1 happens then also E_2 must happen. We may also say that E_2 contains E_1, written $E_2 \supset E_1$. A common technique for showing that two sets E_1, E_2 are the same set is to prove separately that $E_1 \supset E_2$ and $E_2 \supset E_1$. From the definitions, certain results follow immediately. These will be listed in a particular way to bring out a certain feature. Let E_1, E_2, E_3 be events in a sample space S and $\varnothing$ be the empty set.

The operations of forming intersections and unions satisfy

(1) $E_1 \cup (E_2 \cup E_3) = (E_1 \cup E_2) \cup E_3$, $E_1 \cap (E_2 \cap E_3) = (E_1 \cap E_2) \cap E_3$

(2) $E_1 \cup E_2 = E_2 \cup E_1$, $E_1 \cap E_2 = E_2 \cap E_1$

(3) $E_1 \cup E_1 = E_1$, $E_1 \cap E_1 = E_1$

(4) $E_1 \cup E_1' = S$, $E_1 \cap E_1' = \varnothing$

(5) $E_1 \cup \varnothing = E_1$, $E_1 \cap S = E_1$

(6) $E_1 \cup S = S$, $E_1 \cap \varnothing = \varnothing$

(7) $\varnothing' = S$, $S' = \varnothing$

The pairs of statements are duals in the sense that if in any one, we interchange union with intersection and S with $\varnothing$, leaving complementation undisturbed, we obtain a twin statement which is also true.

Other results can be verified with the assistance of a Venn diagram. In a Venn diagram, the sample space is represented by a rectangle and any event by a circle

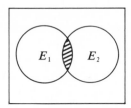

Fig. 2.1

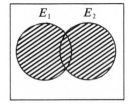

Fig. 2.2

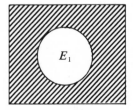

Fig. 2.3

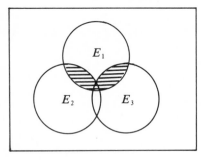

Fig. 2.4

in this rectangle. In the three diagrams using this scheme (Figs. 2.1, 2.2 and 2.3), the shaded areas represent $E_1 \cap E_2$, $E_1 \cup E_2$, E_1' respectively. From such a diagram other results can be 'read off' by identifying certain areas in two different ways. Thus from the fourth diagram (Fig. 2.4) we have at once:

$$E_1 \cap (E_2 \cup E_3) = (E_1 \cap E_2) \cup (E_1 \cap E_3).$$

This distributive law can be heard as well as seen to be true. For a point which is in E_1 and either E_2 or E_3 is certainly either in E_1 and E_2 or E_1 and E_3 (and conversely.

Problem 1
(1)   Show that $(E_1 \cup E_2) \cap (E_1 \cup E_3) = E_1 \cup (E_2 \cap E_3)$. What is the dual of this result?
(2)   Show that $(E_1 \cap E_2)' = E_1' \cup E_2'$.
(3)   Show that $(E_1 \cup E_2)' = E_1' \cap E_2'$. What is the dual of this result? (2) and (3) constitute De Morgan's laws.
(4)   Show that $E_1 \subset (E_1 \cup E_2)$ but $E_1 \supset (E_1 \cap E_2)$.
(5)   If $E_1 \subset E_2$ show that

$$E_1 \cup E_2 = E_2$$
$$E_1 \cap E_2 = E_1$$
$$(E_1 \cap E_3) \subset (E_2 \cap E_3)$$
$$(E_1 \cup E_3) \subset (E_2 \cup E_3)$$

(6)   Use (4) and (5) to deduce that

$$E_1 \cup (E_1 \cap E_2) = E_1$$
$$E_1 \cap (E_1 \cup E_2) = E_1$$

Verify the results using Venn diagrams.
(7)   Sometimes, the information about an event E_1 is provided in terms of whether or not a second event E_2 has occurred. Show that

$$E_1 = (E_1 \cap E_2) \cup (E_1 \cap E_2').$$ ■

If the sets contain a finite number of points then we can find a certain relationship between these numbers and the numbers of points found in related sets formed by set operations. Let $N(E)$ be the number of points in the event E. If E_1, E_2 are disjoint sets or are mutually exclusive events we have at once

$$N(E_1 \cap E_2) = 0,$$
$$N(E_1 \cup E_2) = N(E_1) + N(E_2).$$

If E_1, E_2 are not disjoint, then

$$N(E_1 \cup E_2) = N(E_1) + N(E_2) - N(E_1 \cap E_2)$$

since the number of points in $E_1 \cap E_2$ is counted twice in

$$N(E_1) + N(E_2).$$

Also, $E_1 \cup E_2$ can be expressed as the union of the disjoint sets,

$$E_1 \cap E_2', \quad E_1' \cap E_2, \quad E_1 \cap E_2$$

whence

$$N(E_1 \cup E_2) = N(E_1 \cap E_2') + N(E_1' \cap E_2) + N(E_1 \cap E_2).$$

For any event $E_1, N(E_1) = N(E_1 \cap E_2) + N(E_1 \cap E_2').$

Example 1

An ice-cream firm, before launching three new flavours, conducts a tasting with the assistance of 60 schoolboys. The findings were summarized as:

32 liked A	10 liked A and B	
24 liked B	11 liked A and C	6 liked A and B and C.
31 liked C	14 liked B and C	

Since there are only three flavours, A, B, C to consider, the information provided can easily be grasped through a diagram (Fig. 2.5). [*A* is the set of people who liked flavour A.]

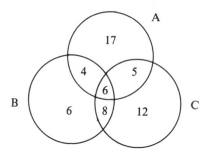

Fig. 2.5

By inserting the 6 who liked all three, we can quickly fill in all the other spaces. Thus these 6 are part of the 10 who liked A and B, hence there must have been $10 - 6 = 4$ who liked A and B but not C. For more complicated cases, a formal approach may be preferred. Thus suppose that we require the number who like A *only*. Since

$$A = (A \cap C) \cup (A \cap C'), \quad N(A) = N(A \cap C) + N(A \cap C'),$$

or

$$32 = 11 + N(A \cap C'),$$

that is

$$N(A \cap C') = 21.$$

Similarly

$$N(A \cap B \cap C') = N(A \cap B) - N(A \cap B \cap C) = 10 - 6 = 4.$$

Finally, the number liking A only must be

$$N(A \cap B' \cap C') = N(A \cap C') - N(A \cap B \cap C') = 21 - 4 = 17.$$

Two boys are unlisted from the information provided, those who liked none of the flavours! If the number who liked B is not reported, we can set bounds to the missing number. From Fig. 2.5, it must exceed 18, since these like B and something else; but it cannot be greater than $60 - 34 = 26$, since 34 did not care for B.

Problem 2. 50 patients suffering from a disease are classified as to the presence or absence of three symptoms A, B, C. The presence of symptom B implies symptom A also, but precludes symptom C. It is noted that 8 have B, 36 have A, and 30 have C. Find limits for the number having both C and A.

Problem 3. In a workforce of 80 men and 95 women, an enquiry is made regarding ownership of a car and possession of a mortgage. It is found that 44 men and 21 women have a mortgage. If 26 men have neither a car nor a mortgage, find the number of women who have a car but no mortgage.

2.5 AXIOMS FOR PROBABILITIES FOR A SAMPLE SPACE WITH A FINITE NUMBER OF POINTS

For every event, E, in the sample space S we assign a non-negative number, called the probability of E denoted by $\Pr(E)$, so that the following axioms are satisfied.

(a)   For every event E, $\Pr(E) \geqslant 0$.
(b)   For the certain event, $\Pr(S) = 1$.
(c)   If E_1, E_2 are mutually exclusive events $\Pr(E_1 \cup E_2) = \Pr(E_1) + \Pr(E_2)$.

There is no unique way of assigning probabilities to the events so that the axioms are satisfied. However, if the assignment is to bear any reasonable relation to the reality represented by the experiment, then the number assigned to an event should be the limiting proportion of times that event occurs in a long series of uniform trials of the experiment. It will be observed that the axioms have the previously noted properties of the relative frequencies of outcomes. If S contains a finite number, k, of points we assign a probability to each point in the sample space, so that the sum of all these probabilities is unity. To find the appropriate probability for any event E, we then merely sum the probabilities attached to the points contained in E. An important example is when equal probability $1/k$ is given to each of the points in the sample space. This **symmetric** case is appropriate to many games of chance. In this case if an event contains r points, the probability of the event will be r/k. Various examples of the symmetric case will be later examined in detail.

Deductions from the axioms

We first show that the probability of the empty set is zero and hence apart from its inclusion as an event, contributes nothing to our calculations. Now the empty set and the whole space S have no points in common, since $\varnothing$ has no points. Hence, by axiom (c),

$$\Pr(\varnothing \cup S) = \Pr(\varnothing) + \Pr(S) = \Pr(\varnothing) + 1, \text{ by axiom (b).} \qquad (2.1)$$

But $\varnothing \cup S$ is S. Hence, $\Pr(\varnothing \cup S) = 1$ and $\Pr(\varnothing) = 0$.

An important situation arises when on event E is included in another event F. This means that every point of E is a point of F. Another way of putting this is to say that E implies F. Let G be the set of points of F not in E, then E and G are mutually exclusive events whose union is F. Hence, by axiom (c)

$$\Pr(F) = \Pr(E \cup G) = \Pr(E) + \Pr(G).$$

Hence

$$\Pr(E) = \Pr(F) - \Pr(G) \leqslant \Pr(F). \tag{2.2}$$

Further, since every event is contained in the sample space S,

$$\Pr(E) \leqslant \Pr(S) = 1. \tag{2.3}$$

If $\Pr(E)$ is known, then $\Pr(E')$ or the probability of not E can immediately be calculated. For E, E' are mutually exclusive events whose union is S, since every point of S is either in E or 'not E' and cannot be in both. Hence,

$$1 = \Pr(S) = \Pr(E \cup E') = \Pr(E) + \Pr(E'),$$

or

$$\Pr(E') = 1 - \Pr(E). \tag{2.4}$$

We shall frequently meet events E and F which are not mutually exclusive. Let G be the set of points of F which are *not* in E, then E and G are mutually exclusive events and $E \cup G$ is the set $E \cup F$.

$$\Pr(E \cup F) = \Pr(E \cup G) = \Pr(E) + \Pr(G).$$

But F is the union of $E \cap F$ and G, and these are mutually exclusive events.

$$\Pr(E \cap F) = \Pr(F) - \Pr(G),$$

or

$$\Pr(G) = \Pr(F) - \Pr(E \cap F)$$

that is,

$$\Pr(E \cup F) = \Pr(E) + \Pr(F) - \Pr(E \cap F). \tag{2.5}$$

That is, the probability of E or F is the probability of E plus the probability of F minus the probability of E and F. Naturally, when $E \cap F$ is empty, $\Pr(E \cap F) = 0$ and (2.5) reduces to axiom (c). The result may also be seen intuitively from the consideration that $\Pr(E) + \Pr(F)$ counts the probability of $E \cap F$ *twice*, and this must be corrected by subtracting it once. In any case

$$\Pr(E \cup F) \leqslant \Pr(E) + \Pr(F). \tag{2.6}$$

Problem 4. Show that, if E, F, G are events,

$$\Pr(E \cup F \cup G) = \Pr(E) + \Pr(F) + \Pr(G) - \Pr(E \cap F)$$

$$- \Pr(E \cap G) - \Pr(F \cap G) + \Pr(E \cap F \cap G). \qquad \blacksquare$$

Suppose $E_1, E_2, \ldots, E_n$ are all mutually exclusive events in the sample space S, then

$$\Pr(E_1 \cup E_2 \cup E_3 \ldots \cup E_n) = \Pr(E_1) + \Pr(E_2) + \ldots \Pr(E_n). \quad (2.7)$$

For we may regard $(E_2 \cup E_3 \ldots \cup E_n)$ as one event and then

$$\Pr[E_1 \cup (E_2 \cup E_3 \ldots \cup E_n)] = \Pr(E_1) + \Pr(E_2 \cup E_3 \ldots \cup E_n)$$

and the result is obtained by continued application.

All the results above refer to a sample space with a finite number of points. If the sample space contains infinitely many points then some modifications are required. For in such spaces it may be possible to define some subsets of S to which it is not possible to assign probabilities, satisfying the axioms. To meet this difficulty it is necessary to redefine those subsets of S which are to be called events. In a more advanced course, this would be done and it would be shown that in this case also, all finite unions, intersections, and complements of a finite number of events are also events. Even so, the result (2.7) above fails for the union of infinitely many events and axiom (c) has to be modified to state that the probability of the union of an infinite sequence of mutually exclusive events is the sum of the infinite series of the probabilities of the separate events.

2.6 SETS AND EVENTS

It has been convenient to discuss the rules for manipulating probabilities using the framework of sets in a sample space. In any particular problem it is more natural to think of an event as a statement which is or is not confirmed by the actual outcome of an experiment. Thus, if a die is rolled, the event 'the number is even' inclines us more to picture the concrete outcomes $2, 4, 6$ than to reflect on the set of three points in a sample space corresponding to these outcomes. It will be clear, from the way they have been defined, that there is an exact matching between operations on sets representing events and the ordinary connectives between statements about actual events, namely:

$$\cup \equiv \text{or}$$

$$\cap \equiv \text{and}$$

$$(\;)' \equiv \text{not}$$

As from now on we shall frequently use the statement language, we here repeat the main results of section 2.5 suitably translated.

$$0 \leqslant \Pr(E) \leqslant 1$$

$$\Pr(\text{not } E) = 1 - \Pr(E)$$

$$\Pr(E_1 \text{ and } E_2) = 0 \text{ if } E_1 \text{ and } E_2 \text{ are inconsistent}$$

$$\Pr(E_1 \text{ or } E_2) = \Pr(E_1) + \Pr(E_2) - \Pr(E_1 \text{ and } E_2).$$

2.7 COUNTING METHODS

Suppose a man has a choice of three different routes from London to Exeter and thence a choice of two different routes from Exeter to Torquay. It is evident that he has $3 \times 2 = 6$ different routes from London to Torquay via Exeter provided we assume that neither choice from Exeter to Torquay is influenced by the route from London to Exeter. In general, if action A_1 may be carried out in a_1 different ways and may be followed by action A_2 in a_2 different ways then the joint action A_1 followed by A_2 may happen in $a_1 a_2$ different ways, assuming that a_2 does not depend on the particular choice of A_1. The result extends to k actions by repetition.

Example 2
How many different numbers of three digits can be formed from the numbers, 1, 2, 3, 4, 5 − (a) if repetitions are allowed (b) if repetitions are not allowed? How many of these numbers are even in either case?

(a)   Each digit can be chosen in five different ways, since repetitions are allowed, and hence there are $5 \times 5 \times 5 = 125$ such numbers. If the number is even, the final digit must be either 2 or 4, hence there are $2 \times 5 \times 5 = 50$ such numbers.

(b)   The final digit can be chosen in five ways, *and regardless of the choice* there are four choices for the next digit and then three for the remaining digit. Hence $5 \times 4 \times 3 = 60$ numbers in all. There are $2 \times 4 \times 3 = 24$ even numbers.

Problem 5. In how many different orders can the letters of the word CINEMA be arranged? How many do not begin with M but end with C?

Example 3
In how many different ways can five men stand in a row if two particular men must be next to each other? In how many ways can this be done in a circle? The two men can be paired in two ways. The pair and the remaining three men can be arranged in $2(4 \times 3 \times 2 \times 1) = 48$ different orders in a row. In a circle, only the orders relative to a fixed man are different. Having fixed a single man, there are $2(3 \times 2 \times 1) = 12$ different orders in the circle for the remaining two men, and the pair (clockwise and anticlockwise counting as different).

Problem 6. Each of four questions on a multiple choice test has three possible answers. How many candidates must sit the test to ensure that at least two candidates give the same answers?

Problem 7. If there are n_i counters of colour i and k colours in all, show that the number of selections which can be made taking any number at a time is

$$\prod_{i=1}^{k} (1 + n_i) - 1$$

Example 4

Consider the placing of r different balls in n different boxes. We are not concerned with the order in which the balls are picked up or the order in which they sit in the boxes. If the balls are placed one at a time, then each ball may be placed in one of n (provided there is ample room) and hence there are n^r different distributions. Now, if each ball is placed at random, the probability of each distribution is $1/n^r$. If there are restrictions, say each box can only take one ball, then the first ball may be placed in n ways but the second only $n - 1$ ways and in all there are $n(n - 1) \ldots (n - r + 1)$ different distributions. Yet another problem is posed if the balls cannot be distinguished from each other. In this case, two distributions are the same if corresponding boxes contain the same *number* of balls. The various assumptions possible about the balls, boxes and capacity restrictions can be used as simple models for real experiments. (For an interesting discussion of such models in physics see ref. [1]).

Problem 8. Four different balls are placed at random in four different boxes. Calculate the probability that each box contains just one ball.

2.8 PERMUTATIONS

In how many different ways can n people form a single-file queue at a bus-stop? The first position can be filled in n ways, then the second in $(n - 1)$ ways since one person is not now available, the ith place in $n - (i - 1)$ ways and so on until there is just one person to fill the last place. Thus there are $n(n - 1) \ldots 3.2.1$ ways in which the queue may be formed. Such products are continually appearing in counting processes, and a standard notation for such a product is $n!$, read as n factorial. The number $n!$ is thus defined for all positive integers and, by convention, $0!$ is 1. The people in the queue are always the same, but an observer may detect $n!$ distinct orders in which they may stand. It is sometimes objected that if someone is chosen to be first then he is no longer available for the second place. This is quite true, but it should be realized that every person is in front in turn for some of the arrangements; in fact each person is front for $(n - 1)!$ of the different orders.

Now suppose a bus arrives, in how many different *orders* can just r people mount the bus one at a time? To answer this, we do not *need* to know the order in the queue or even if they are in a queue at all, provided they mount the bus one at a time. The first person to get on can be any one of n persons, the second any one of $n - 1$ and the rth person any one of $n - (r - 1)$, *since $r - 1$ persons have already boarded*. Hence the total number of distinct ways is $n(n - 1)(n - 2) \ldots (n - r + 1)$. Now

$$n(n - 1) \ldots (n - r + 1) = \frac{n(n - 1) \ldots (n - r + 1)(n - r)(n - r - 1) \ldots 3.2.1}{(n - r)(n - r - 1) \ldots 3.2.1}$$

$$= \frac{n!}{(n - r)!}. \tag{2.8}$$

The number of different arrangements of n distinct things taken r at a time is also called the number of **permutations** of n things r at a time and is written $(n)_r$

$$(n)_r = \frac{n!}{(n-r)!} \tag{2.9}$$

2.9 COMBINATIONS

We next find the number of different samples of r elements that can be drawn from a population of n distinct elements. 'Element' may mean person, institution, or, say, car registration number. We have still to settle how the sample is to be drawn. We may take r elements all at once or make up the sample by drawing one element at a time until r have been obtained. Further, if we draw one at a time, we may either put each one aside until we have collected r elements (*sampling without replacement*) or may make a note of which element it is and then put it back into the population before drawing another element (*sampling with replacement*). Without replacement means that no two elements in the sample may be the same, while with replacement means that an element may be recorded as appearing several times in the sample. In this section we shall consider sampling without replacement. Taking the elements one at a time differs from drawing the sample all at once in one important respect, namely that we can record the order in which the elements appeared in the sample. Thus we should distinguish between ordered samples and unordered samples. Two ordered samples will be the same when they have the same elements and these were drawn in the same order. The number of different ordered samples is the number of distinct arrangements or permutations of n distinct elements, r at a time and this is $n!/(n-r)!$. Two unordered samples on the other hand, will be the same, if they contain the same elements, in whatever order they appear. We can find the number of distinct ordered samples by taking a particular unordered sample and finding all the distinct ordered samples which its elements can form. Since there are $r!$ distinct arrangements of r different elements, each unordered sample yields $r!$ ordered samples. Thus $r!$ times the number of distinct unordered samples must equal the number of distinct ordered samples, namely $n!/(n-r)!$. Hence the number of distinct unordered samples is $n!/[r!(n-r)!]$. This number, also known as the number of **selections** or of **combinations** of n different elements taken r at a time, is denoted ^nC_r, or (more usually), $\binom{n}{r}$.

Example 5

The number of different selections of three from the five letters A, B, C, D, E is

$$\frac{5!}{3!2!} = 10$$

The number of selections of three which contain A is $\binom{4}{2}$, since having taken A we select two more from the remaining four.

Example 6

A college offers four courses in statistics and five in mathematics to students in their first year. A student must take two courses in statistics and three in mathematics. How many different first-year programmes may be devised? A student may select two courses in statistics from four in $\binom{4}{2}$ ways and with each such selection he may choose three courses from five in mathematics in $\binom{5}{3}$ ways. Hence the number of programmes is $\binom{4}{2}\binom{5}{3} = 60$. If later it is found that one particular course in statistics involves a timetable clash with one particular course in mathematics, then we must subtract the number of programmes which contain both these courses. There are $\binom{3}{1}\binom{4}{2} = 18$ of these and hence 42 different programmes possible. Verify this answer by finding the number of programmes which include just one of the classing courses or neither of them.

Problem 9. Find the maximum number of points of intersection of m straight lines and n circles.

Problem 10. Show that the number of distinct ways in which r different balls can be placed in n different boxes so that one particular box contains exactly k balls is $\binom{r}{k}(n-1)^{r-k}$. What is the sum of this expression over all values of k?

Problem 11. A bag contains $m + k$ discs. On m of these is inscribed a different non-zero number, while the remaining k bear the value zero. If n discs are drawn one at a time without replacement, calculate the probability that the ith disc is a zero.

2.10 ARRANGEMENTS IN A ROW

We next consider the number of distinct arrangements of a_1 similar white balls and a_2 similar red balls in a row. Here the $(a_1 + a_2)!$ arrangements are not all distinct. Suppose there are $a_1 + a_2$ vacant spaces in a row to be filled by the balls. Each distinct arrangement of the balls may be found by stating which places are to be filled by red balls. We have to select a_2 places from the $a_1 + a_2$ available and this can be done in

$$\binom{a_1 + a_2}{a_2}$$

ways. Alternatively, for any arrangement of the balls, the whites can be interchanged among themselves in $a_1!$ ways and the red balls in $a_2!$ ways without yielding an arrangement of different appearance. Hence the number of distinct arrangements multiplied by $a_1!a_2!$ gives the total number of rearrangements, which is $(a_1 + a_2)!$. Hence the number of distinct arrangements is

$$\frac{(a_1 + a_2)!}{a_1!a_2!} = \binom{a_1 + a_2}{a_2}$$

as before. Using this second argument, it can be easily shown that with a_i balls of colour i and k colours in all, the number of distinct arrangements of the

$$\sum_{i=1}^{k} a_i = N$$

balls is $N!(a_1!a_2!\ldots a_k!)$. Show by the argument using the selection of places, that the number of distinct distributions is also

$$\binom{N}{a_1}\binom{N-a_1}{a_2}\binom{N-a_1-a_2}{a_3}\ldots\binom{N-a_1-a_2\ldots-a_{k-1}}{a_k}.$$

Hence the number of distinguishable ways of arranging 3 white, 4 black and 2 red balls in a row is

$$\frac{(3+4+2)!}{3!4!2!} = 1260.$$

Problem 12. How many different 10-digit numbers can be formed from the numbers $1, 1, 1, 2, 2, 3, 4, 4, 6, 9$ so that no multiples of three are adjacent?

Problem 13. In how many distinguisable ways can four statistics books, three psychology books, and five novels be arranged on a shelf so that books of the same type are together when (a) books of the same type are different; (b) books of the same type are identical?

Example 7
Two similar packs of n different cards are laid out in two rows side by side. In how many ways can this be done so that no pair of cards is the same? This is the basic situation in the problem of derangements, though it is often jokingly phrased in terms of 'nobody getting his own hat from a cloakroom' or 'no letter being put in the right envelope'. We begin by supposing that the two packs are laid out so that every pair is the same (or is matching). We then leave one row fixed and derange all the cards in the other row. Let there be $\phi(n)$ derangements so that no pair is matching. If we concentrate on a particular card A then we see that another card, chosen in $n - 1$ ways, takes the place of A in two exclusive and only possible ways. Either this card merely changes places with A and the remaining $n - 2$ cards suffer total derangement in $\phi(n - 2)$ ways or A is not permitted to rest in the place of the other card and the $(n - 1)$ cards including A are deranged in $\phi(n - 1)$ ways. Thus altogether we have,

$$\phi(n) = (n-1)\,[\phi(n-2) + \phi(n-1)].$$

This equation looks intractable but yields to rewriting in the form

$$\phi(n) - n\phi(n-1) = -1[\phi(n-1) - (n-1)\,\phi(n-2)]$$
$$= (-1)^2\,[\phi(n-2) - (n-2)\,\phi(n-3)]$$
$$= (-1)^{n-2}\,[\phi(2) - 2\phi(1)].$$

Now there is only one way of deranging two matched pairs, hence, $\phi(2) = 1$ and $\phi(1) = 0$.

Thus $\phi(n) = n\phi(n-1) + (-1)^n$. Applying the same result to $\phi(n-1)$

$$\phi(n) = n[(n-1)\,\phi(n-2) + (-1)^{n-1}] + (-1)^n$$
$$= n(n-1)\,\phi(n-2) + n(-1)^{n-1} + (-1)^n$$
$$= \frac{n!}{2!} - \frac{n!}{3!} + \frac{n!}{4!} - \ldots + \frac{n!}{n!}\,(-1)^n$$
$$= n!\left(\frac{1}{2!} - \frac{1}{3!} + \ldots \frac{(-1)^n}{n!}\right).$$

For large n, $\phi(n)/n!$ is approximately e^{-1}.

Problem 14. If n letters are placed one in each of n addressed envelopes, find the number of ways in which just r letters are in their correct envelopes. Hence deduce that,

$$\sum_{r=0}^{n} \binom{n}{r}\,\phi(n-r) = n!,$$

where $\phi(n)$ is defined as in Example 7.   ∎

Suppose that an urn contains r red and $m-r$ white balls. A random sample of n balls is drawn without replacement and we require the probability that the sample contains just k red balls. Though all the red balls are similar, since each selection of n is to have the same change of being drawn, we may think of them as bearing the numbers $1, 2, \ldots, r$, while the white balls are numbered $r+1$, $r+2, \ldots, m$.

The red balls can be selected in $\binom{r}{k}$ ways and with each such way we can select $n-k$ whites in $\binom{m-r}{n-k}$ ways. Hence there are $\binom{r}{k}\binom{m-r}{n-k}$ samples which have equal probability. The probability of just k reds is thus

$$\binom{r}{k}\binom{m-r}{n-k}\bigg/\binom{m}{n}.$$

The formula obtained is applicable to many practical situations. The 'population' may be of machines and 'red' may stand for defective or the population may be of adults in a particular town and 'red' may mean 'owns or rents a television set'.

The reader will have detected various side restrictions on the possible value of k. Thus $k \leqslant r$ and $k \leqslant n$ imply $k \leqslant \min (r, n)$ while $n - k \leqslant m - r$ with $k \geqslant 0$ demands $k \geqslant \max [n - (m - r), 0]$. However, values of k which do not satisfy these bounds will have zero probability.

2.11 RANDOM SAMPLING

To make forward planning possible, governments must have reliable statistics concerning the characteristics and composition of society. Periodically a census is carried out to obtain this information. Such is the labour and cost of compiling the results, that only a short list of key questions is administered to all the reachable and relevant citizens. For information on other matters, the survey office must rely on the responses of a sample of the population to a longer list of questions. There is always a danger that some feature of the process used to select this sample will lead to persistent and uncorrectable bias in the conclusions drawn for the population. Therefore it would be unwise to select the sample by taking names from telephone directories when estimating earnings. One precaution we can take is to give all possible samples, of the required number of elements, an equal chance of being chosen. The resulting process is called random sampling.

Thus suppose there are m elements in the population and we wish to draw a random sample of n elements. In survey work there is no point in examining the same element more than once so that the results of section 2.10 apply. Therefore there are $\binom{m}{n}$ different unordered samples which can be drawn without replacement and each of these is given probability $1 / \binom{m}{n}$.

For example, suppose we wish to draw a random sample of two from four elements, without replacement. The four elements may be labelled A, B, C, D and there are $\binom{4}{2} = 6$ selections of the elements. These are $\{A, B\}$, $\{A, C\}$, $\{A, D\}$, $\{B, C\}$, $\{B, D\}$, $\{C, D\}$ which we number 1, 2, 3, 4, 5, 6 respectively. If a fair die is rolled, the number obtained can be used to indicate which sample is to be taken.

A consequence of drawing a random sample of n from m, is that all sets of $k(\leqslant n)$ elements have the same probability of appearing in the sample. For if a particular set of k elements be included, then $n - k$ further elements can be selected from the $m - k$ remaining in $\binom{m-k}{n-k}$ ways. Hence the probability that k particular elements appear in the sample is $\binom{m-k}{n-k} / \binom{m}{n}$. This probability, although a function of k, does not depend on the elements involved. When $k = 1$, the probability that any individual element is included is

$$\binom{m-1}{n-1} / \binom{m}{n} = \frac{n}{m}.$$

It is worth pointing out that although random sampling without replacement ensures that each element in the population has the same chance of appearing in the sample, the converse need not be true. For example, if the population has four elements A, B, C, D and the pairs $\{A, C\}$, $\{B, D\}$ are selected each with probability 1/2, then each of A, B, C, D appears with probability 1/2 but A and B cannot appear together.

Example 8

A box contains ten articles, of which just three are defective. If a random sample of five is drawn, without replacement, calculate the probabilities that the sample contains (a) just one defective, (b) at most one defective, (c) at least one defective.

(a)   There are $\binom{10}{5}$ samples with equal probability. Of these $\binom{3}{1}\binom{7}{4}$ contain just one defective and four non-defective. Hence rquired probability is,

$$\frac{\binom{3}{1}\binom{7}{4}}{\binom{10}{5}} = \frac{5}{12}.$$

(b)   Similarly, the probability of no defective is

$$\frac{\binom{3}{0}\binom{7}{5}}{\binom{10}{5}} = \frac{1}{12}.$$

Hence the probability of at most one defective is

$$\frac{1}{12} + \frac{5}{12} = \frac{1}{2}.$$

(c)   Pr(at least on defective)

$$= 1 - \text{Pr(no defective)}$$

$$= 1 - \frac{1}{12} = \frac{11}{12}.$$

Example 9

A hand of 13 cards is drawn at random without replacement from a full pack of playing cards. Find the probability that it contains 4 cards in each of three suits and a singleton. We can draw 4 cards from 13 in $\binom{13}{4}$ ways. Hence we can draw 4 from 13 in three particular suits and a singleton from the remaining suit in $\binom{13}{4}^3\binom{13}{1}$ ways. But the suit to provide the singleton may be nominated in $\binom{4}{1}$ ways. Hence the probability of the stipulated hand is

$$\frac{\binom{4}{1}\binom{13}{4}^3\binom{13}{1}}{\binom{52}{13}}.$$

Problem 15. A bag contains 3 red discs, 3 green discs, and 3 white discs. A random sample of two is drawn without replacement. Calculate the probability that the discs have different colours.

Problem 16. A bag contains n white discs and n black discs. Pairs of discs are drawn without replacement until the bag is empty. Show that the probability that every pair consists of one white and one black disc is $2^n \Big/ \binom{2n}{n}$.

Problem 17. A number is composed from k different pairs of digits. If r digits are chosen at random, what is the probability that they are all different?

2.12 COMBINATORIAL IDENTITIES

We next derive some of the elementary properties of the quantities $\binom{n}{r}$, some of which are of frequent application in the evaluation of probabilities. By definition if n, r are positive integers and $r \leqslant n$, then

$$\binom{n}{r} = \frac{n!}{r!(n-r)!}.$$

If in this formula we put $r = 0$, we obtain $1/0!$ but we have defined $0!$ as 1, hence we define $\binom{n}{0}$ as 1. Further, we shall agree that $\binom{n}{r} = 0$ if $r > n$, which is reasonable, since there are no selections of more than n from n things.

We have at once,

$$\binom{n}{r} = \frac{n!}{r!(n-r)!} = \binom{n}{n-r}.$$

This illustrates that if a selection of r elements is made from n, then $n - r$ elements are left behind and, of course, the number of different remainders must equal the number of selections. (Part of the charm of the formulae used in counting processes is the possibility of finding a representation which makes a formula seem obvious.) Next

$$\binom{n+1}{r} = \binom{n}{r-1} + \binom{n}{r}. \tag{2.10}$$

This is readily verified, for the right-hand side is

$$\frac{n!}{(r-1)!(n-r+1)!} + \frac{n!}{r!(n-r)!} = \frac{n!}{(r-1)!(n-r)!}\left[\frac{1}{n-r+1} + \frac{1}{r}\right]$$

$$= \frac{n!}{(r-1)!(n-r)!}\left[\frac{n+1}{r(n-r+1)}\right]$$

$$= \frac{(n+1)!}{r!(n-r+1)!} = \binom{n+1}{r}.$$

We may also argue as follows: every selection of r elements from $n + 1$ either does or does not include a particular element A. If A is included, the remaining $r - 1$ elements must be selected from n elements which may be done in $\binom{n}{r-1}$ ways. If A is not included, then we may select all r elements from the remaining n elements in $\binom{n}{r}$ ways. Hence, by addition, the result.

Example 10
This last result is the key to many identities. Writing it in the form

$$\binom{n}{r-1} = \binom{n+1}{r} - \binom{n}{r}$$

$$\sum_{n=j}^{k}\binom{n}{r-1} = \sum_{n=j}^{k}\left[\binom{n+1}{r} - \binom{n}{r}\right]$$

$$= \binom{j+1}{r} - \binom{j}{r} + \binom{j+2}{r} - \binom{j+1}{r}\cdots$$

$$\binom{k+1}{r} - \binom{k}{r},$$

the only remaining terms are those involving the lowest and highest values for n. Hence

$$\sum_{n=j}^{k} \binom{n}{r-1} = \binom{k+1}{r} - \binom{j}{r}.$$

Problem 18. Show that

$$\sum_{k=0}^{m} \binom{n-k}{r} = \binom{n+1}{r+1} - \binom{n-m}{r+1}, \quad n \geqslant m \geqslant r.$$

Problem 19. Show that

$$\sum_{r=0}^{m} (-1)^r \binom{n}{r} = (-1)^m \binom{n-1}{m}, \quad n > m.$$

Problem 20. Show that

$$\sum_{r=0}^{m} \binom{r+k}{k} = \binom{m+k+1}{k+1}. \qquad\qquad \blacksquare$$

The number of selections of r elements from n is $\binom{n}{r}$, hence the total number of selections is

$$\sum_{r=1}^{n} \binom{n}{r}.$$

But we have already shown that the total number of ways of selecting a sample from n elements is $2^n - 1$, hence since $\binom{n}{0} = 1$,

$$\sum_{r=0}^{n} \binom{n}{r} = 2^n.$$

In this formular $\binom{n}{0}$ corresponds to selecting no elements.

Finally, we obtain a result derived from drawing a sample of n from m balls, r of which are red. The number of selections (involving different balls) so that the sample contains just k reds is $\binom{r}{k}\binom{m-r}{n-k}$. But every selection of n must contain either 0, 1, 2, ..., or n reds $(n \leqslant r)$. Hence

$$\sum_{k=0}^{n} \binom{r}{k}\binom{m-r}{n-k}$$

must be the total number of selections of n balls from m, that is, $\binom{m}{n}$. Hence

$$\sum_{k=0}^{n} \binom{r}{k}\binom{m-r}{n-k} = \binom{m}{n}. \tag{2.11}$$

Problem 21. Discuss (2.11) for the case $n > r$.

2.13 THE QUANTITIES $\binom{n}{r}$ AND THE BINOMIAL THEOREM

The expansion

$$(1+x)^n = 1 + \binom{n}{1}x + \binom{n}{2}x^2 + \ldots + \binom{n}{r}x^r + \ldots + \binom{n}{n}x^n$$

where x is any real number and n a positive integer will already have been met. It is in this connection that the numbers $\binom{n}{r}$ are called the binomial coefficients. The theorem may readily be proved by the method of induction, but it is enlightening to relate it to a certain selection process. Since

$$(1+x)^n = (1+x)(1+x)\ldots(1+x)$$

to n factors, each term in the expansion is the produce of n symbols, one from each of the n factors. If the product is x^r, then x has been selected r times and 1 has been selected $n - r$ times. But we may select r out of n factors in $\binom{n}{r}$ ways, hence $\binom{n}{r}$ is the coefficient of x^r. From this expansion we can obtain some of the results already found in the last section. For example, if we put $x = 1$, then

$$2^n = 1 + \binom{n}{1} + \binom{n}{2} + \ldots + \binom{n}{n}.$$

Consider also the identity $(1+x)^m = (1+x)^r(1+x)^{m-r}$ for $r \leqslant m$. The coefficient of x^n in $(1+x)^m$ is $\binom{m}{n}$. The term in x^n in $(1+x)^r(1+x)^{m-r}$ is found by taking all terms like x^k from $(1+x)^r$, with coefficient $\binom{r}{k}$, and multiplying by x^{n-k}, with coefficient $\binom{m-r}{n-k}$ from $(1+x)^{m-r}$ and adding them together. Thus we have

$$\binom{m}{n} = \sum_{k=0}^{n} \binom{r}{k}\binom{m-r}{n-k}.$$

Problem 22. In the binomial expansion of $(1 + x)^n$, show that the sum of the binomial coefficients of the odd powers of x equals the sum of the binomial coefficients of the even powers of x and each has sum 2^{n-1}. *Hint*: show that

$$\sum_{r=0}^{n} (-1)^r \binom{n}{r} = 0.$$

Problem 23. Use the identity $\phi(n) = n\phi(n-1) + (-1)^n$ to prove that

$$\sum_{r=0}^{n} \binom{n}{r} \phi(n-r) = n \sum_{r=0}^{n-1} \binom{n-1}{r} \phi(n-r-1) = n!$$

Problem 24. Show that,

$$\sum_{i=0}^{j} (-1)^i \binom{a}{i} \binom{a-i}{j-i} = 0.$$

Problem 25. Prove that

(a) $\displaystyle \sum_{r=0}^{n} \binom{n}{r}^2 = \binom{2n}{n}$

(b) $\displaystyle \sum_{r=0}^{n-1} \binom{n}{r} \binom{n}{r+1} = \binom{2n}{n-1}.$

2.14 MULTINOMIAL EXPANSION

As another example of the application of the selection method, we consider the expansion of $(1 + x + x^2 + \ldots + x^r)^n$, which is a polynomial of the form

$$\sum_{m=0}^{rn} A_m x^m$$

and we seek the general form of the coefficients A_m. Now x^m is obtained by taking one symbol from each of the n factors $(1 + x + x^2 + \ldots + x^r)$ and there may be repetitions in that x^i may be taken from several factors. Suppose in fact that the term x^m is to be made up from the product of α_0 selections of 1, α_1 selections of $x^1, \ldots, \alpha_i$ selections of x^i, and so on. Then, since there are n factors,

$$\alpha_0 + \alpha_1 + \alpha_2 + \ldots + \alpha_r = n \tag{2.12}$$

where we note that some of the α_i can be zero. Also, the degree of the product is required to be m, that is

$$1^{\alpha_0}(x)^{\alpha_1}(x^2)^{\alpha_2}\ldots(x^r)^{\alpha_r} = x^{\alpha_1+2\alpha_2+\ldots+r\alpha_r} = x^m,$$

or

$$\alpha_1 + 2\alpha_2 + \ldots + r\alpha_r = m. \tag{2.13}$$

In how many ways can we make such a selection? The α_0 symbols 1 can be taken from n factors in $\binom{n}{\alpha_0}$ ways, *then* the α_1 symbols x from the *remaining* $n - \alpha_0$ factors in $\binom{n - \alpha_0}{\alpha_1}$ ways and so on. Thus there are

$$\binom{n}{\alpha_0}\binom{n-\alpha_0}{\alpha_1}\binom{n-\alpha_0-\alpha_1}{\alpha_2}\ldots\binom{n-\alpha_0-\alpha_1\ldots-\alpha_{r-1}}{\alpha_r} = \frac{n!}{\alpha_0!\alpha_1!\ldots\alpha_r!} \tag{2.14}$$

ways of selecting the α_i. There are of course, several sets of α_i and A_m is the sum of all terms like (2.14) for those sets which satisfy (2.12) and (2.13). For numerical work, we see that $1 + x + x^2 + \ldots + x^r$ is the sum of $r + 1$ terms of a G.P. and is

$$\frac{1 - x^{r+1}}{1 - x},$$

hence using the binomial theorem for a negative index, which is valid if $|x| < 1$,

$$\left(\frac{1 - x^{r+1}}{1 - x}\right)^n = (1 - x^{r+1})^n (1 - x)^{-n}$$

$$= \left(1 - \binom{n}{1}x^{r+1} + \binom{n}{2}x^{2r+2} \ldots\right) \times$$

$$\left(1 + nx + \frac{n(n+1)x^2}{2!} + \ldots\right). \tag{2.15}$$

In particular, the coefficient of x^r involves only the term in x^r from the second bracket in (2.15) which has coefficient

$$\frac{n(n+1)(n+2)\ldots(n+r-1)}{r!} = \frac{(n+r-1)!}{(n-1)!\,r!} = \binom{n+r-1}{r}. \tag{2.16}$$

We can find a model for this bit of algebra. Consider the placing of r *similar* balls in n *distinct* boxes. A box contains $0, 1, 2, \ldots, r$ balls. Suppose α_i boxes contain i balls. Then since the total number of boxes is n,

$$\sum_{i=0}^{r} \alpha_i = n$$

and the number of balls is r,

$$\sum_{i=0}^{r} i\alpha_i = r.$$

Thus $n!/(\alpha_0!\alpha_1!\ldots\alpha_r!)$ corresponds to the number of different ways in which we can select α_0 boxes to be empty, α_1 boxes to contain one ball, and so on. It does not matter at any stage which of the available balls goes into a box as these are all alike. Hence the number of distinguishable ways in which r similar balls may be placed in n distinct boxes is $\binom{n+r-1}{r}$. This is not the most concise way of finding this result, but it does lend itself readily to modifications which solve allied problems.

Distribution of r similar balls in n distinct boxes

We now obtain the number of distinguishable distributions by a combinatorial argument. Suppose we have $n-1$ parallel strokes in a row, and that the spaces between and outside these strokes represent the n boxes. Next put down r crosses in some order. Thus for $n = 5, r = 4$, we could have

$$\times\,|\times\times\,|\,\,|\times\,|$$

which represents 1 ball in box one, 2 balls in box two, 0 balls in box three, one ball in box four and 0 balls in box five. There are $n + r - 1$ symbols in a row, $n - 1$ of one kind and r of another kind. There are $\binom{n+r-1}{r}$ distinguisable permutations of these symbols, each one of which corresponds to a distinguishable arrangement of the r similar balls into n different boxes.

Formulae of this type, which depend on positive integers may be verified by the method of mathematical induction. Since *two* variables, r and n are involved, a little care is needed. Suppose $P(n, r)$ is a proposition that depends on n and r then it does *not* suffice to show that (a) $P(n + 1, r + 1)$ is true if $P(n, r)$ is true, (b) $P(1, 1)$ is true. This will merely prove the result for $P(1, 1), P(2, 2), P(3, 3)$, etc. Several adequate procedures are available, for example,

(a)   show that $P(1, r)$ is true for all r,
(b)   show that if $P(n, r)$ is true for all r, then $P(n + 1, r)$ is true for all r.

The induction now 'works' as follows: since (a) is true, put $n = 1$ in (b), then $P(2, r)$ is true for all r – and so on.

We can also find the number of distinguishable ways in which r similar balls can be placed in n distinct boxes *so that no box is empty*, though we must have $r \geq n$. We first take any n balls and put one in each box, thus ensuring that no box is empty. This leaves $r - n$ balls and these may be put in n boxes in

$$\binom{n + (r - n) - 1}{r - n}$$

distinguishable ways, that is $\binom{r-1}{n-1}$ distinguishable ways. The first stage of this argument would fail entirely if the balls were not similar.

Example 11

Find the coefficient of x^{10} in the expansion of $(x + x^2 + x^3)^4$. We may write this as

$$\left(\frac{x - x^4}{1 - x}\right)^4 = x^4(1 - x^3)^4 (1 - x)^{-4}$$

$$= x^4(1 - 4x^3 + 6x^6 - 4x^9 + x^{12})$$

$$\left(1 + 4x + \frac{4.5x^2}{2!} + \frac{4.5.6x^3}{2!} + \dots\right.$$

$$\left. + \frac{4.5.6.7.8.9 x^6}{6!} \dots\right).$$

The total coefficient of x^{10} is

$$6 - \frac{4.4.5.6}{3!} + \frac{4.5.6.7.8.9}{6!} = 10.$$

This is also the number of ways in which ten single pound coins can be distributed among four men so that each gets at least one and at most three pounds. There are four distributions of type $(3, 3, 3, 1)$ and six of type $(3, 3, 2, 2)$.

Problem 26. Find the number of distinct ways in which r similar balls may be placed in n different boxes so that no box is empty by calculating the coefficient of x^r in the expansion of $(x + x^2 + \dots + x^r)^n$.

Show further that the number of different distributions in which exactly m cells are empty is $\binom{n}{m}\binom{r-1}{n-m-1}$, and evaluate

$$\sum_{m=0}^{n-1} \binom{n}{m}\binom{r-1}{n-m-1}.$$

Problem 27. If $\phi_n^{(r)}$ is the number of distinguishable ways in which r similar balls can be placed in n different boxes, show that

$$\phi_n^{(r)} = \phi_n^{(r-1)} + \phi_{n-1}^{(r)}.$$

Verify that this equation is satisfied by

$$\phi_n^{(r)} = \binom{n+r-1}{r} .$$

Problem 28. Show that the number of distinct ways in which r different balls may be put into n *different* boxes is the coefficient of x^r in

$$r! \left(1 + \frac{x}{1!} + \frac{x^2}{2!} + \ldots + \frac{x^r}{r!} \right)^n$$

or in $r!e^{nx}$ and verify that this is n^r. By considering the coefficient of x^r in

$$r! \left(\frac{x}{1!} + \frac{x^2}{2!} + \ldots + \frac{x^r}{r!} \right)^n ,$$

prove that the number of distributions in which no box is empty is

$$n^r - \binom{n}{1} (n-1)^r + \binom{n}{2} (n-2)^r \ldots .$$

2.15 RUNS

Suppose letters of the alphabet are placed in a row and each letter may appear more than once. A sequence of r similar letters of one kind, which is not preceded or followed by a letter of the same kind, is said to be a *run* of length r. Thus the arrangement ABBABBBAA consists of runs of length 1, 2, 1, 3, and 2, five runs in all. Suppose a similar letters A and b similar letters B are placed at random in row, we seek the probability that the sequence contains just k runs of letter A. The number of permutations of $(a + b)$ things is $(a + b)!$ and since the letters are placed at random, all these are equally likely and the probability of each is $1/(a + b)!$. However, there are $a!b!$ permutations which give the same distinguishable order. Hence the probability of each distinguishable order is $a!b!/(a + b)! = 1/\binom{a+b}{a}$. If there are k runs of 'A's then there must be $k - 1$, k, or $k + 1$ runs of 'B's, since A-runs and B-runs must alternate. Suppose in fact there are $(k - 1)$ B-runs, then the spacings between and outside these B-runs may be regarded as k distinct boxes in which we have to place a letters which correspond to similar balls. Each box must contain at least one 'ball' or there would be no A-run. The number of distinguishable ways of placing a balls in k boxes so that no box is empty is $\binom{a-1}{k-1}$. Now the b letters B may be arranged in *their* $k - 1$ boxes in $\binom{b-1}{k-2}$ distinguishable arrangements which yield k A-runs and $k - 1$ B-runs. On taking account of the other three possibilities, the total number of distinguishable ways of obtaining k A-runs is

$$\binom{a-1}{k-1}\binom{b-1}{k-2} + 2\binom{a-1}{k-1}\binom{b-1}{k-1} + \binom{a-1}{k-1}\binom{b-1}{k}$$

$$= \binom{a-1}{k-1} \left\{ \left[\binom{b-1}{k-2} + \binom{b-1}{k-1} \right] + \left[\binom{b-1}{k-1} + \binom{b-1}{k} \right] \right\}$$

$$= \binom{a-1}{k-1} \left[\binom{b}{k-1} + \binom{b}{k} \right]$$

$$= \binom{a-1}{k-1}\binom{b+1}{k}.$$

The probability of just k A-runs is

$$\binom{a-1}{k-1}\binom{b+1}{k} \bigg/ \binom{a+b}{a},$$

which is zero if either $k > a$ or $k > b + 1$, as it should be. Run theory is a source of methods for testing whether or not such a sequence of symbols is 'sufficiently random'. Thus if the examination results for a women and b men are put in order of increasing score, then similar results of the sexes would tend to give a large number of runs. The above formula can be used to compute for given a, b, the probability of at least the observed number of runs of woman's marks. It should be noted that actual marks gained are not used and thus some information has been lost. On the other hand, little has been asserted about the population from which the marks have been drawn.

Problem 29. Show that the probability of just k runs in all is

$$2\binom{a-1}{j-1}\binom{b-1}{j-1} \bigg/ \binom{a+b}{a}$$

where $k = 2j$ is even and is

$$\left[\binom{a-1}{j}\binom{b-1}{j-1} + \binom{a-1}{j-1}\binom{b-1}{j} \right] \bigg/ \binom{a+b}{a}$$

where $k = 2j +$ is odd.

Problem 30. Show the probability of starting with an A run of length k is

$$\binom{a+b-k-1}{b-1} \bigg/ \binom{a+b}{a}.$$

REFERENCE

[1]   W. Feller, *An Introduction to Probability Theory and its Applications*, John Wiley, New York. 1957.

BRIEF SOLUTIONS AND COMMENTS ON THE PROBLEMS

Problem 1. The dual of (1) is

$$(E_1 \cap E_2) \cup (E_1 \cap E_3) = E_1 \cap (E_2 \cup E_3).$$

The dual of (3) is result (2)!

Problem 2. The number with A but not B is $36 - 8 = 28$. Hence

$$N(C \cap A) \leqslant \min(28, 30) = 28,$$

since $C \cap B$ is empty. Since the total number of patients is 50,

$$30 - N(C \cap A) + 36 \leqslant 50, \text{ or } N(C \cap A) \geqslant 16.$$

Problem 3. For the men, $80 = N(M') + N(M)$

$$= N(M' \cap C') + N(M' \cap C) + N(M) = 26 + N(M' \cap C) + 44.$$

Hence $N(M' \cap C) = 80 - 70 = 10$, and the required number of women is $25 - 10 = 15$. The total number of women is here redundant information.

Problem 4.

$$\Pr[E \cup (F \cup G)] = \Pr(E) + \Pr(F \cup G) - \Pr[E \cap (F \cup G)]$$
$$\Pr(F \cup G) = \Pr(F) + \Pr(G) - \Pr(F \cap G)$$
$$\Pr[E \cap (F \cup G)] = \Pr[(E \cap F) \cup (E \cap G)]$$
$$= \Pr(E \cap F) + \Pr(E \cap G) - \Pr(E \cap F \cap G).$$

For the last line, consult Problem 1, part (1).

Problem 5.

$$6.5.4.3.2.1. = 720.$$
$$4.4.3.2. = 96.$$

Problem 6.

$$3^4 + 1 = 82.$$

Problem 7.

$n_i + 1$ of colour i (including none). Subtract 1 from product since 0 overall is not a selection.

Problem 8.

$$\frac{4.3.2.1}{4^4} = \frac{3}{32}.$$

Problem 9.

$$2\binom{n}{2} + 2mn + \binom{m}{2}.$$

Problem 10.

$$\sum_{k=0}^{r} \binom{r}{k}(n-1)^{r-k} = (1+n-1)^r = n^r.$$

n^r is the total number of ways of placing the balls.

Problem 11. If the n discs include s zeros, the probability that a zero appears in the ith place is $s(n-1)!/n! = s/n$. The probability that a selection of n contains s zeros is

$$\binom{m}{n-s}\binom{k}{s} \bigg/ \binom{m+k}{n}.$$

Hence the probability that the ith disc drawn is a zero is

$$\frac{1}{n}\sum_{s=1}^{k} s\binom{m}{n-s}\binom{k}{s} \bigg/ \binom{m+k}{n}$$

$$= \frac{k}{n}\sum_{s=1}^{k}\binom{m}{n-s}\binom{k-1}{s-1} \bigg/ \binom{m+k}{n}$$

$$= \frac{k}{n}\binom{m+k-1}{n-1} \bigg/ \binom{m+k}{n} = \frac{k}{k+m}.$$

Notice that the answer depends only on the proportion of zeros.

Problem 12. There are $7!/(3!2!2!) = 210$ different arrangements of 1112244. For each such, there are 8.7.6 ways of inserting 3, 6, 9 so that no two are adjacent. Ans. 70 560.

Problem 13. (a) $(4!3!5!)3!$. (b) $3!$.

Problem 14. Choose the r in $\binom{n}{r}$ ways then the remaining $n-r$ deranged in $\phi(n-r)$ ways. But the sum must be the total number of permutations.

Problem 15.

$$\frac{\binom{3}{1}\binom{3}{1}\binom{3}{0}}{\binom{9}{2}} \times \binom{3}{2} = \frac{3}{4}.$$

Problem 16. The first members of each pair are a sample of n from $2n$. But the members of each pair can be W, B or B, W, i.e. 2^n exchanges. Alternatively

$$\frac{n^2(n-1)^2 \dots 1^2}{\binom{2n}{2}\binom{2n-2}{2}\dots} = \frac{2^n}{\binom{2n}{n}}.$$

Problem 17.

(a)   Choose r pairs from k in $\binom{k}{r}$ ways and then from each pair one digit in two ways. Hence the probability is

$$2^r \binom{k}{r} \Big/ \binom{2k}{r}.$$

(b)   Choose the first digit in $2k$ ways, the next in $2k - 2$ which are different from the first choice – and so on. The required probability is

$$\frac{2k(2k-2)\dots 2k-2(r-1)}{2k(2k-1)\dots 2k-(r-1)},$$

which can be converted into the answer (a).

Problem 18.

$$\sum_{k=0}^{m}\binom{n-k}{r} = \sum_{k=0}^{m}\left[\binom{n-k+1}{r+1} - \binom{n-k}{r+1}\right] = \binom{n+1}{r+1} - \binom{n-m}{r+1}$$

Problem 19. Use $\binom{n}{r} = \binom{n-1}{r-1} + \binom{n-1}{r}.$

Write sum as

$$\binom{n}{0} - \left[\binom{n-1}{0} + \binom{n-1}{1}\right]\dots + (-1)^m\left[\binom{n-1}{m-1} + \binom{n-1}{m}\right].$$

Collect terms.

Problem 20.

$$\sum_{r=0}^{m} \binom{r+k}{k} = \sum_{r=0}^{m} \left[\binom{r+k+1}{k+1} - \binom{r+k}{k+1} \right]$$

$$= \binom{m+k+1}{k+1}.$$

Problem 21. $\binom{r}{k}$ is zero for $k > r$.

Problem 22.

$$(1+x)^n = \sum_{r=0}^{n} \binom{n}{r} x^r.$$

Set $x = -1$ to obtain result.

Problem 23. Apply the identity to $\phi(n-r)$

$$\sum_{r=0}^{n} \binom{n}{r} \left[(n-r)\,\phi(n-r-1) + (-1)^{n-r} \right],$$

$$= n \sum_{r=0}^{n-1} \binom{n-1}{r} \phi(n-r-1) + \sum_{r=0}^{n} \binom{n}{r} (-1)^{n-r}$$

$$= n \sum_{r=0}^{n-1} \binom{n-1}{r} \phi(n-r-1) + (1-1)^n$$

$$= n(n-1) \sum_{r=0}^{n-2} \binom{n-2}{r} \phi(n-r-2) + 0$$

$$= n!, \text{ by repeated application and using } \phi(0) = 1.$$

Note that $\phi(1) = \phi(0) - 1$, but $\phi(1)$ is zero.

Problem 24. After cancelling $(a-i)!$, the l.h.s. can be written

$$\sum_{i=0}^{j} (-1)^i \frac{a!}{i!(j-i)!\,(a-j)!} = \binom{a}{j} \sum_{i=0}^{j} (-1)^i \binom{j}{i}$$

$$= \binom{a}{j} \sum_{i=0}^{j} (1-1)^j = 0.$$

Problem 25.

(a) $\displaystyle\sum_{r=0}^{n} \binom{n}{r}\binom{n}{r} = \sum \binom{n}{r}\binom{n}{n-r} = \binom{2n}{n}.$

(b) $\displaystyle\sum_{r=0}^{n-1} \binom{n}{r}\binom{n}{r+1} = \sum \binom{n}{r}\binom{n}{n-1-r} = \binom{2n}{n-1}.$

Problem 26.

$$(x + x^2 + \ldots + x^r)^n = [(x - x^{r+1})/(1 - x)]^n$$
$$= x^n(1 - x^r)^n (1 - x)^{-n}.$$

We require the coefficient of x^{r-n} in $(1 - x)^{-n}$, which is

$$\frac{n(n + 1) \ldots (n + r - n - 1)}{(r - n)!} = \binom{r-1}{n-1}.$$

The empty boxes can be chosen in $\binom{n}{m}$ ways. All r balls can be placed in $\binom{r-1}{n-m-1}$ ways in $n - m$ boxes so that none are empty. The sum of the products must be the total number of ways of placing the balls, that is $\binom{n+r-1}{r}$.

Problem 27. Either one particular box is empty, and all the r balls are dispersed among the remaining $n - 1$ boxes, or that same particular box can have one ball placed in it and the remaining $r - 1$ balls are dispersed over all n boxes.

Problem 28. First part follows from remarks on the multinomial expansion. Second part requires the coefficient of x^r in

$$r!(e^x - 1)^n = r! \sum_{k=0}^{n} \binom{n}{k} e^{kx} (-1)^{n-k}.$$

But the coefficient of x^r in e^{kx} is $k^r/r!$ and the result follows. The introduction of e^{kx} contributes additional terms, which do not alter the required coefficients!

Problem 29. If there are $2j$ runs there must be j of each type. If $2j + 1$ either $j + 1$ of type A and j of type B or j of A and $j + 1$ of B.

Problem 30. A run of k letters A followed by a B uses $k + 1$ letters. There are $\binom{a+b-k-1}{b-1}$ distinct arrangements of the remaining letters.

3

Conditional Probability and Independence

3.1 INTRODUCTION

We have now considered several methods which will help us to count the points in various kinds of events when the sample space consists of a finite number of points. The impression may have been given that 'equally likely outcomes' are somehow always self-evident. Difficulties arising involve either hidden or unwarranted assumptions about the population being sampled. In a much-quoted example, it is held that if a fair coin is tossed twice, the probability that heads will appear at least once is 2/3. The basis of the argument is that there are three cases to consider — heads on the first throw, heads on the second throw, and heads on neither; two of these are favourable, hence the probability is 2/3. The objection is that these cases are neither equally likely nor mutually exclusive since heads on the first throw does not preclude heads on the second throw. E. Parzen [1] discusses in detail calculating the probability that for a randomly chosen month the thirteenth is a Friday. At first sight, every day of the week seems equally likely, and hence the probability is 1/7. However, a count of all cases over a stipulated long period of time gives a different result. This question of the actual relative frequency of an event is important if useful predictions are to be made in real situations. An unborn child may be either a boy or a girl. The probability that it will be a boy is not 1/2, but nearer 0.52 if the actual proportion of boys over several decades is used as an estimate.

3.2 EVALUATING PROBABILITIES

Example 1
Two cards are drawn without replacement from a well-shuffled pack. Calculate the probability that at least one card is a club.

A pack contains 52 cards, which are divided into four suits of 13 cards each, named spades, hearts, diamonds, and clubs. Since the pack is well shuffled, all pairs are to be given equal probability. There are $\binom{52}{2}$ such pairs. It often pays to calculate some easier probability. In this case we notice that the complement of 'at least one card is a club' is 'neither card is a club'. The number of samples of two that can be selected from the 39 cards which are not clubs is $\binom{39}{2}$. Hence the probability of at least one club is $1 - \binom{39}{2} / \binom{52}{2} = 15/34$. Alternatively, if the cards are drawn one at a time then there are 52×51 different ordered pairs altogether, and 39×38 of these are such that neither is a club. Again, the probability of at least one club is

$$1 - \frac{39 \times 38}{52 \times 51} = \frac{15}{34}.$$

We are not always fortunate enough to find a simplifying tactic of this kind, and so we shall also solve the problem directly in more than one way. There are $\binom{13}{1}\binom{39}{1}$ pairs in which just one of the cards is a club card and $\binom{13}{2}\binom{39}{0}$ pairs for which both cards are clubs. Hence the required probability that at least one is a club is

$$\frac{\binom{13}{1}\binom{39}{1} + \binom{13}{2}\binom{39}{0}}{\binom{52}{2}} = \frac{15}{34}.$$

If once again the cards are drawn one at a time, there are 52×51 ordered pairs, each with equal probability. Of these, there are 13×39 pairs in which a club is drawn first and the second is not a club, while there is the same number in which the first card is not a club and the second is a club. There are also 13×12 pairs in which both are clubs. Hence the probability that at least one card is a club is $(2 \times 13 \times 39 + 13 \times 12)/(52 \times 51) = 15/34$. There is yet another method possible on the basis of ordered pairs, namely to calculate, using eqn. (2.5), Pr(club on first card or club on second card) = Pr(club on first card) + Pr (club on second card) − Pr(club on first card and club on second card). Thus we have 13×15 ordered pairs in which the first card is club and the second is any one of the remaining cards. Hence the probability of a club first is $(13 \times 51)/(52 \times 51) = 1/4$. And we notice that this is the same result as would be obtained by considering drawing one card from the pack at random. For the probability that both are clubs we have $(13 \times 12)/(52 \times 51) = 1/17$. Finding the probability that the *second* card is a club poses a slight difficulty. It is true that there are 52 cards for the first draw, but the number of ways in which the second may be a club certainly depends on whether the first card was itself a club or not. For if a club came first there are only 12 cards from which to choose, and if not, there are 13. For the moment we shall split up the event 'club second' into the mutually exclusive events 'club first and second' for which there are 13 × 12 pairs and 'not club first and club second' for which there are 39 × 13 pairs. This implied that the required probability of 'club second' is (13 × 12 + 39 × 13)/(52 × 51) = 1/4. This result is thoroughly comforting. It means that provided that nothing is known about the first card drawn, then the probability that the second card is a club is just the same as for the first card. Thus the probability that at least one card is a club is

$$\frac{1}{4} + \frac{1}{4} - \frac{1}{17} = \frac{15}{34}.$$

* Further discussion

A population contains N elements $a_1, a_2, \ldots, a_N$. Elements are to be drawn one at a time without replacement. If one of the elements $a_1, a_2, \ldots, a_n$ is drawn at the kth draw, then event A is said to have occurred. Then *before any element is drawn*, the probability of A on draw k is a constant and equal to the probability of A on the first draw. Once some elements have been drawn then we should have to revise this probability on the basis of the information yielded. If, however, we are not informed as to which elements have been withdrawn, then the result still stands. That the result is true is shown by recalling that among the $(N)_k$ different ordered samples of k elements drawn without replacement, each of the elements $a_1, a_2, \ldots, a_n$ appears equally often in each of the draws, $1, 2, \ldots, k$. Thus, if there are 10 identical boxes, only one of which contains treasure, and 10 persons select a box in turn, each person has the same probability of obtaining the treasure.

If may be complained that the argument advanced takes no account of the number of times that A may have occurred before drawing the kth element. If we think about the first draw only, then we have a sample space with N points, each with equal probability, the event A contains n points and hence $\Pr(A) = n/N$. If in fact k elements are to be drawn, then we construct a different sample space, but the $\Pr(A$ on first draw$)$ remains n/N. For the sample space now consists of $(N)_k$ points of equal probability, one for each of the permutations of N, taken k at a time. There are n ways of choosing a first elements so that A happens, the remaining $(k-1)$ members of the ordered sample appearing in $(N-1)_{k-1}$ ways.

$$\Pr(A \text{ on first draw}) = n(N-1)_{k-1}/(N)_k$$

$$= \frac{n(N-1)!}{(N-k)!} \frac{(N-k)!}{N!} = \frac{n}{N}.$$

Next, let us consider the kth element drawn and suppose that r of the n elements favouring A have already appeared. These r elements can be *selected* in $\binom{n}{r}$ ways, then $(k-1-r)$ from $N-n$ not implying A in $\binom{N-n}{k-1-r}$ ways. We thus have $\binom{n}{r}\binom{N-n}{k-1-r}$ different such selections and each yields $(k-1)!$ arrangements. Finally there are $n-r$ ways in which A may happen on the kth draw. Hence, the total number of arrangments in which A may appear on the kth draw is

$$\sum_{r=0}^{k-1}\left[(k-1)!\binom{n}{r}\binom{N-n}{k-1-r}(n-r)\right]$$

$$= n(k-1)!\sum_{r=0}^{k-1}\left[\binom{n-1}{r}\binom{N-1-(n-1)}{k-1-r}\right]$$

$$= n(k-1)!\binom{N-1}{k-1} \quad \text{by eqn (2.11)}$$

$$= n(k-1)!\,\frac{(N-1)!}{(k-1)!(N-k)!} = \frac{n(N-1)!}{(N-k)!}$$

and again

$$\Pr(A \text{ on the } k\text{th draw}) = \frac{n(N-1)!/(N-k)!}{(N)_k} = \frac{n}{N}. \qquad \blacksquare$$

Problem 1. Among M balls, just K are white. A random sample of n is drawn without replacement and from this sample another random sample of m is drawn without replacement. Calculate the probability that the second sample contains just j white balls.

Problem 2. Packets of coffee each contain a coupon, which is equally likely to bear the letter A, B or C. If n packets are bought, what is the probability that the coupons cannot be used to spell CAB?

Problem 3. A bag contains a white balls and b black balls. Two balls are drawn at random. These are replaced and another two are drawn at random. Calculate the probability that i white balls are seen ($i = 0, 1, 2, 3, 4$).

3.3 APPLICATIONS

Example 2
A psychologist wishes to test the ability of a clairvoyant. He uses cards, each of which is blue on one side and either black or red on the other. His intention is to show the blue side only and to ask the clairvoyant to state the colour of the

other side. The clairvoyant makes a statement for each card and the psychologist wishes to assess his performance on the basis that the clairvoyant has in fact no special ability and is only guessing. An assessment may be made using an argument based on the rules of probability, due attention being paid to the way in which the experiment is designed. By guessing we shall mean that each of the possible sets of responses is to have equal probability.

Suppose that the psychologist offers ten cards and the information that five of them are red and five are black. To behave in a way consistent with this information the clairvoyant divides the cards into two groups of five, one of which he calls 'red', the other 'black'. The psychologist makes no comment while the cards are being sorted, but at the end finds that eight cards have been correctly classified. In such a situation, the more cards that are correctly classified, the more impressed we are, for it seems to indicate that the clairvoyant can indeed tell the cards. But should we be impressed? We need to know the probability of doing as well as this even if he were only guessing. In fact we wish to find the probability of doing *at least* as well as this for we want to divide the possible results of such an experiment into two sets, those which favour the clairvoyant and those which do not. How we perform this division in part depends on the risks we are prepared to take. If the clairvoyant is only guessing, then all $\binom{10}{5}$ ways of dividing the ten cards into two groups of five are assigned equal probability. We notice that only an even number of correct classifications are possible, for if a black card is called 'red' then a red card has to be called 'black', since there are to be five cards in each group. There is one way of classifying all the cards correctly. To obtain just eight correction classifications we can choose four 'reds' from five reds in $\binom{5}{4}$ ways, with each of which we can select one 'red' from five blacks in $\binom{5}{1}$ ways, the remaining cards being called 'black'. Thus there are $\binom{5}{4}\binom{5}{1}$ ways of classifying eight cards correctly and the probability of classifying at least eight correctly is

$$\frac{1+\binom{5}{4}\binom{5}{1}}{\binom{10}{5}} = \frac{13}{126}.$$

We *interpret* the calculation as follows. If this whole experiment were repeated a large number of times, in a proportion of about 13/126 of them (rather more than 1 in 10), the clairvoyant would classify at least eight correctly even if he were guessing. We are still left with the problem of asking ourselves whether we are impressed or not. It is usual in experimental work to propose some artificial level, by which such a probability is to be judged. This level may be fixed differently according to the field of application. A common level employed is

5%, or 1 in 20. Our calculated probability is greater than this, and hence is declared *not significant* at the 5% level. There is no question of proving or disproving the existence of the clairvoyant's talent.

The psychologist could have organized the test differently. For example, he might have presented the cards two at a time, together with the (true) information that one card was red while the other was black. The clairvoyant makes a response to each pair and he either classifies both correctly (R) or incorrectly (W) and these are assigned equal probability since one of each pair is known to be red. His five responses are scored in the form *RRWRW*, there being $2^5 = 32$ such sequences in all, each with equal probability 1/32. There is only one sequence *RRRRR* and altogether $\binom{5}{4} = 5$ sequences in which just four pairs are called correctly. Hence the probability that at least eight cards are classified correctly is $(1 + 5)/32 = 3/16$. Such a result is thus more likely in this *design* of the experiment and this makes sense since after all he is provided with *more* information namely that one of each pair is red, whereas previously he knew only that five of the ten were red.

The amount of information given is the key to modifying the design so that to classify at least eight cards correctly becomes more difficult. Suppose now that the psychologist merely gives the cards one at a time but does not announce how many reds and blacks there are. As far as the clairvoyant is concerned, each card, as it is presented, has equal probability of being black or red. The clairvoyant makes a sequence of ten responses of the type red, black, black . . . and there are 2^{10} such sequences, which if he is guessing, are assigned equal probability. Now the cards have been dealt in some order and there is just one sequence in which he classifies all ten cards correctly, just nine correctly in $\binom{10}{9} = 10$ sequences and eight correctly in $\binom{10}{8} = 45$ sequences. Hence the probability of at least eight correct classifications is $(1 + 10 + 45)/2^{10} = 7/128$. This is, naturally, less than the first design, and reflects the fact that he does not know that he should restrict himself to sequences containing five reds and five blacks.

Example 3

Suppose we wish to test whether a new drug aids the recovery of patients from a mental illness, when it is known that some patients recover spontaneously without any treatment at all. Starting with $m + n$ patients diagnosed as having the illness, we form them into two groups. One group of m patients gets no treatment and acts as a *control* or *comparison* group for the *treatment* group, which contains n patients. At the end of the experiment, the numbers of patients who recovered in the control and treatment groups are j and k respectively. If in fact the drug has no effect, then it is as if there were $j + k$ among the $n + m$ patients who were bound to recover anyway and k appear in the treatment group in virtue only of drawing a random sample of n from $n + m$. The probability of this is

$$\frac{\binom{j+k}{k}\binom{n+m-(j+k)}{n-k}}{\binom{n+m}{n}} = \frac{(j+k)!\,(n+m-j-k)!\,n!\,m!}{k!\,j!\,(n-k)!\,(m-j)!\,(n+m)!}.$$

This formula can be used to find the probability that at least k recover in the treatment group, subject to $j+k$ being a constant.

The information in the experiment is displayed in Table 3.1. From the table it can be seen that the probability of k recovering in the treatment group can be written as the product of the factorials of the row and column totals divided by the produce of the factorials of the table entries and the total number in the experiment.

Table 3.1

	Recovered	Not recovered	
Treatment group	k	$n-k$	n
Control group	j	$m-j$	m
	$k+j$	$n+m-k-j$	$n+m$

Such medical trials are full of difficilties. How, for instance, are the groups to be made up? There may be sources of *bias* which distort the results of the experiment. This could happen if, say, the illness varied in intensity and on moral grounds, the n worst cases received the drug or if the chance of recovery is strongly related to age and for administrative reasons, n younger patients from one ward are put *en bloc* into the treatment group. For the previous calculation to be valid, and to guard against bias, the n patients for the treatment group should be chosen at random from the $n+m$ available. Other matters which have been ignored are the possibility that patients drop out, the necessity for a time limit so that some patients are recorded as 'not recovered' though they might have if more time had been available, and the problem of measuring recovery as though it were an all-or-nothing condition.

3.4 CONDITIONAL PROBABILITIES

In an earlier example we showed that if two cards are drawn one at a time without replacement, then the probability that the second card is a club is 13/52, an answer that took account of the first card not being restricted. If, however, a first card is drawn and seen to be a club, the probability that the second card is a club becomes 12/51. This probability is called the *conditional probability* that the second card is a club *given* that the first card is a club. For two events E_2, E_1, the conditional probability of E_2 given E_1 is written $\Pr(E_2 | E_1)$. The above calculation could also have been carried out on the sample space containing 52×51 points of equal probability corresponding to all possible ordered pairs. The event E_1, 'a club is drawn first' contains 13×51 points and when we ask for the probability of a club on the second draw (E_2), we are in fact regarding E_1 as a sample space containing 13×51 points of equal probability. There are 13×12 points in this space, for which a club is second hence $\Pr(E_2 | E_1) = (13 \times 12)/(13 \times 51) = 12/51$, as before. The last ratio, $(13 \times 12)/(13 \times 51)$, can also be written

$$\frac{(13 \times 12)/(52 \times 51)}{(13 \times 51)/(52 \times 51)} = \frac{\Pr(E_2 \text{ and } E_1)}{\Pr(E_2)},$$

where the probabilities are evaluated on the original sample space. We take $\Pr(E_2 | E_1) = \Pr(E_2 \text{ and } E_1)/\Pr(E_1)$ as the definition of the conditional probability of E_2 given E_1 and applicable to sample spaces other than those which contain a finite number of points. From the frequency point of view such a definition is supported by the following consideration. If an experiment is repeated n times and on r_2 occasions E_2 is observed, on r_1 occasions E_1 is observed and there are r_{12} occasions when both E_1 and E_2 are noted, then the relative frequencies of E_1, E_2 are r_1/n, r_2/n respectively. But we may also consider the *relative* frequency of E_2 among the occasions when E_1 occurred, and this is

$$\frac{r_{12}}{r_1} = \frac{r_{12}/n}{r_1/n},$$

which under statistical regularity will be ultimately $\Pr(E_1 \text{ and } E_2)/\Pr(E_1)$ as n increases while r_{12}/r_1 corresponds to $\Pr(E_2 | E_1)$, since r_1 increases as n increases. Further, E_1 and E_2, regarded as an event in S, and E_2 given E_1, regarded as an event in E_1, are the same set of points, and hence we can reasonably require $\Pr(E_2 | E_1)$ to be proportional to $\Pr(E_2 \text{ and } E_1)$. That is, $\Pr(E_2 | E_1) = k \Pr(E_2 \text{ and } E_1)$ and by choosing E_2 as E_1,

$$1 = \Pr(E_1 | E_1) = k \Pr(E_1 \text{ and } E_1) = k \Pr(E_1), \quad k = 1/\Pr(E_1).$$

Definition. Provided $\Pr(E_1) \neq 0$, then

$$\Pr(E_2 | E_1) = \Pr(E_2 \text{ and } E_1)/\Pr(E_1), \tag{3.1}$$

while if $\Pr(E_1) = 0$, then also $\Pr(E_2 | E_1) = 0$. We may interchange E_1 with E_2 throughout. ∎

We immediately use the definition to obtain

$$\Pr(E_2 | E_1) = \frac{\Pr(E_2 \text{ and } E_1)}{\Pr(E_1)} = \frac{\Pr(E_1 | E_2)\Pr(E_2)}{\Pr(E_1)}. \tag{3.2}$$

Further,

$$\Pr(E_3 \text{ and } E_2 \text{ and } E_1) = \Pr[E_3 \text{ and } (E_2 \text{ and } E_1)]$$

$$= \Pr(E_3|E_2 \text{ and } E_1)\Pr(E_2 \text{ and } E_1)$$

$$= \Pr(E_3|E_2 \text{ and } E_1)\Pr(E_2|E_1)\Pr(E_1), \quad (3.3)$$

this last being a kind of 'chain rule' which is easily generalized. Lastly, if $E_1 \ldots E_k$ are k mutually exclusive and exhaustive events in a sample space, then every event A is the union of the intersections of A with the E_i. Hence,

$$\Pr(A) = \sum_{i=1}^{k} \Pr(A \text{ and } E_i) = \sum_{i=1}^{k} \Pr(A|E_i)\Pr(E_i).$$

Now,

$$\Pr(A|E_i) = \frac{\Pr(E_i|A)\Pr(A)}{\Pr(E_i)},$$

from (3.2), thus

$$\Pr(E_i|A) = \frac{\Pr(A|E_i)\Pr(E_i)}{\Pr(A)} = \frac{\Pr(A|E_i)\Pr(E_i)}{\sum_{i=1}^{k} \Pr(A|E_i)\Pr(E_i)}. \quad (3.4)$$

The formula for $\Pr(E_1 \text{ or } E_2)$ can be written, using a conditional probability

$$\Pr(E_1 \text{ or } E_2) = \Pr(E_1) + \Pr(E_2) - \Pr(E_1 \text{ and } E_2)$$

$$= \Pr(E_1) + \Pr(E_2) - \Pr(E_1|E_2)\Pr(E_2)$$

$$= \Pr(E_1) + \Pr(E_2) - \Pr(E_2|E_1)\Pr(E_1).$$

We emphasize that conditional probabilities are but probabilities, and obey the same rules, thus

$$\Pr(E_2 \text{ or } E_3|E_1) = \Pr(E_2|E_1) + \Pr(E_3|E_1) - \Pr(E_2 \text{ and } E_3|E_1)$$

$$= \Pr(E_2|E_1) + \Pr(E_3|E_1) - \Pr(E_2|E_3 \text{ and } E_1)\Pr(E_3|E_1).$$

Example 4

A bag contains one white ball and two red balls. A ball is drawn at random which if white is replaced together with an extra white ball, but which if red is replaced together with two extra red balls. A second ball is then drawn at random; find the probability that it is red. This calculation has to be carried out on the basis that the result of the first draw is not known. Let R_i, W_i be the events that the colour of the ball on the ith draw was red or white respectively. Then the probability of the second ball being red is made up of the two exclusive and only possible cases, when the first ball is white and when it is red.

$$\Pr(R_2) = \Pr(R_2 \text{ and } R_1 \text{ or } R_2 \text{ and } W_1)$$

$$= \Pr(R_2 \text{ and } R_1) + \Pr(R_2 \text{ and } W_1)$$

$$= \Pr(R_2 | R_1) \Pr(R_1) + \Pr(R_2 | W_1) \Pr(W_1)$$

$$= \frac{4}{5} \times \frac{2}{3} + \frac{2}{4} \times \frac{1}{3}$$

$$= \frac{7}{10}.$$

This is greater than $\Pr(R_1) = 2/3$, which is not surprising since there are more red balls in the bag to start with and hence it is more likely that the number of reds will increase. We have also $\Pr(W_2) = 1 - \Pr(R_2) = 3/10$, without any other intermediate calculation.

Suppose next that the result of the first draw remains unknown but the second ball drawn is found to be red. We wish to find $\Pr(R_1 | R_2)$. Now

$$\Pr(R_1 | R_2) = \frac{\Pr(R_1 \text{ and } R_2)}{\Pr(R_2)} = \frac{\Pr(R_2 | R_1) \Pr(R_1)}{\Pr(R_2)}$$

$$= \frac{\dfrac{4}{5} \times \dfrac{2}{3}}{\dfrac{7}{10}} = \frac{16}{21}$$

and this is greater than $\Pr(R_1)$, and again

$$\Pr(W_1 | R_2) = 1 - \Pr(R_1 | R_2) = 1 - \frac{16}{21} = \frac{5}{21}.$$

Neither $\Pr(R_1 | R_2)$ nor $\Pr(R_2 | R_1)$ can be found without knowing the composition of the balls at the start. Lacking this information, other, perhaps unwarranted, assumptions must be made. Our next example brings this out.

Example 5

A bag contains two balls, each of which may be either red or white. A ball is drawn at random and found to be red. What is the probability that the other ball is red? To start with, the bag could contain 0, 1 or 2 red balls. Everything turns on what we are prepared to assume about the probabilities of these conjectures. Before *any* ball is drawn, it might be argued that these hypotheses H_0, H_1, H_2 are equally likely to be true and each should have probability of 1/3. In which case, if $\Pr(R)$ means probability of drawing a red

$$\Pr(R) = \sum_{i=0}^{2} \Pr(R \text{ and } H_i) = \sum_{i=0}^{2} \Pr(R|H_i) \Pr(H_i)$$

$$= 0 \times \frac{1}{3} + \frac{1}{2} \times \frac{1}{3} + 1 \times \frac{1}{3} = \frac{1}{2}.$$

If in fact a red is drawn, we can recalculate the probabilities of H_0, H_1, H_2 conditional on that event.

$$\Pr(H_0|R) = \frac{\Pr(R|H_0)\Pr(H_0)}{\Pr(R)} = 0,$$

which reflects that if a red is drawn then there could not have been two whites in the bag.

$$\Pr(H_1|R) = \frac{\Pr(R|H_1)\Pr(H_1)}{\Pr(R)} = \frac{\dfrac{1}{2} \times \dfrac{1}{3}}{\dfrac{1}{2}} = \frac{1}{3}$$

$$\Pr(H_2|R) = \frac{\Pr(R|H_2)\Pr(H_2)}{\Pr(R)} = \frac{1 \times \dfrac{1}{3}}{\dfrac{1}{2}} = \frac{2}{3}.$$

If $1/3, 1/3, 1/3$ are regarded as the *prior probabilities* of H_0, H_1, H_2, then having drawn a red, $0, 1/3, 2/3$ may be called the *posterior probabilities* of H_0, H_1, H_2.

[It may be objected that if a red is drawn, then the prior probabilities of H_1, H_2 should be revised. Suppose these are taken to be $1/2, 1/2$ respectively. In this case

$$\Pr(R) = \frac{1}{2} \times \frac{1}{2} + 1 \times \frac{1}{2} = \frac{3}{4}.$$

We still calculate the posterior probabilities since taking a ball at random and finding it to be red contains more information than that the two balls are not both white.

$$\Pr(H_1|R) = \frac{\dfrac{1}{2} \times \dfrac{1}{2}}{\dfrac{3}{4}} = \frac{1}{3}, \quad \Pr(H_2|R) = \frac{1 \times \dfrac{1}{2}}{\dfrac{3}{4}} = \frac{2}{3}$$

and the posterior probabilities are as before.]

The assumption that the prior probabilities should be $1/3$ for each of the H_i corresponds to all proportions of red being equally likely. There are plenty of other assumptions one might make, a plausible one being that each ball has equal probability of being white or red. This means that *WW, RW, WR, RR* are to have equal probability. The prior probabilities of H_0, H_1, H_2 are then 1/4, 1/2, 1/4 and

$$\Pr(R) = 0 \times \frac{1}{4} + \frac{1}{2} \times \frac{1}{2} + 1 \times \frac{1}{4} = \frac{1}{2}$$

$$\Pr(H_0|R) = 0, \quad \Pr(H_1|R) = \frac{1}{2}, \quad \Pr(H_2|R) = 1 - \Pr(H_1|R) = \frac{1}{2}.$$

When so applied, (3.4) is known as Bayes' formula

Example 6

We now return to Example 4, and take it one stage further. That is, after the first draw has been completed, a second ball is drawn at random, if white it is replaced together with another white, if red it is replaced together with two more reds. A third ball is now drawn at random. What is the probability that it is red? The advantages of splitting up the event R_3 into mutually exclusive and only possible cases will now be apparent.

$$\Pr(R_3) = \Pr(R_3 R_2 R_1) + \Pr(R_3 R_2 W_1) + \Pr(R_3 W_2 R_1)$$
$$+ \Pr(R_3 W_2 W_1)$$
$$= \Pr(R_3|R_2 R_1) \Pr(R_2|R_1) \Pr(R_1)$$
$$+ \Pr(R_3|R_2 W_1) \Pr(R_2|W_1) \Pr(W_1)$$
$$+ \Pr(R_3|W_2 R_1) \Pr(W_2|R_1) \Pr(R_1)$$
$$+ \Pr(R_3|W_2 W_1) \Pr(W_2|W_1) \Pr(W_1).$$

Working out the probabilities for such an expression is made easier by the construction of a simple tree diagram showing the possible stages (Fig. 3.1). (The first number in each bracket shows the number of whites, the second the number of reds.)

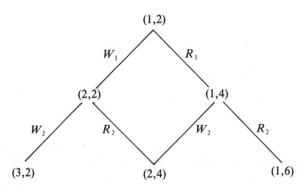

Fig. 3.1

$$\Pr(R_3) = \left(\frac{6}{7} \times \frac{4}{5} \times \frac{2}{3}\right) + \left(\frac{4}{6} \times \frac{2}{4} \times \frac{1}{3}\right) + \left(\frac{4}{6} \times \frac{1}{5} \times \frac{2}{3}\right)$$

$$+ \left(\frac{2}{5} \times \frac{2}{4} \times \frac{1}{3}\right)$$

$$= \frac{16}{35} + \frac{1}{9} + \frac{4}{45} + \frac{1}{15} = \frac{76}{105}.$$

Two comments seem worth making. While it is true that W_1 followed by R_2 gives the same composition as R_1 followed by W_2, these paths do not have the same probability. Also, since R_3 must be preceded either by R_2 or W_2, there is a temptation to state

$$\Pr(R_3) = \Pr(R_3 R_2) + \Pr(R_3 W_2) = \Pr(R_3|R_2)\,\Pr(R_2)$$

$$+ \Pr(R_3|W_2)\,\Pr(W_2).$$

However, $\Pr(R_3|R_2)$ is not uniquely defined, and in fact depends on whether R_1 or W_1 preceded R_2.

3.5 INDEPENDENT EVENTS

We have defined the conditional probability of A given B as $\Pr(A|B) = \Pr(A$ and $B)/\Pr(B)$, when $\Pr(B) \neq 0$. Now, it may happen that $\Pr(A|B) = \Pr(A)$, in which case we say that A is **independent** of B. Since

$$\Pr(A|B) = \Pr(B|A)\,\Pr(A)/\Pr(B),$$

if A is independent of B, then $\Pr(B|A) = \Pr(B)$, or B is independent of A. We need therefore only speak of A and B being independent. The sense of this concept is that the information 'B has happened' does not alter the probability of A occurring. In other words, B is redundant as far as A is concerned. This is not generally the case, extreme examples being when B precludes A or $\Pr(A|B) = 0$ and when B always implies A when $\Pr(A|B) = 1$. Should A and B happen to be independent

$$\frac{\Pr(A \text{ and } B)}{\Pr(B)} = \Pr(A|B) = \Pr(A)$$

that is,

$$\Pr(A \text{ and } B) = \Pr(A)\,\Pr(B), \tag{3.5}$$

a result which is extremely useful and could be the starting-point for the definition of independence.

Problem 4. If A is independent of B, show that A is also independent of not B.

Example 7

If two fair dice are rolled, then the outcome on one cannot influence the outcome on the other. For instance

$$\text{Pr(both show six)} = \text{Pr(first shows six)} \times \text{Pr(second shows six)}$$

$$= \frac{1}{6} \times \frac{1}{6} = \frac{1}{36}$$

which checks with the 36 possible pairs of results being equiprobable. Also,

$$\text{Pr(either shows a six)} = \frac{1}{6} + \frac{1}{6} - \frac{1}{36}.$$

Example 8

If two balls are drawn *with replacement* from a bag containing three white and seven red balls, what is the probability that both are red? Since the ball is replaced, the system returns to its starting point and the event 'the ball is red' is independent of the draw number.

$$\text{Pr(both red)} = \frac{7}{10} \times \frac{7}{10} = \frac{49}{100}.$$

This solution may be contrasted with one involving the 10^2 ordered equally likely outcomes for which the event 'both are red' contains 7^2 points — a method which becomes increasingly cumbersome.

Example 9

Suppose a card is drawn at random from a pack, then the probability that it is black is 26/52. Now the probability that it is an ace is 4/52 and that it is a black ace is 2/52. Since $4/52 \times 26/52 = 2/52$, 'black' and 'ace' are independent events. In fact, the proportion of black among the aces is the same as the proportion of black cards.

   If we are dealing with three events A, B, C, then it could happen that these events are independent by pairs, and yet A fails to be independent of the event (B and C). Examples of this do not seem to appear in practice, though artificial ones may be constructed. If in fact A is also independent of (B and C) then it may be verified that so is B of (A and C) and is C of (A and B) and

$$\text{Pr}(A \text{ and } B \text{ and } C) = \text{Pr}(A)\,\text{Pr}(B)\,\text{Pr}(C). \tag{3.6}$$

The terminology of conditional probabilities becomes so cumbersome as the number of events increases that although they appeared in the original definition we state an alternative definition.

Definition. The events $E_1, E_2, \ldots, E_n$ are said to be independent if for every selection of k of them ($k = 1, \ldots, n$), the probability of all the members of the selection happening is the product of the probabilities of the members of the selection happening separately.

3.6 SAMPLING WITH REPLACEMENT

A bag contains m balls, k of which are red, the remainder being white. A random sample of n balls is drawn one at a time *with replacement*. To find the probability that the sample contains just r red balls, we first argue in terms of a sample space generaged by all possible sequences of drawings. To this end, it is helpful to think of the balls being numbered. There are m^n different ordered samples with equal probability and $k^r(m-k)^{n-r}$ of these for which there are just r reds and $n-r$ whites in some particular order. But there are $\binom{n}{r}$ distinguishable arrangements of r reds and $n-r$ whites, hence the probability of just r reds is

$$\frac{\binom{n}{r}k^r(m-k)^{n-r}}{m^n} = \binom{n}{r}\left(\frac{k}{m}\right)^r\left(\frac{m-k}{m}\right)^{n-r} = \binom{n}{r}p^r(1-p)^{n-r},$$

where $k/m = p$. Now the probability that the ith ball drawn is red can be obtained as follows. There are $m^{n-1}k$ ordered samples of size n in which the ith ball is red, regardless of the outcomes of the other drawings. Hence $\Pr(R_i) = m^{n-1}k/m^n = k/m = p$, and $\Pr(W_i) = 1 - k/m = 1 - p = q$, say. Consider any particular sequence of draws which contains just r reds and $n-r$ whites, then we have already shown that the probability of such a sequence is $p^r q^{n-r}$. But this is the product of r probabilities p that there will be a red on r particular draws and $n-r$ probabilities q that there will be a white on the remaining draws.

We now obtain the same formula from another point of view. By the way the experiment is carried out, the events 'red', 'white' on different draws are independent. Hence the probability of a particular sequence of r reds and $n-r$ whites is $p^r q^{n-r}$. But there are $\binom{n}{r}$ distinguishable sequences of r reds and $n-r$ whites. Hence the probability of just r reds is $\binom{n}{r}p^r q^{n-r}$. Thus for independent and uniform trials the construction of a sample space for the whole series is not necessary.

Example 10

A box contains five balls, two of which are red and three are white. Two balls are drawn one at a time with replacement and their colour noted. Since the drawings are independent, $\Pr(R_1) = \Pr(R_2) = 2/5$, $\Pr(W_1) = \Pr(W_2) = 3/5$. From this information we find (a) Pr(both red); (b) Pr(just one red); and (c) Pr(at least one red).

(a) $\Pr(R_1 \text{ and } R_2) = \Pr(R_1)\Pr(R_2) = \dfrac{2}{5} \times \dfrac{2}{5} = \dfrac{4}{25}$

(b) $\Pr(\text{just one red}) = \Pr(R_1 \text{ and } W_2) + \Pr(W_1 \text{ and } R_2)$

$$= \left(\frac{2}{5} \times \frac{3}{5}\right) + \left(\frac{3}{5} \times \frac{2}{5}\right) = \frac{12}{25}$$

(c) $\Pr(\text{at least one red}) = 1 - \Pr(\text{no red}) = 1 - \Pr(\text{both white})$

$$= 1 - \left(\frac{3}{5} \times \frac{3}{5}\right) = \frac{16}{25}.$$

Alternatively

$$\Pr(\text{at least one red}) = \Pr(R_1 \text{ or } R_2)$$

$$= \Pr(R_1) + \Pr(R_2) - \Pr(R_1 \text{ and } R_2)$$

$$= \frac{2}{5} + \frac{2}{5} - \left(\frac{2}{5} \times \frac{2}{5}\right) = \frac{16}{25}.$$

If the sequence of drawing pairs is repeated three times

$$\Pr(\text{just two red pairs}) = \binom{3}{2}\left(\frac{4}{25}\right)^2\left(1 - \frac{4}{25}\right).$$

While

$$\Pr(\text{just two red pairs and one white pair}) = \binom{3}{2}\left(\frac{4}{25}\right)^2\left(\frac{9}{25}\right).$$

Example 11
A fair die is tossed eight times; calculate the probability that on the eighth throw the third 'six' is observed.

On the previous seven throws, just two sixes are obtained with probability

$$\binom{7}{2}\left(\frac{1}{6}\right)^2\left(\frac{5}{6}\right)^5.$$

The probability of another 'six' on the eighth throw is 1/6 independently of previous outcomes. Hence probability is

$$\binom{7}{2}\left(\frac{1}{6}\right)^3\left(\frac{5}{6}\right)^5.$$

Problem 5. Show that if a random sample of n articles is drawn from a large number of articles of which a proportion p is defective, the chance of the sample containing r defective articles is

$$\binom{n}{r} p^r (1-p)^{n-r}$$

A large batch of manufactured articles is accepted if either of the following conditions is satisfied:

(a)   A random sample of 10 articles contains no defective articles.
(b)   A random sample of 10 contains one defective article and a second random sample of 10 is then drawn which contains no defective articles.

Otherwise the batch is rejected.

If in fact, 5% of the articles in a batch to be examined are defective find the chance of the batch being accepted. $(0.95)^{10} = 0.5987$.

Oxford & Cambridge, A.L., 1964.

Problem 6. Two integers, i, j, are drawn one at a time, without replacement from the set $1, 2, 3, \ldots, n$. Calculate $\Pr[i \leqslant k | j \leqslant k]$ and $\Pr[j \leqslant k | i \leqslant k]$.

Problem 7
(i) A, B and C are events with A and C exclusive, B and C independent, and probabilities $P(A) = 1/12$, $P(B) = 1/3$, $P(C) = 1/4$ and $P(A \cap \bar{B}) = 0$. Calculate $P(A \cap B \cap C)$, $P(A \cap B)$, $P(A \cup B)$, $P(A|B)$, $P(B \cap C)$ and $P(A \cup B \cup C)$.
(ii)   Box A contains 3 blue beads, 2 green and 2 white. Box B contains 2 blue beads, 3 green and 4 white. One bead is drawn from each box. Given that one of the beads is blue, what is the probability that the other is white?

University of Surrey, 1983.

Problem 8. It is known from past experience of carrying out surveys in a certain area that one-fifth of the houses will be unoccupied during the day. In a proposed survey it is planned that houses unoccupied will be visited a second time and a member of the survey team produces the following rough calculations.

The proportion of homes that will be occupied at either the first or second visit $= 1 - (1/5)^2 = 24/25$. Another member of the team argues that the probability of no one being in at the first or second visit is $2(2/5)$ and that the answer should have been $1 - (2/5) = 3/5$. Examine each of the calculations and explain carefully the assumptions that have been made and the laws of probability that have been used. Where you disagree with the assumptions explain why and indicate if necessary what further information you would require to calculate the desired proportion.

London University, B.Sc. Gen., pt. I, 1967.

Problem 9. Discuss the following incorrect arguments, explaining carefully in each case the nature of the mistake and giving the correct analysis:

(a)   In a population of (two-eyed) men the probability that a man's left eye is brown is p, and the probability that a man's right eye is brown is also p. The probability that a man has at least one brown eye is, by the addition law of probability,

prob(left eye brown or right eye brown)

= prob(left eye brown) + prob(right eye brown)

= $2p$.

(b)   To be eligible for a certain type of work, a man must be above both a certain minimum weight and a certain minimum height. The separate probabilities that these conditions are satisfied are p_w and p_h respectively. Therefore for a man selected at random the probability that he is eligible is, by the multiplication law of probability,

prob(weight satisfactory and height satisfactory)

= prob(weight satisfactory) prob(height satisfactory)

= $p_w p_h$.

(c)   A subject guesses the answers to three 'true–false' questions, and has independently for each question an equal chance of being right or wrong. There are four possibilities for the total number of correct answers, namely 0, 1, 2, 3 right and therefore, since all possibilities are equally likely, the probability is 1/4 that all three questions are answered correctly.

London University, B.Sc. Gen., pt. I, 1965.

Problem 10. Three players x, y, z play a game and at each round of the game they have an equal probability of winning that round. The player who wins each round scores one point and the game is won by the *first* person to score a total of 3 points. x wins the first round; calculate the probability that y wins the game.

Oxford & Cambridge (S.P.), 1968.

Problem 11. In a game, a player wins a game if he is the first to score 9 points, provided that the score has not been 8 points each. If the score reaches 8–8, then a subsequent difference of two clear points decides the winner. For a particular player, the probability of winning any point is p, regardless of any other circumstance. Calculate the probability that n points have to be decided for him to win the game. Show that the probability of the *game* remaining undecided is zero.

Problem 12. Show that the number of ways in which a set of n elements can be divided into r labelled subsets containing $n_1, \ldots, n_r$ elements where $n_1 + \ldots + n_r = n$, is the multinomial coefficient $\begin{pmatrix} n \\ n_1, n_2, \ldots, n_r \end{pmatrix}$. State and prove the multinomial theorem. Six fair dice are rolled independently. Show that

(i)   the probability of seeing two pairs and two singletons (two scores appearing twice each, two others one each) is $16\,200/6^6$ ($= 0.347$),

(ii) the probability of seeing one pair and four singletons is $10\,800/6^6$ ($= 0.231$).

Westfield College, 1983.

Problem 13. State Bayes theorem.

A pattern recognition machine outputs one of a set of responses $R_1, R_2, \ldots, R_N$ as a result of receiving one of a set of input signals $S_1, S_2, \ldots, S_N$. In theory, R_i should be output whenever signal S_i is received, $i = 1, 2, \ldots, N$, but, in fact, this only occurs with probability $p(0 < p < 1)$. It may be assumed that each signal is processed independently by the machine and that if an error is made the wrong response is selected at random.

One of the signals is chosen at random by an experimenter and fed through the machine on two separate occasions. If the response R_i resulted in both cases, show that the probability of the signal being S_i is given by

$$\frac{(N-1)p^2}{(1 - 2p + Np^2)} \, .$$

What would this probability be if the machine selected R_i on the first occasion but $R_j (j \neq i)$ on the second? Comment on these two probabilities in the case where N is large.

University College, 1977.

3.7 THE PROBABILITY OF AT LEAST ONE EVENT

We have frequently used the following for two events, E_1, E_2

$$\Pr(E_1 \text{ or } E_2) = \Pr(E_1) + \Pr(E_2) - \Pr(E_1 \text{ and } E_2).$$

The corresponding result for n events $E_1, E_2, \ldots, E_n$ is

$$\Pr(E_1 \text{ or } E_2 \text{ or } E_3, \ldots, \text{ or } E_n) = \Sigma \Pr(E_i) - \Sigma \Pr(E_i \text{ and } E_j)$$

$$+ \Sigma \Pr(E_i \text{ and } E_j \text{ and } E_k) \ldots$$

$$(-1)^{n-1} \Pr(E_1 \text{ and } E_2 \ldots E_n),$$

where the summation for m simultaneous events is only over the $\binom{n}{m}$ selection of m different suffices from n. This may be proved by induction.

A simplification is noted when the probability of the simultaneous happening of m events is the same for every set of m different events, say P_m. Since there are $\binom{n}{m}$ selections of m events from n we have

$$\Pr(E_1 \text{ or } E_2 \ldots \text{ or } E_n) = \binom{n}{1} P_1 - \binom{n}{2} P_2 \ldots$$

$$+ (-1)^{n-1} \binom{n}{n} P_n.$$

Using

$$\Pr(E_1 \text{ or } E_2 \ldots \text{ or } E_n) = \Pr(\text{at least one of } E_1, E_2, \ldots, E_n)$$

this provides one way of computing $\Pr(\text{none of the events})$ as $1 - \Pr(\text{at least one of the events})$.

Example 12

If r different balls are placed at random in n different cells, find the probability that no cell is empty $(r > n)$. All distributions have equal probability $1/n^r$. Any particular set of m cells will be empty in $(n - m)^r$ different ways since the r balls are distributed in all possible ways in the remaining $(n - m)$ cells. Hence the probability that these m cells at least are empty is $(n - m)^r/n^r$.

$$\Pr(\text{no cell empty}) = 1 - \Pr(\text{at least one cell empty})$$

$$= 1 - \binom{n}{1}\left(\frac{n-1}{n}\right)^r + \binom{n}{2}\left(\frac{n-2}{n}\right)^r + \ldots$$

$$(-1)^{n-1} \binom{n}{n-1}\left(\frac{1}{n}\right)^r.$$

Problem 14. Two similar packs of n different cards are randomly shuffled and laid out side by side. Show that the probability of no matching pairs is

$$\sum_{k=0}^{n} \frac{(-1)^k}{k!}.$$

Show also that for three similar packs, the probability of no matching triple is

$$\sum_{k=0}^{n} (-1)^k \frac{(n-k)!}{n!k!}.$$

3.8 INFINITE SEQUENCE OF INDEPENDENT TRIALS

We have managed to avoid, so far, discussing what happens if the sequence of independent trials may be infinite. Consider a sequence of independent ordered trials on each of which there is a constant probability of success equal to p, where $0 < p < 1$. Suppose now we wish to calculate the probability that the kth trial is the *first* success. At first sight we seem to have an immediate and simple answer. For the kth trial will be the first success when it is preceded by $k-1$ failures, but the trials are independent and hence the required probability is $q^{k-1}p$, and indeed this sensible answer is the one we wish to keep. Thence we may calculate the probability that the first success occurs at or before the kth trial as

$$\sum_{j=1}^{k} q^{j-1}p = p \frac{(1-q^k)}{1-q} = 1 - q^k.$$

Furthermore, we calculate, formally anyhow, the probability that the first success occurs *after* the kth trial as

$$\sum_{j=k+1}^{\infty} q^{j-1}p = \frac{pq^k}{1-q} = q^k$$

since $0 < q < 1$.

We note with satisfaction that since $(1 - q^k) + q^k = 1$, the probability of getting a success on some trial is 1. This agreeable feeling diminishes somewhat when we ask ourselves what we mean by the event 'first success comes after the kth trial'. Did we then after all envisage an infinite sequence of trials for which such an event has meaning? The calculation above involved the multiplication of the probabilities of independent events although the case when this was justified involved only a finite number of trials. Now, for any fixed integer k we may choose another integer j greater than k and consider a sample space in connection with j independent trials. We consider the space for which the label for a point consists of a sequence of j symbols each of which may be S, for success, or F for failure. Since the trials are independent, to such a point, we assign probability $p^r q^{j-r}$, where r is the number of Ss in the sequence. The event 'the kth trial is the first success' consists of those points for which the sequence begins with $k-1$ Fs followed by S followed by any distinguishable sequence of $j-k$ letters F or S. Hence the probability of the event is

$$q^{k-1}p\sum_{r=0}^{j-k}\binom{j-k}{r}p^r q^{j-k-r} = q^{k-1}p(p+q)^{j-k} = q^{k-1}p$$

This is the same result as obtained above and does not depend on j, *provided* $j > k$. The event 'the first success happens after the kth trial' has probability

$$\sum_{i=k+1}^{j}q^{i-1}p = p\,\frac{(q^k - q^j)}{1-q} = q^k - q^j$$

and this does depend on j. However, since $0 < q < 1$, the limit as j tends to infinity of this probability exists and is q^k. This result allows us, in the present problem, to speak of the probability of the event 'first success after the kth trial', in the sense of 'the limit of the probability of the first success after the kth trial in j trials as j tends to infinity'. In this and other cases like it, this is what we *shall* mean and it is not intended to examine situations where the ambiguity involved leads to any difficulty. We discuss briefly what has been evaded by taking the limit just considered. The event 'first success happens after trial k' can only be defined on a sample space generated by an infinite sequence of trials. The sequence which is the label or description for a point in this space now consists of an infinite sequence of symbols. If, however, the probability assigned to such a sequence is the product of the probabilities of the constituent successes and failures, then the product has limit *zero*, since each term in it is less than one. Thus by using the independence rule mechanically, we end up by assigning probability zero to each point in the sample space. In fact there is no way of assigning positive probability to each point in the space because there is more than a countable infinity of such points. This last point may be seen as follows. Each sequence of symbols may be regarded as a decimal in the binary scale by replacing S by 1 and F by 0. Thus .010101 ... is an alternating sequence of F with S and in the scale of ten is the number

$$\frac{1}{2^2} + \frac{1}{2^4} + \ldots = \frac{1}{3}.$$

Every such sequence represents a point in the closed interval $[0, 1]$. Moreover, every real number x in the interval has such a representation by the method of bisection of intervals. That is, we bisect the interval $[0, 1]$, then x is either in the left or right half of the interval. If the right half, write down one, if the left write down zero. Now bisect the half in which x lies and repeat. Eventually x will either be the only point contained in the intersection of all such intervals or x lies on one of the points of subdivision. In the latter case, a choice of representation is available. For example, the number 1/4 may be represented either as .0100... or as .00111 .... In any case there are at least as many points in our sample space as there are in the closed interval $[0, 1]$ and this number is known to be uncountable. This suggests that though we can't assign a positive probability to each point, perhaps we can to 'sets' of these sequences. We sketch a method suitable for p rational (Fig. 3.2). Divide the interval $[0, 1]$ in the ratio $p : q$ then divide each of the intervals so formed in the ratio $p : q$, and so on. Any infinite sequence of successes and failures is assigned a point in the interval $[0, 1]$ according to the following rule. If the first trial results in a success choose the left of the first two intervals, if a failure choose the right. Whichever interval we are in is itself divided into two intervals in the ratio $p : q$. Now take the second trial and repeat the choice. Eventually the intersection of all such intervals will define a point which is to be the one assigned to the particular sequence of trials. Every infinite sequence that commences with a success will be assigned a point in the interval $[0, p]$ and it therefore seems not unreasonable to assign to the collection of all such sequences, a probability equal to the length of the interval, namely p. In the same way, any sequence that begins failure, success will be assigned a point in the interval $[p, p + qp]$ and the length of this interval is qp.

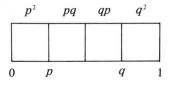

Fig. 3.2

We conclude this section with a few remarks on the notion of a countable set. A set is said to be countable if its elements can be counted in the sense that one element may be called the first, another the second, and so on, such that every element must appear somewhere in the listing in a place which can be identified. Clearly, every set which contains a finite number of elements is countable. The set of integers $1, 3, 5, \ldots$ is countable, since $2r + 1$ can be put in place r. Moreover, the set of all integers is countable since 0 may be put first, $+r$ in position $2r$ and $-r$ in position $2|r| + 1$. More surprisingly, the set of all fractions between zero and one is countable. On the other hand, the set of all numbers between zero and one is not countable.

Problem 15. Player A is about to serve in a game of badminton against B. If A wins this rally, he will score 1 point and can serve again. This continues until A loses a rally, when neither player scores, but B gains the right to serve. The game then continues in the same way but with B serving, and so on. The outcome of each rally depends only on which player serves and is otherwise independent of all previous rallies.

The probability that either player wins a rally in which he has served is $p(\neq 0, 1)$ while the probability that the non-server wins is $q(= 1 - p)$. Find the probability that A wins exactly r rallies before losing the right to serve. Find also the average number of points he will have gained before losing the right to serve.

The probability that A, serving first, scores the first point is f. By considering the result of the first rally, or otherwise, show that

$$f = \frac{1}{1+q}.$$

Show that the probability that B scores exactly 2 of the first 3 points scores is

$$\frac{q(1+q+q^2)}{(1+q)^3}.$$

Oxford & Cambridge, A.L., 1979.

Problem 16. A chess match between two grandmasters A and B is won by whomever first wins a total of 2 games. A's chances of winning, drawing or losing any particular game are p, q, r respectively. The games are independent and $p + q + r = 1$. Show that the probability that A wins the match after $(n+1)$ games $(n \geq 1)$ is

$$p^2 [nq^{n-1} + n(n-1) rq^{n-2}].$$

By considering suitable operations on the infinite geometric series $(q + q^2 + \ldots)$, or otherwise, show that the probability that A wins the match is $p^2(p+3r)/(p+r)^3$.

Find the probability that there is no winner and comment on your result.

Oxford & Cambridge, M.E.1, 1977.

REFERENCE

[1] E. Parzen *Modern Probability Theory and its Applications*, John Wiley & Sons Inc., New York, 1960.

BRIEF SOLUTIONS AND COMMENTS ON THE PROBLEMS

Problem 1. The probability that the first sample contains i white balls is

$$\binom{K}{i}\binom{M-K}{n-i} \Big/ \binom{M}{n}.$$

If there are i whites, the probability that the second sample contains j whites is

$$\binom{i}{j}\binom{n-i}{m-j} \Big/ \binom{n}{m}.$$

After unwinding the binomial coefficients and a little cancelling the total probability is, when $K > n$,

$$\sum_{i=j}^{K} \left[\frac{K!(M-K)!}{(K-i)!(M-k-n+i)!} \times \right.$$

$$\left. \times \frac{1}{j!(i-j)!(m-j)!(n-i-m+j)!} \right] / \binom{M}{n}\binom{n}{m}.$$

The summation over i involves

$$\sum_{i=j}^{K} \binom{M-n}{K-i}\binom{n-m}{i-j} = \binom{M-m}{K-j}.$$

The final probability can be shown to be

$$\binom{K}{j}\binom{M-K}{m-j} / \binom{M}{m}$$

just as though a sample of m had been drawn immediately from the M available. On reflection, the first sample is seen to have no part to play — hence there is unnecessary complication here.

Problem 2. There is no letter A with probability $(2/3)^n$.
There is no letter A or B $\Rightarrow$ C, with probability $(1/3)^n$.
The packets must contain a coupon

$$\Pr(\bar{A} \cup \bar{B} \cup \bar{C}) = 3(2/3)^n - 3(1/3)^n.$$

Problem 3. The total number of selections is $\binom{a+b}{2}\binom{a+b}{2}$. Number of ways yielding no whites is $\binom{b}{2}^2$. For one white is $2\binom{a}{1}\binom{b}{1}\binom{b}{2}$, since the solitary white may appear either on the first or second draw. There may be two whites either if there is one at each draw or two in one and none in the other. Hence the total number of ways, of selecting two whites is

$$2\binom{a}{2}\binom{b}{2} + \left[\binom{a}{1}\binom{b}{1}\right]^2.$$

For three whites, $2\binom{a}{1}\binom{b}{1}\binom{a}{2}$ and finally for four, $\binom{a}{2}^2$. Hence the required probabilities can be immediately stated.

Problem 4.

$$\Pr(A) = \Pr(A \ \& \ B) + \Pr(A \ \& \ \text{not} \ B)$$

$$= \Pr(A|B)\Pr(B) + \Pr(A \ \& \ \text{not} \ B)$$

$$= \Pr(A)\Pr(B) + \Pr(A \ \& \ \text{not} \ B).$$

$$\Pr(A)[1 - \Pr(B)] = \Pr(A \& \text{ not } B)$$

$$\Pr(A)\Pr(\text{not } B) = \Pr(A \& \text{ not } B).$$

Problem 5. Probability of acceptance is

$$(0.95)^{10} + 10(0.95)^9 \, (0.05) \, (0.95)^{10} = 0.79.$$

Problem 6.

$$\Pr(j \leqslant k | i \leqslant k) = \Pr(i \leqslant k \text{ and } j \leqslant k)/\Pr(i \leqslant k)$$

$$= \frac{\binom{k}{2}/\binom{n}{2}}{\binom{k}{1}/\binom{n}{1}} = \frac{k-1}{n-1},$$

which, on reflection, should occasion no surprise. Indeed, given j is an integer not exceeding k, i may be one of $k - 1$ values from the remaining $n - 1$. Hence

$$\Pr(i \leqslant k | j \leqslant k) = (k - 1)/(n - 1).$$

Problem 7

(i) $\Pr(A \cap B \cap C) = 0$, since A and C exclusive

$\Pr(A) = \Pr(A \cap \bar{B}) + \Pr(A \cap B)$, $\Pr(A \cap B) = 1/12 \, (A \subset B)$.

$\Pr(A \cup B) = \Pr(A) + \Pr(B) - \Pr(A \cap B) = \Pr(B) = 1/3$.

$\Pr(A|B) = \Pr(A \cap B)/\Pr(B) = \Pr(A)/\Pr(B) = 1/4$.

$\Pr(B \cap C) = \Pr(B) \Pr(C) = 1/12$, since B, C independent.

$\Pr(A \cup B \cup C) = \Pr(A \cup B) + \Pr(C) - \Pr[(A \cup B) \cap C]$

$\qquad = 1/3 + 1/4 - 1/12 = 1/2.$

(ii) $\Pr(\text{least one blue}) = \dfrac{3}{7} + \dfrac{2}{9} - \dfrac{3}{7} \times \dfrac{2}{9} = \dfrac{35}{63} = \dfrac{5}{9}$

$\Pr(\text{one blue and one white}) = \dfrac{3}{7} \times \dfrac{4}{9} + \dfrac{2}{9} \times \dfrac{2}{7} = \dfrac{16}{63}$

$\Pr(\text{one blue and one white}|\text{least one blue}) = \dfrac{16}{63} \Big/ \dfrac{5}{9} = \dfrac{16}{35}.$

Problem 8. First calculation assumes that being unoccupied is an independent event for different visits. The second assumes that not being at home on first and second events are exclusive. People who go to work are often out, while others are nearly invariably at home.

Problem 9

(a)   Both eyes can be brown. If left brown then almost certainly right also brown. $\Pr(\text{both brown}) = p$. Hence $p + p - p = p$.

(b)   Height and weight not independent — indeed they are positively associated. Hence we require

Pr(weight satisfactory|height satisfactory) Pr(height satisfactory).

(c)   The possibilities are not equally likely. Ans. $(1/2)^3 = 1/8$.

Problem 10. y wins in three more games with probability $(1/3)^3$. y wins on the fourth game with probability

$$[3(1/3)^3 + 3(1/3)^3]/3.$$

y wins on the fifth game with probability

$$[6(1/3)^4 + 12(1/3)^4]/3.$$

y wins on the sixth game with probability

$$[30(1/3)^5]/3.$$

The total probability is $55/243$, which is also the result for z. As a matter of self-respect, the reader should verify *directly* that the probability that x wins is $133/243$.

Problem 11. If $9 \leqslant n \leqslant 16$, the player must win the nth point and 8 out of the previous $n - 1$ points. Required probability is

$$\binom{n-1}{8} p^8 q^{n-9}, \quad q = 1 - p.$$

If $n > 16$, it must be even, say $2m$. Also the players must have shared the first 16 points, then won alternately except that the last two points were won by the player in question. This sequence has probability

$$\binom{16}{8} p^8 q^8 (2pq)^{m-9} p^2, \quad m = 9, 10, \ldots \; .$$

The probability of an unending sequence includes a factor

$$(2pq)^{m-9} \to 0 \text{ as } m \to \infty.$$

Problem 12. $\binom{6}{4}$ ways of choosing 4 different numbers to appear and $\binom{4}{2}$ ways of choosing 2 to be doubletons. $6!/2!2!1!1!$ ways in which such a selection appears on different dice, the product is 16 200. Similarly

$$\binom{6}{5}\binom{5}{1}\frac{6!}{2!1!1!1!1!} = 10\,800.$$

Problem 13. R_{i1} on first transmission if either

(a) S_i chosen with probability $1/N$ then R_{i1} with probability p, or
(b) S_j with probability $1/N$, not R_{j1} with probability q but R_{i1} with probability $1/(N-1), j \neq i$.

Hence

$$\Pr(R_{i1}) = \frac{p}{N} + \frac{(N-1)\,q}{N(N-1)}.$$

Similarly $\Pr(R_{i2}, R_{i1} | S_i) = p^2$,

$$\Pr(R_{i2}, R_{i1} | S_j) = \left(\frac{q}{N-1}\right)^2, \quad j \neq i.$$

Hence

$$\Pr(S_i | R_{i2}, R_{i1}) = \frac{\Pr(R_{i2}, R_{i1} | S_i) \Pr(S_i)}{\sum_j \Pr(R_{i2}, R_{i1} | S_j) \Pr(S_j)}$$

$$= \frac{p^2/N}{\dfrac{p^2}{N} + \dfrac{(N-1)\,q^2}{N(N-1)^2}}$$

$$= \frac{(N-1)p^2}{(1-2p+Np^2}.$$

Answer to the second part,

$$\frac{\dfrac{pq}{N(N-1)}}{\left(\dfrac{pq}{N(N-1)} + \dfrac{q^2(N-2)}{N(N-1)^2} + \dfrac{pq}{N(N-1)}\right)}.$$

Problem 14. Suppose k pairs are matched, one set fixed. Since there are $(n-k)!$ arrangements of the remaining cards,

$$P_k = \frac{(n-k)!}{n!}.$$

Hence

$$\Pr(\text{at least one match}) = \sum_{k=1}^{n} (-1)^{k-1} \binom{n}{k} \frac{(n-k)!}{n!}$$

$$= \sum_{k=1}^{n} (-1)^{k-1}/k!.$$

Hence result.

For three packs, if k triples are matched, one set fixed, since there are $[(n-k)!]^2$ arrangement of the other cards in two sets,

$$P_k = \left[\frac{(n-k)!}{n!}\right]^2 .$$

$$\Pr[\text{at least one match}] = \sum_{1}^{n} \binom{n}{k} \left[\frac{(n-k)!}{n!}\right]^2 (-1)^{k-1} .$$

$$= \sum_{1}^{n} \frac{(n-k)!}{k!n!} (-1)^{k-1} .$$

Hence result.

Problem 15. There can be many losses of service between points scored. One possible sequence BBA, with probability $(1-f)f(1-f)$.

Problem 16. A wins on the $(n+1)$th game if either

(i)   he has won one of the previous n, lost one and drawn $n-2$, or
(ii)   he has won one of the previous games and drawn $n-1$.

Probability of no winner is zero.

4

Random Variables

4.1 INFINITE SAMPLE SPACES

We have restricted ourselves to the situation when the sample space contains only a finite number of points. Without giving a rigorous treatment we briefly discuss some of the situations where this is not the case. We shall look at some situations where the outcomes are numbers. If the outcome of the experiment is a number, then the sample space can conveniently be taken as a set of points on the real line. If the number of flashes from radioactive material in a fixed time is recorded by a Geiger counter or the number of organisms in a drop of fluid counted under a microscope, then though there is some practical limit to the number that can be recorded, we may for theoretical reasons assume that the number can be any positive integer. If every finite interval of the real line contains only a finite number of points of the sample space, then we shall say that the sample space is discrete. The property stated guarantees that there are at most a countable infinity of points in the sample space. If again, an event is any set of these points, countability ensures that we shall have no difficulty in listing them and determining which points are in which event. Furthermore we are able to assign a non-zero probability to every point, subject to their sum being unity. For example, if the sample space consists of points listed as 0, 1, 2, . . . , we can assign probability $\lambda^r \exp(-\lambda)/r!$ to the point having label r. However, if there be not a finite number of points, we cannot give equal probability to each, however small, for the sum of the probabilties would then be infinity! The axiom stating that the probability of an event which is the countable union of exclusive events is to be the sum of the probabilities of the events, again means that the probability of an event can be found by summing the probabilities of the points in the set defining the event. Not that this will be an easy task unless the probability assigned to a point is a function of its position in the list of points.

It is also possible that a completely contrasting property to discreteness is shown by numerical outcomes. When the 'experiment' is to measure the height of a man chosen at random from a city, then any height between certain limits is *possible*, though only a certain number of values can actually occur, since there is a finite number of men in any city. Similar remarks apply to weights of plants, densities of acids and times to failure of electric bulbs. In such cases, a reasonable model to adopt is that the sample space consists of the positive part of the real line. In the extreme case no point has positive probability, and only intervals of the real line are assigned positive probability. Thus, if a person is chosen at random from a city, in this model we would assign a probability to his being between 5 feet and 6 feet high, but zero probability to his having height 6 feet. We shall require all intervals, unions of intervals, and complements of intervals in the real line to be events, though at this level we cannot decide whether any other kinds of sets ought to be included as events. To assign probabilities, we need a non-negative function of an interval whose value for the whole real line is unity and such that the value for the union of disjoint intervals is the sum of the values for the separate intervals.

4.2 RANDOM VARIABLES

Definition. A random variable is a function on a sample space with two properties:

(1)   the values it assumes are real numbers; and
(2)   for every real number x, the probability that the value of the function is less than or equal to x can be calculated.   ∎

If the points in the sample space are themselves numerical outcomes, then the outcome is itself a random variable. 'Random variable' is a wider concept than 'numerical outcome of a random experiment', since a random variable can be defined on a sample space for which the simple events are not numbers. At the moment we are only dealing with those random variables which yield a single number at each performance of the experiment. Random variables are denoted by capital letters such as X, Y, Z or X_1, X_2, ..., X_n, while the values they take are denoted by small letters x, y, z or x_1, x_2, ..., x_n respectively.

If six cards are drawn at random without replacement from a pack, then the number of clubs is a random variable say X, which takes the values 0, 1, 2, 3, 4, 5, 6. Also, X^2 is a random variable which takes the values 0, 1, 4, 9, 16, 25, 36. If each card is given its face value, except that ace counts as 1 and any picture card counts as 0, then the sum of the values of the cards, Y, is a random variable which takes the values 0, 1, 2, ..., 58. Also, $X + Y$, XY are random variables.

If an electric light bulb is selected at random from a batch, its length of life T_1 is a random variable, as are T_1^3 and $\log_e T_1$. If on failure it is replaced by another bulb, then the sum of their lives $T_1 + T_2$ is a random variable.

The previous remarks will have given the impression that any function on the sample space will qualify as a random variable. To qualify, the function must be sufficiently 'well behaved'. As the adjective 'random' indicates, the value of X, until the outcome of the experiment, is unknown. Nevertheless, unless we can compute $\Pr(X \leqslant x)$, for any real number x, the function is useless. How then should we calculate such a probability? We must find the set E_x of all those sample points which are mapped by X into numbers not exceeding x, then $\Pr(X \leqslant x)$ is the probability that the experiment yielded a point in E_x. Now this probability may be calculated provided E_x is an event. In sum, the extra requirement is that E_x be an event for every x.

Example 1

Two fair dice are rolled, the sample space containing 36 points labelled by the ordered paris (s, t), $s = 1, 2, 3, 4, 5, 6$, $t = 1, 2, 3, 4, 5, 6$. Suppose each point is given probability $1/36$ and the random variable X is defined as the sum of the values showing on the faces of the dice. Then $X \leqslant 4$, provided the outcome belongs to the set $(1, 1), (1, 2), (2, 1), (1, 3), (3, 1), (2, 2)$. This set is an event, and has probability $6/36 = 1/6$. Hence, $\Pr(X \leqslant 4) = 1/6$.

To verify such a property in a general way would be exceedingly tiresome and it is not intended to consider any cases in which the requirement is in doubt. Indeed we shall restrict our attention to those random variables for which $\Pr(X \leqslant x)$ can be computed via a function known as the probability density function of X. This function allows the sample space to be dispensed with and roughly speaking describes how the probability is smeared out in terms of x.

One further aspect of example 1 deserves our attention. It is plain that the total score X is composed of the sum of two other random variables S, T which are related to the first and second die respectively. On the sample space, S is the function which assumes the value of the first component of each ordered pair (s, t). Thus $S = s$ at the six points $(s, 1), (s, 2), (s, 3), (s, 4), (s, 5), (s, 6)$ and for fair dice, has total probability $6/36$. Similarly T is the second component and $\Pr(T = t) = 6/36$. There is one point for which $S = s$ *and* $T = s$, with probability $1/36$. Thus we have

$$\Pr(S = s \text{ and } T = t) = \Pr(S = s) \Pr(T = t) \text{ for all } (s, t).$$

In the spirit of the terminology used for events in Chapter 3, it is appropriate to say that S, T are independent random variables. In this example, it is apparent that the result reflects the physical independence of the outcomes for the two dice. More generally, if for the discrete random variables X, Y we have

$$\Pr(X = x \text{ and } Y = y) = \Pr(X = x) \Pr(Y = y) \text{ for all } x, y.$$

then we say X, Y are independent. This important concept will be considered in greater detail in Chapter 8. For such independent random variables we can, for example, calculate $\Pr(X + Y = k)$ by decomposition as

$$\Pr(X + Y = k) = \sum_x \Pr(X = x \text{ and } Y = k - x)$$

$$= \sum_x \Pr(X = x) \Pr(Y = k - x).$$

Note that some values of X may make $X + Y = k$ impossible, in which case $\Pr(Y = k - x)$ will be zero.

4.3 DISCRETE AND CONTINUOUS RANDOM VARIABLES

Definition. Suppose the random variable X take either a finite or a countable number of values and that there is a function $f(x)$ which satisfies the following conditions:

(a) $f(x) \geqslant 0$.

(b) $\sum f(x) = 1$, where the summation is over all values of x assumed by X.

(c) For every interval $[a, b]$,

$$\Pr(a \leqslant X \leqslant b) = \sum_{x=a}^{b} f(x).$$

Then X is said to be a **discrete random variable** with **probability density function** $f(x)$ (abbreviation p.d.f.). The basic idea is that, for a discrete random variable, $f(x)$ gives $\Pr(X = x)$. The values assumed by X in applications are usually in some set of the integers.

Example 2
X is a discrete random variable with p.d.f.

$$f(x) = \binom{3}{x}\left(\frac{1}{2}\right)^3, \quad x = 0, 1, 2, 3$$

$$= 0 \text{ otherwise.}$$

Then, certainly $f(x) \geqslant 0$ and

$$\sum_{x=0}^{3} f(x) = \sum_{x=0}^{3} \binom{3}{x}\left(\frac{1}{2}\right)^3 = \left(\frac{1}{2} + \frac{1}{2}\right)^3 = 1.$$

This might be the p.d.f. suggested for the random variable which counts the number of heads when three fair pennies are tossed at once. From the p.d.f. we can calculate $\Pr(2 \leqslant X \leqslant 3)$ as $\Pr(X = 2) + \Pr(X = 3)$.

Definition. Suppose that X is a random variable and there is a function $f(x)$ such that

(a) $f(x) \geqslant 0$.

(b) $f(x)$ has at most a finite number of discontinuities in every finite interval of the real line.

(c) $\int_{-\infty}^{+\infty} f(x)\,dx = 1$.

(d)   For every interval $[a, b]$,

$$\Pr(a \leqslant X \leqslant b) = \int_a^b f(x)\,dx.$$

Then X is said to be a **continuous random variable** with **probability density function** $f(x)$.

Example 3

X is a continuous random variable with p.d.f.

$$f(x) = \frac{1}{k}, \quad 0 \leqslant x \leqslant k$$

$$= 0 \text{ otherwise.}$$

Then, for $k > 0$,

$$f(x) \geqslant 0, \quad \int_{-\infty}^{+\infty} f(x)\,dx = \int_0^k \frac{1}{k}\,dx = 1$$

and if $0 \leqslant a < b \leqslant k$

$$\Pr(a \leqslant X \leqslant b) = \int_a^b f(x)\,dx = \int_a^b \frac{1}{k}\,dx = \frac{b - a}{k}.$$

This p.d.f. would be suitable for a continous random variable X defined as the distance from the origin of a point chosen at random on a segment of length k.

A random variable which has a p.d.f. $f(x)$ is also said to have the **distribution** $f(x)$. Any constants in a probability density function are called the **parameters** of the distribution.

We now consider the probability density functions of some random variables which are of frequent use in statistics.

4.4 THE BERNOULLI DISTRIBUTION

The discrete random variable X is said to have the **Bernoulli distribution** with parameter p, $0 < p < 1$, if its p.d.f. satisfies

$$f(x) = p \text{ if } x = 1$$
$$= 1 - p \text{ if } x = 0$$
$$= 0 \text{ otherwise.}$$

We have

$$\sum_{x=0}^{1} f(x) = p + 1 - p = 1.$$

This p.d.f. applies when in a single trial there is a probability p of a success, X being the number of successes; it is of theoretical interest mainly.

4.5 THE BINOMIAL DISTRIBUTION

The discrete random variable X is said to have the **binomial distribution** with parameters n and p where n is a positive integer and $0 < p < 1$, if its p.d.f. satisfies

$$f(x) = \binom{n}{x} p^x (1-p)^{n-x}, \quad x = 0, 1, 2, \ldots, n$$

$$= 0 \text{ otherwise.} \qquad (4.1)$$

This distribution is so named because $f(x)$ is the $(x + 1)$th term in the binomial expansion of $[p + (1 - p)]^n$, whence

$$\sum_{x=0}^{n} f(x)$$

is identically 1, whatever the values of n and p. This is the appropriate p.d.f. when X is the number of successes in the course of n independent trials on each of which the probability of success is p. From this we see that the binomial is the 'sum' of n independent Bernoulli distributions. The applicability of the binomial distribution is wide and ranges from the number of defectives in a sample drawn with replacement from a batch, to situations more complex but from which a binomial distribution can be constructed. Suppose it is wished to determine whether a new chemical has any effect on the growth of a particular type of plant. Ten pairs of matched seedlings can be planted in similar positions, so that one of each pair receives the chemical. At the end of a fixed time, some appropriate measure, say the height of each plant is taken. If the chemical has no effect then, *other things equal*, the number of pairs in which the treated plant is taller than the untreated should follow a binomial distribution with $n = 10$ and $p = 1/2$. It will be appreciated that by making the test on this basis, information is discarded, for example, how much taller one plant is than its paired plant.

A useful computing formula

Consider the ratio $f(x + 1)/f(x)$

$$\frac{f(x + 1)}{f(x)} = \frac{\binom{n}{x + 1} p^{x+1}(1 - p)^{n-x-1}}{\binom{n}{x} p^{x}(1 - p)^{n-x}}$$

$$= \frac{n!x!(n - x)!}{(x + 1)!(n - x - 1)!n!} \frac{p}{1 - p} = \left(\frac{n - x}{x + 1}\right) \frac{p}{q}, \quad (4.2)$$

where $q = 1 - p$.

Hence, if $f(0)$ be calculated as $(1 - p)^{n}$, we may obtain all other values in succession. Such a method is of little use if n is large, and extensive tables such as those mentioned below are available:

(a)   For $n \leqslant 50$, National Bureau of Standards, *Tables of the Binomial Probability Distribution*.

(b)   For $50 \leqslant n \leqslant 100$, H. Romig's *50–100 Binomial Tables*.

Both for practical and theoretical reasons, it is useful to have an approximation for $n!$. One such is provided by Stirling's formula:

$$n! \sim \sqrt{(2\pi)}n^{n+1/2} e^{-n}.$$

From (4.2) we have

$$\frac{f(x + 1) - f(x)}{f(x)} = \left(\frac{n - x}{x + 1}\right) \frac{p}{q} - 1 = \frac{p(n + 1) - (x + 1)}{(x + 1)q}.$$

From this we observe that if $(x + 1) < (n + 1)p$, then $f(x + 1) > f(x)$, while if $(x + 1) > (n + 1)p$, we have $f(x + 1) < f(x)$. That is, the p.d.f. is at first increasing and then decreasing. There is one special case when $(n + 1)p$ is an integer, say m. In this case for $x + 1 = m$, we have $f(m) = f(m - 1)$. Thus, unless $(n + 1)p$ is an integer, the number of successes equal to the largest integer less than $(n + 1)p$ is the most probable.

Example 4

If a fair coin is tossed four times, what is the probability of at least two heads?

The number of heads has the binomial distribution with parameters $n = 4$, $p = 1/2$.

The probability of at least two heads is

$$\sum_{x=2}^{4} \binom{4}{x} \left(\frac{1}{2}\right)^{x} \left(\frac{1}{2}\right)^{4-x} = \frac{6}{16} + \frac{4}{16} + \frac{1}{16} = \frac{11}{16}.$$

However the probability of at least four heads in eight tosses is

$$\sum_{x=4}^{8} \binom{8}{x} \left(\frac{1}{2}\right)^x \left(\frac{1}{2}\right)^{8-x} = \frac{163}{256} < \frac{11}{16}.$$

Example 5
A proportion p of a large number of items in a batch is defective. A sample of n items is drawn and if it contains no defective items the batch is accepted, while if it contains more than two defective items, the batch is rejected. If, on the other hand, it contains one or two defectives, an independent sample of m is drawn, and if the combined number of defectives in the sample does not exceed two, the batch is accepted. Calculate the probability of accepting the batch.

The probability of no defectives in the first sample is q^n.

The probability of one defective in the first sample, and at most one in the second sample, is

$$\binom{n}{1} pq^{n-1} \left[q^m + \binom{m}{1} pq^{m-1} \right].$$

The probability of two defectives in the first sample and none in the second is

$$\binom{n}{2} p^2 q^{n-2} (q^m).$$

Hence the probability of acceptance is the sum of the probabilities for these three cases. It is assumed that either the sampling is with replacement or that the number of items in a batch is so large that the proportion of defectives is scarcely affected by sampling without replacement.

Problem 1. Samples, each of eight articles, are taken at random from a large consignment in which 20% of the articles are defective. Find the most likely number of defective articles in a single sample and the chance of obtaining precisely this number.

If 100 samples of eight are to be examined, calculate the number of samples in which you would expect to find three or more defective articles.

Oxford & Cambridge, A.L., 1963.

Problem 2. In a certain inspection scheme a sample of 50 items is selected at random from a very large batch and the number of defectives is recorded. If this number is more than three the batch is rejected; if it is less than three the batch is accepted. If the number of defectives is exactly three a further sample, this time of 25 items is taken and the batch is rejected if there is more than one defective in the second sample but accepted otherwise.

If the proportion of defective items in the batch is 1%, determine the values of the following probabilities:

(a)   That the batch is accepted as a result of inspection of the first sample.
(b)   That a further sample has to be taken and the batch is accepted as a result of inspection of that sample.
(c)   That the batch is rejected.

(Take $(0.99)^{48}$ as 0.6173.)

Oxford & Cambridge (S.P.), 1968.

Problem 3. The probability that a seed germinates is p independently for each seed. If n rows of m seeds are planted, find the probability that:

(a)   Two rows each contain one seed that fails to germinate, all other seeds germinating.
(b)   One row contains two seeds failing to germinate, all other seeds germinating.

Verify that the sum of the probabilities in (a) and (b) is the probability that just two of the mn seeds fail to germinate.

4.6 THE HYPERGEOMETRIC DISTRIBUTION

The discrete random variable X is said to have the **hypergeometric distribution** with parameters N, n, and k (all positive integers) if its p.d.f. satisfies

$$f(x) = \frac{\binom{k}{x}\binom{N-k}{n-x}}{\binom{N}{n}}, \quad x = 0, 1, 2, \ldots, n$$

$$= 0 \text{ otherwise.} \tag{4.3}$$

We have previously shown that

$$\sum_{x=0}^{n} \binom{k}{x}\binom{N-k}{n-x} = \binom{N}{n}$$

so that we have

$$\sum_{x=0}^{n} f(x) = 1.$$

It should be noted that if $k < n$, then $\binom{k}{x}$ is zero for $k < x \leqslant n$. This distribution would apply if X is the number of defectives in a sample drawn without replacement from a batch of N articles, there being k defectives in the batch. The hypergeometric distribution is related to the binomial distribution with $p = k/N$. Indeed if n is small compared to N, sampling without replacement must be approximately equivalent to sampling with replacement.

4.7 THE GEOMETRIC DISTRIBUTION

The discrete random variable X is said to have the **geometric distribution** with parameter p, $0 < p < 1$ if its p.d.f. satisfies

$$f(x) = p(1-p)^{x-1}, \quad x = 1, 2, \ldots$$

$$= 0 \text{ otherwise.} \tag{4.4}$$

$$\sum_{x=1}^{\infty} f(x) = p \sum_{x=1}^{\infty} (1-p)^{x-1} = \frac{p}{1-(1-p)} = 1.$$

We can also find explicitly

$$\Pr(X \leqslant x) = p \sum_{r=1}^{x} q^{r-1} = \frac{p(1-q^x)}{1-q} = 1 - q^x.$$

This distribution applies to a series of independent trials, in each of which there is a constant probability of success p, the random variable X then being the number of trials *up to and including the first success*. This is the case if items are drawn one at a time with replacement from a batch when X is the number of draws to the first defective item discovered. In this sense, the number of trials is often called the waiting time. If $0 < p < 1$, then

$$f(x+1)/f(x) = 1 - p < 1,$$

successive terms are decreasing and 1 is the most probable waiting time. This does not mean that this waiting time is more probable than any other conceivable event, for $\Pr(X \geqslant 1)$ is $1 - p$ and this is greater than $\Pr(X = 1)$ if $p < 1/2$.

The geometric distribution has a property which seems intuitively 'almost obvious' in terms of independent trials. If the first success has not appeard by trial x_1, then the probability that it appears on trial $x_2 (> x_1)$ depends only on $x_2 - x_1$. That is, no account is taken of what happened on the first x_1 trials. The distribution is said to 'lack memory'. More formally,

$$\Pr(X = x_2 | X > x_1) = \frac{\Pr(X = x_2 \text{ and } X > x_1)}{\Pr(X > x_1)}$$

$$= \frac{\Pr(X = x_2)}{\Pr(X > x_1)}, \quad \text{since } x_2 > x_1,$$

$$= pq^{x_2 - 1}/q^{x_1}$$

$$= pq^{x_2 - x_1 - 1},$$

and the conditional probability has the form of another geometric distribution with parameter p.

We may also find the approximate average number of trials needed to obtain a first success. Let the trials proceed until n successes have been obtained, the number of trials being x_n. Then $x_1, x_2 - x_1, x_3 - x_2, \dots, x_n - x_{n-1}$ are the result of n attempts to obtain a first succes. These have average

$$[x_1 + x_2 - x_1 + \dots + x_n - x_{n-1}]/n = x_n/n.$$

But the proportion of successes up to trial x_n is n/x_n, and, for large n, this must be approximately $1/p$. Thus, if $p = 1/4$, the average number of trials required to obtain a first success is approximately 4. In Chapter 9 the idea of 'average' will be given a more extended discussion.

Example 6
Suppose a computer is reputedly printing a long stream of random binary digits in which 0, 1 are equiprobable. The number of digits after each zero until the next zero is encountered must have a geometric distribution with parameter 1/2. The probability of a waiting time of x digits is $1/2^x$, $x = 1, 2, \dots$. Thus an output of the form 0, 0, 0, 1, 1, 1, 0, 0, 0, 1, 1, 1, ..., bears the correct proportion of zeros but is clearly not random.

Problem 4. X_1, X_2 have independent geometric distributions with a common parameter p. Calculate $\Pr(X_1 + X_2 = t)$ and hence $\Pr(X_1 = x_1 | X_1 + X_2 = t)$.

Problem 5. Three players A, B, C, alternatively throw a die each in that order, the first player to throw a 6 being deemed the winner. A's die is fair, whereas B and C throw biased dice with probabilities p_1, p_2 respectively of throwing a 6. Show that A wins the game with probability

$$\frac{1}{6(1 - u)},$$

where $u = 5(1 - p_1)(1 - p_2)/6$ and find the probabilities of B and C winning. For what values of p_1 and p_2 is the game fair?

University College of Wales, Aberystwyth (part question), 1979

4.8 THE NEGATIVE BINOMIAL DISTRIBUTION

The discrete random variable X is said to have the **negative binomial distribution** with parameters r and p, where r is a positive integer and $0 < p < 1$, if its p.d.f. satisfies

$$f(x) = \binom{r+x-1}{x} p^r (1-p)^x, \quad x = 0, 1, 2, \ldots \tag{4.5}$$

$$= 0 \text{ otherwise.}$$

In order to check that the total probability is 1, and to throw some light on how this distribution obtains its name (also known as the Pascal distribution), we first observe that if $|y| < 1$,

$$\frac{1}{(1-y)^r} = (1-y)^{-r} = 1 + ry + \frac{r(r+1)}{2!} y^2 \ldots$$

$$+ \frac{r(r+1) \ldots (r+k-1)}{k!} y^k \ldots$$

$$= 1 + \binom{r}{1} y + \binom{r+1}{2} y^2 \ldots$$

$$+ \binom{r+k-1}{k} y^k \ldots$$

$$= \sum_{i=0}^{\infty} \binom{r+i-1}{i} y^i.$$

Hence

$$\sum_{x=0}^{\infty} f(x) = p^r \sum_{x=0}^{\infty} \binom{r+x-1}{x} (1-p)^x = p^r [1 - (1-p)]^{-r} = 1.$$

Now, the binomial distribution consists of the terms $\binom{n}{x} p^x (1-p)^{n-x}$, found from the expansion of $(q+p)^n$. In the present distribution we may write

$$1 = \frac{p^r}{p^r} = \frac{p^r}{(1-q)^r} = \left[\frac{1}{p} + \left(-\frac{q}{p} \right) \right]^{-r} = (Q+P)^m$$

with $Q + P = 1/p - q/p = 1$, and $m = -r$. The $(x+1)$th term, written formally, is

$$\binom{m}{x} P^x Q^{m-x} = \binom{-r}{x} \left(-\frac{q}{p} \right)^x \left(\frac{1}{p} \right)^{-r-x}$$

and this suggests a binomial distribution with 'parameters' $-r$ and $-q/p$. This artificial relation to the binomial distribution is less enlightening than a consideration of how it may be generated in the course of independent trials in each of which the probability of success is p. Now consider the random variable X which is equal to the number of failures before the rth success. If there are x failures, then the $(r + x)$th trial is the rth success. In which case the first $(r + x - 1)$ trials must include just $r - 1$ successes. The probability that in $(r + x - 1)$ trials we have just $r - 1$ successes is

$$\binom{r + x - 1}{r - 1} p^{r-1}(1 - p)^x = \binom{r + x - 1}{x} p^{r-1}(1 - p)^x.$$

The probability that such a sequence is followed by a success is found by multiplying this last probability by p, giving

$$\binom{r + x - 1}{x} p^r(1 - p)^x.$$

Hence the number of failures before the rth success has the negative binomial distribution. The number of trials up to and including the rth success, that is, the waiting time to the rth success, can be found from the p.d.f. of the number of failures. If $Y = X + r$, the p.d.f. of Y is

$$\binom{y - 1}{y - r} p^r(1 - p)^{y-r}, \tag{4.6}$$

by putting $x = y - r$. We notice that the waiting time to the rth success is the sum of r independent waiting times to a first success.

Example 7

In a sequence of independent trials, the probability of a success on each trial is $1/4$. It is required to find the probability that the second success occurs on the fourth trial or later. From equation (4.6), we need

$$\sum_{y=4}^{\infty} \binom{y - 1}{y - 2} \left(\frac{1}{4}\right)^2 \left(\frac{3}{4}\right)^{y-2}$$

but, it is clearly more convenient to find

$$1 - \Pr(\text{second success on or before the third trial})$$

$$= 1 - \sum_{y=2}^{3} \binom{y - 1}{1} \left(\frac{1}{4}\right)^2 \left(\frac{3}{4}\right)^{y-2}$$

$$= 1 - \frac{1}{16} - \frac{6}{64} = \frac{54}{64}.$$

The reader may have spotted an alternative calculation based on the binomial distribution. For, equivalently, we require the probability of at most one success in three trials when $p = 1/4$. This is

$$\binom{3}{0} \left(\frac{1}{4}\right)^0 \left(\frac{3}{4}\right)^3 + \binom{3}{1} \left(\frac{1}{4}\right)^1 \left(\frac{3}{4}\right)^2 = 54/64.$$

More generally, in a series of trials the rth success will appear on or before the nth trial, if and only if there are at least r successes in n trials. Even when the trials are independent and the probability of success is constant on each, this apparently innocent result is surprisingly awkward to verify when disguised in algebraic form. This is the subject of Problem 6. (optional)

Problem 6. Prove that

$$\sum_{x=r}^{n} \binom{n}{x} p^x (1-p)^{n-x} = \sum_{y=r}^{n} \binom{y-1}{r-1} p^r (1-p)^{y-r}.$$

(Hint, compare the coefficients of p^k, $r \leqslant k \leqslant n$.)

Problem 7. Prove that

$$\sum_{y=r}^{z-1} \binom{y-1}{r-1} p^r (1-p)^{y-r} p(1-p)^{z-y-1}$$

$$= \binom{z-1}{r} p^{r+1} (1-p)^{z-(r+1)}.$$

By considering the waiting time to the $(r+1)$th success in a series of independent trials, obtain a probabilistic proof. ∎

The reader is warned that other texts may employ slightly different definitions for the geometric and negative binomial distributions. The commonest variant is to base the geometric distribution on the number of failures *before* the first success, the p.d.f. is $f(x) = p(1-p)^x$, $x = 0, 1, 2, \ldots$. This choice has the merit of viewing the geometric distribution as a particular negative binomial distribution. The waiting time may be defined as the number of failures before the rth success in the negative binomial distribution.

4.9 THE POISSON DISTRIBUTION

The discrete random variable X is said to have the **Poisson distribution** with positive parameter λ, if its probability density function $f(x)$ satisfies

$$f(x) = \frac{\lambda^x e^{-\lambda}}{x!}, \quad x = 0, 1, 2, \ldots$$

$$= 0 \text{ otherwise.} \tag{4.7}$$

For positive λ,

$$f(x) > 0 \quad \text{and} \quad \sum_{x=0}^{\infty} f(x) = e^{-\lambda} \sum_{x=0}^{\infty} \frac{\lambda^x}{x!} = e^{-\lambda} \cdot e^{+\lambda} = 1.$$

The relation between two values of $f(x)$ for the successive values of x is useful for computing purposes.

$$\frac{f(x+1)}{f(x)} = \frac{\lambda^{x+1} e^{-\lambda}}{(x+1)!} \Big/ \frac{\lambda^x e^{-\lambda}}{x!} = \frac{\lambda}{x+1}$$

$$f(x+1) = \frac{\lambda}{x+1} f(x).$$

Thus when $x = 0$, $f(0) = e^{-\lambda}$ is found from tables of the exponential function, and $f(1)$ is found from the above expression and hence $f(2)$, and so on. Furthermore, $f(x+1) > f(x)$ if

$$\frac{\lambda}{x+1} > 1 \quad \text{or} \quad x < \lambda - 1$$

and $f(x+1) < f(x)$ if $x > \lambda - 1$. However, if $\lambda < 1$, there is no value of x with positive probability $< \lambda - 1$ and $f(x)$ has a maximum at $x = 0$. If λ is an integer then $f(\lambda - 1) = f(\lambda)$. Tables for this useful and important distribution are available in E. C. Molina's *Poisson's Exponential Binomial Limit* (Van Nostrand), which tabulates

$$f(x) = \frac{\lambda^x e^{-\lambda}}{x!} \quad \text{and} \quad \sum_{x=x_0}^{\infty} f(x)$$

for $0 \leqslant x_0 \leqslant 100$.

Problem 8. Prove that for the Poisson distribution with parameter λ,

$$\sum_{x=0}^{n} f(x) = \frac{1}{n!} \int_{\lambda}^{\infty} e^{-x} x^n \, dx \qquad \blacksquare$$

As the name suggests, the distribution is due to S. Poisson, who first studied it as an approximation to the binomial distribution, when n is large and p is small. It will be appreciated that the evaluation of binomial probabilities for large n is tedious. The p.d.f. for the binomial distribution is

$$f(x) = \binom{n}{x} p^x (1-p)^{n-x}, \quad x = 0, 1, 2, \ldots, n.$$

Now consider limit $f(x)$ as $n \to \infty$ and $p \to 0$ in such a way that np remains constant, say k. Then $p = k/n$. Substituting in $f(x)$ for p, we have

$$f(x) = \frac{n!}{x!(n-x)!} \left(\frac{k}{n}\right)^x \left(1-\frac{k}{n}\right)^{n-x}$$

$$= \frac{k^x}{x!} \left(1-\frac{k}{n}\right)^{-x} \left(1-\frac{k}{n}\right)^n \frac{n!}{n^x(n-x)!}.$$

Now, for fixed x,

$$\lim_{n\to\infty} \left(1-\frac{k}{n}\right)^{-x} = 1$$

and

$$\lim_{n\to\infty} \frac{n!}{(n-x)!\,n^x} = \lim_{n\to\infty} \frac{n(n-1)\ldots(n-x+1)}{n^x}$$

$$= \lim_{n\to\infty} \left[1 \left(1-\frac{1}{n}\right) \left(1-\frac{2}{n}\right) \ldots \right.$$

$$\left. \left(1-\frac{x-1}{n}\right) \right] = 1,$$

while

$$\lim_{n\to\infty} \left(1-\frac{k}{n}\right)^n = e^{-k}.$$

Hence

$$\lim_{n\to\infty} f(x) = \frac{k^x e^{-k}}{x!}$$

for fixed x. But this is the p.d.f. of a Poisson distribution with parameter $k = np$. For example, if X has the binomial distribution with parameters $n = 100$, $p = 1/100$, then $\Pr(X = 0) \approx 0.366$, $\Pr(X = 1) \approx 0.370$, $\Pr(X = 2) \approx 0.185$. The corresponding values using the approximating Poisson distribution are 0.368, 0.368, and 0.184. Even for small values of n, the Poisson distribution is not outrageously bad. The values in Table 4.1 give the probabilities that $X = 0, 1, 2, 3$, when X has the binomial distribution with parameters $n = m$, $p = 1/m$ for $m = 3, 4, 5, 6, 10$ and the corresponding values for the Poisson distribution with $\lambda = np = 1$.

Table 4.1 Binomial probabilities

	$m = 3$	4	5	6	10	Poisson $\lambda = 1$
$\Pr(X = 0)$	0.296	0.316	0.328	0.335	0.349	0.368
$\Pr(X = 1)$	0.444	0.422	0.410	0.401	0.387	0.368
$\Pr(X = 2)$	0.222	0.211	0.205	0.201	0.194	0.184
$\Pr(X = 3)$	0.037	0.047	0.051	0.053	0.057	0.061

Values extracted by permission from W. Feller, *An Introduction to Probability Theory and its Applications*, Vol. I, John Wiley, New York.

A variety of practical applications have been made of the Poisson distribution as an approximation to the binomial distribution, including such topics as quality control, suicide rates, misprints, radioactivity, flying bomb hits on London, and bacteria counts. There seem to be two types of application, those in which n and p are known and those in which only their product, that is λ, is known or can be estimated. If $x_1, x_2, \ldots, x_n$ are reputedly from the same Poisson distribution, then an estimate of its parameter is formed from $(x_1 + x_2 + \ldots + x_n)/n$.

Example 8
An industrial process produces items of which 1% are defective. If a random sample of 100 of these are drawn from a large consignment, calculate the probability that the sample contains no defectives. Since the consignment is large, we may suppose that the number of defectives in the sample obeys the binomial distribution with parameter $n = 100$ and $p = 1/100$. On this basis, the probability that the sample contains no defectives is

$$\left(1 - \frac{1}{100}\right)^{100} = 0.366.$$

The corresponding Poisson approximation with $\lambda = np = 1$, is 0.368. In this case it is reasonably clear that the Poisson distribution is the limit of a binomial distribution.

Example 9
Radioactive material emits particles which can be detected by a Geiger counter. However, the information available is in the form of the number of particles emitted in a fixed time. It is not known how many particles are available for release in the material, thus n and p are unknown. Suppose, however, that during N different non-overlapping periods of time of duration t_0, a total of M particles are counted, then the average number of particles per period is M/N. This, for

reasons which will be clearer later, is a satisfactory estimate for the parameter in the Poisson distribution. We are in fact in a position to check roughly whether the number of particles emitted in time t_0 appears to obey a Poisson distribution. We can calculate the probability of x particles being emitted as

$$f(x) = \frac{(M/N)^x \, e^{-M/N}}{x!}, \quad x = 0, 1, 2, \ldots$$

Now, if N is large and N_x is the *observed* number of periods in which x particles *were* emitted, then N_x/N should be approximately $f(x)$. Thus the calculated frequencies $Nf(x)$ can be compared with the observed N_x. There will of course be discrepancies and a test would then be carried out to see whether these are acceptable or not.

Example 10

X_1, X_2 have independent Poisson distributions with parameters $\lambda_1 = 1$ and $\lambda_2 = 2$ respectively. Evaluate (a) $\Pr[(X_1 \text{ and } X_2) \leqslant 1]$, (b) $\Pr[(X_1 \text{ or } X_2) \leqslant 1]$, and (c) $\Pr[(X_1 + X_2) \leqslant 1]$.

(a) $\Pr(X_1 \leqslant 1) = e^{-1} + 1 e^{-1} = 2 e^{-1}$

      $\Pr(X_2 \leqslant 1) = e^{-2} + 2 e^{-2} = 3 e^{-2}$.

Since X_1 and X_2 are independent,

    $\Pr[(X_1 \text{ and } X_2) \leqslant 1] = \Pr(X_1 \leqslant 1) \Pr(X_2 \leqslant 1) = (2 e^{-1})(3 e^{-2}) = 6 e^{-3}$

(b) $\Pr[(X_1 \text{ or } X_2) \leqslant 1] = \Pr(X_1 \leqslant 1) + \Pr(X_2 \leqslant 1) - \Pr[(X_1 \text{ and } X_2) \leqslant 1]$

                    $= 2 e^{-1} + 3 e^{-2} - 6 e^{-3}$

(c) $\Pr[(X_1 + X_2) \leqslant 1] = \Pr(X_1 = 0 \text{ and } X_2 = 0) + \Pr(X_1 = 0 \text{ and } X_2 = 1)$

                      $+ \Pr(X_1 = 1 \text{ and } X_2 = 0)$

                      $= (e^{-1})(e^{-2}) + (e^{-1})(2 e^{-2}) + (e^{-1})(e^{-2})$

                      $= 4 e^{-3}$.

The same result can be obtained by treating $X_1 + X_2$ as having a Poisson distribution with parameter $\lambda = \lambda_1 + \lambda_2 = 3$.

Problem 9. If X_1, X_2 have independent Poisson distributions with parameters λ_1, λ_2, show that

$$\Pr[(X_1 + X_2) = k] = \frac{(\lambda_1 + \lambda_2)^k \, e^{-(\lambda_1 + \lambda_2)}}{k!}$$

Problem 10. A flat piece of ground is staked out into 120 plots each of one square yard and the number of plants of a particular variety counted in each square. It is conjectured that the number of plants per plot has a Poisson distribution. If this conjecture is correct, how many plots does it suggest should have 0, 1, 2, 3, 4, 5, 6 plants?

[Estimate the parameter from the average number of plants per plot and give the number of plots to the nearest whole number.]

Number of plants	0	1	2	3	4	5	6	7	343
Number of plots	6	20	27	27	20	12	8	0	120

Problem 11. For Problem 10, the probability of 0, 1, 2, plants per plot should be $e^{-\lambda}$, $\lambda e^{-\lambda}$, $\lambda e^{-\lambda}/2$ respectively. Consider four adjacent plots. By considering all possible placings of just two plants among the four, calculate the probability of just two plants falling in such a grouping.

Problem 12. A survey shows that the average daily rate of absence in a work force of 200 is 2%. The management has a reserve of 4 additional staff on stand-by. Assume that the number absent has a Poisson distribution. On what proportion of days will the reserve be inadequate? The pay for a man on reserve is £20 per day, and this is regarded as waste if he is not called out. What is the average cost of this waste per thousand days? $[e^{(-4)} = 0.01832]$.

Problem 13. A volume V of fluid contains n bacteria. What is the probability that a small volume u will contain k bacteria?

BRIEF SOLUTIONS AND COMMENTS ON THE PROBLEMS

Problem 1. One defective, with probability 0.34. About twenty.

Problem 2. (a) 0.9862, (b) 0.0119, (c) 0.0019.

Problem 3. The probability that for just two rows, one seed fails to germinate is

$$\binom{n}{2} \left[\binom{m}{1} p^{m-1}(1-p) \right]^2 (p^m)^{n-2}.$$

The probability that for just one row, two seeds fail to germinate is

$$\binom{n}{1} \left[\binom{m}{2} p^{m-2}(1-p)^2 \right] (p^m)^{n-1}.$$

The sum of the two probabilities is

$$\binom{mn}{2} p^{mn-2} q^2,$$

which is the probability that just two of the mn seeds fail to germinate.

Problem 4. Bearing in mind that $X_1 \geqslant 1$ and $X_2 \geqslant 1$,

$$\Pr(X_1 + X_2 = t) = \sum_{x_1 = 1}^{t-1} \Pr(X_1 = x_1 \text{ and } X_2 = t - x_1)$$

$$= \sum_{x_1 = 1}^{t-1} \Pr(X_1 = x_1) \Pr(X_2 = t - x_1)$$

$$= \sum_{x_1 = 1}^{t-1} pq^{x_1 - 1} \, pq^{t - x_1 - 1} = p^2 \sum_{x_1 = 1}^{t-1} q^{t-2}$$

$$= (t - 1) p^2 q^{t-2}, \quad t \geqslant 2.$$

$$\Pr(X_1 = x_1 | X_1 + X_2 = t) = \frac{\Pr(X_1 = x_1) \Pr(X_2 = t - x_1)}{\Pr(X_1 + X_2 = t)}$$

$$= \frac{pq^{x_1 - 1} \, pq^{t - x_1 - 1}}{(t - 1) p^2 q^{t-2}}$$

$$= \frac{1}{t - 1}, \quad 1 \leqslant x_1 \leqslant t - 1.$$

We conclude that the conditional probability is shared out equally at the points $1, 2, \ldots, t - 1$.

Problem 5. A may win on the *first* trial, with probability $1/6$. He may win on the *fourth* trial, after A, B, C lose trials, with probability $(5/6) (1 - p_1) (1 - p_2) (1/6) = u/6$. Total probability that A wins is

$$\sum_{r=0}^{\infty} u^r/6 = 1/[6(1 - u)].$$

Similarly, the probabilities that B C, win are

$$5p_1/[6(1 - u)], \quad 5(1 - p_1) p_2/[6(1 - u)].$$

By equating the probabilities, the game will be fair if $p_1 = 1/5, p_2 = 1/4$.

Problems 6. Since

$$(1 - p)^{n-x} = \sum_{j=0}^{n-x} \binom{n - x}{j} (-p)^j,$$

the coefficient of p^k in the left-hand side is

$$\sum_{x=r}^{k} \binom{n}{x}\binom{n-x}{k-x}(-1)^{k-x} = \sum_{x=r}^{k} \frac{n!}{x!(n-x)!}$$

$$\cdot \frac{(n-x)!}{(k-x)!(n-k)!}(-1)^{k-x}$$

$$= \frac{n!}{k!(n-k)!} \sum_{x=r}^{k} \frac{k!}{x!(k-x)!}(-1)^{k-x}$$

$$= (-1)^k \binom{n}{k} \sum_{x=r}^{k} \binom{k}{x}(-1)^x$$

$$= (-1)^k \binom{n}{k}(-1)^r \binom{k-1}{r-1},$$

from Problem 19, Chapter 2.

Since

$$(1-p)^{y-r} = \sum_{i=0}^{y-r} \binom{y-r}{i}(-p)^i,$$

the coefficient of p^k in the right-hand side is

$$\sum_{y=k}^{n} \binom{y-1}{r-1}\binom{y-r}{k-r}(-1)^{k-r}$$

$$= \sum_{y=k}^{n} \frac{(y-1)!}{(r-1)!(y-r)!} \cdot \frac{(y-r)!}{(k-r)!(y-k)!}(-1)^{k-r}$$

$$= \frac{(k-1)!(-1)^{k-r}}{(r-1)!(k-r)!} \sum_{y=k}^{n} \frac{(y-1)!}{(y-k)!(k-1)!}$$

$$= \binom{k-1}{r-1}(-1)^{k-r} \sum_{y=k}^{n} \binom{y-1}{k-1}$$

$$= \binom{k-1}{r-1}(-1)^{k-r} \binom{n}{k},$$

from Example 10, Chapter 2.

Problem 7. After collecting terms, we have

$$p^{r+1}(1-p)^{z-(r+1)} \sum_{y=r}^{z-1} \binom{y-1}{r-1} = p^{r+1}(1-p)^{z-(r+1)} \binom{z-1}{r},$$

using the result in Example 10, Chapter 2.

The waiting time to the $(r + 1)$th success is the sum of the waiting time to the rth success, y, and the waiting time to one further success, $z - y$ such that $r \leqslant y \leqslant z - 1$.

Problem 8. Integrate by parts.

$$\frac{1}{n!} \int_\lambda^\infty e^{-x} x^n \, dx = \left(- \frac{e^{-x} x^n}{n!} \right)_\lambda^\infty + \frac{1}{(n-1)!} \int_\lambda^\infty e^{-x} x^{n-1} \, dx$$

$$= \frac{\lambda^n e^{-\lambda}}{n!} + \frac{\lambda^{n-1} e^{-\lambda}}{(n-1)!} + \frac{1}{(n-2)!} \int_\lambda^\infty e^{-x} x^{n-2} \, dx,$$

and the result follows by continued application.

Problem 9.

$$\Pr(X_1 + X_2 = k) = \sum_{i=0}^{k} \Pr(X_1 = i \text{ and } X_2 = k - i),$$

$$= \sum_{i=0}^{k} \Pr(X_1 = i) \Pr(X_2 = k - i),$$

since X_1, X_2 independent,

$$= \sum_{i=0}^{k} \left[\frac{\lambda_1^i}{i!} e^{-\lambda_1} \frac{\lambda_2^{k-i}}{(k-i)!} e^{-\lambda_2} \right],$$

$$= \frac{e^{-\lambda_1 - \lambda_2}}{k!} \sum_{i=0}^{k} \binom{k}{i} \lambda_1^i \lambda_2^{k-i}$$

$$= \frac{e^{-\lambda_1 - \lambda_2}}{k!} (\lambda_1 + \lambda_2)^k.$$

Problem 10. Estimate of λ is $343/120 = 2.858$.

Probability of i plants per plot is $\lambda^i e^{-\lambda}/i! = f(i)$. $e^{-2.858} = 0.0574$.

$120 f(0) \sim 7, \ 120 f(1) \sim 20, \ 120 f(2) \sim 28, \ 120 f(3) \sim 27, \ 120 f(4) \sim 19,$

$120 f(5) \sim 11, \ 120 \ f(6) \sim 5.$

This uses up 117 plots. Actual data is likely to produce few results for the extreme values of a Poisson distribution. It would be appropriate to allocate the remaining 3 plots for 7 or more plants per plot.

Problem 11. 2 plants in one subplot, none in the others, with probability

$$\binom{4}{1} \lambda^2 \frac{e^{-\lambda}}{2} \cdot (e^{-\lambda})^3 = 2\lambda^2\, e^{-4\lambda}.$$

1 plant in each of two subplots and none in the other, with probability

$$\binom{4}{2} (\lambda 1 \quad (\lambda e^{-\lambda})^2 = 6\lambda^2\, e^{-4\lambda}.$$

The total probability is $(4\lambda)^2\, e^{-4\lambda}/2!$, which corresponds to a Poisson distribution with parameter 4λ [an illustration of the result in Problem 9].

Problem 12.

(a) $p = 2\%, \quad n = 200, \quad \lambda = 4$ *Waste*

$f(0) = \quad e^{-4} = 0.01832$ $18.32 \times 4 \times 20 = \quad 1\,465.6$

$f(1) = \quad 4e^{-4} = 0.07328$ $73.28 \times 3 \times 20 = \quad 4\,396.8$

$f(2) = \dfrac{4^2}{2!} e^{-4} = 0.14656$ $146.56 \times 2 \times 20 = \quad 5\,862.4$

                         $195.41 \times 1 \times 20 = \quad 3\,908.8$

$f(3) = \dfrac{4^3}{3!} e^{-4} = 0.19541$ £15 633

$f(4) = \dfrac{4^4}{4!} e^{-4} = 0.19541$

                     0.62898

$\Pr(> 4) = 1 - \Pr(\leqslant 4) \sim 0.37.$

Problem 13. The probability that any particular organism is found in a volume u is u/V. If n is large and u is small, the number has approximately a Poisson distribution with parameter nu/V.

5

Continuous Distributions

5.1 RECTANGULAR DISTRIBUTION

The continuous random variable X is said to have the **rectangular distribution** with parameters a, b (with $a < b$), if its probability density function $f(x)$ satisfies,

$$f(x) = \frac{1}{b-a}, \quad a \leqslant x \leqslant b$$

$$= 0 \text{ otherwise.}$$

With the restriction $a < b$, we have $f(x) > 0$ and

$$\int_a^b f(x) \, dx = 1.$$

Also, provided $a \leqslant x_0 \leqslant b$, then

$$\Pr(X \leqslant x_0) = \int_a^{x_0} f(x) \, dx = \frac{x_0 - a}{b - a}.$$

The rectangular distribution would be appropriate for the 'experiment' in which a point is chosen at random in a segment of length $b - a = l$, when the probability that the point falls in any sub-interval of length l_1 is l_1/l. X is also said to be **uniformly** distributed over (a, b).

Example 1

Two points A, B, are chosen on the circumference of a circle (Fig. 5.1) so that the angle at the centre is uniformly distributed over $(0, \pi)$. What is the probability that the length of AB exceeds the radius of the circle? Fix one end of the chord, then by elementary geometry, the chord exceeds the radius when the angle at the centre exceeds $\pi/3$. Hence the required probability is

$$\frac{2\pi/3}{\pi} = \frac{2}{3}.$$

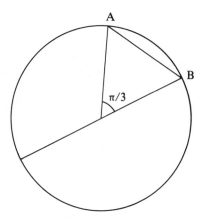

Fig. 5.1

We evaluate this probability more formally to show how to proceed when geometrical insight deserts us. The length of the chord $AB = 2a \sin (\phi/2)$, where a is the radius of the circle and ϕ the angle subtended at the centre by the chord.

$$\Pr(AB \geqslant a) = \Pr\left(2a \sin \frac{\Phi}{2} \geqslant a\right)$$

$$= \Pr\left(\sin \frac{\Phi}{2} \geqslant \frac{1}{2}\right)$$

But in the interval $(0, \pi)$, $\sin (\phi/2)$ is an increasing function of ϕ and $\sin (\pi/6) = 1/2$, hence

$$\Pr(AB \geqslant a) = \Pr\left(\frac{\Phi}{2} \geqslant \frac{\pi}{6}\right)$$

$$= \Pr\left(\Phi \geqslant \frac{\pi}{3}\right)$$

$$= \frac{2}{3}.$$

In this case, the prescription for choosing a chord is equivalent to choosing two points at random on the circumference in the sense that the distance round the circle of a point from some fixed point is uniformly distributed over $(0, 2\pi a)$.

Problem 1. Suppose that two points A, B are chosen on the circumference of a circle, radius a, so that the perpendicular distance from the centre to the chord AB is uniformly distributed over $(0, a)$. Find $\Pr(AB \geqslant a)$.

Problem 2. Two numbers are chosen at random between 0 and 1. Calculate the probability that their sum is less than 1.5. [Hint: Consider a unit square at the origin and the line $x + y = 1.5$.]

Problem 3. A number is chosen at random in the interval $(0, 1)$. Calculate the probability that the square root of x has the integer k in the first decimal place. [Hint: $(k/10) \leqslant \sqrt{x} < (k + 1)/10$.]

5.2 EXPONENTIAL DISTRIBUTION

The continuous random variable X is said to have the **exponential distribution** with positive parameter λ if its probability density function $f(x)$ satisfies

$$f(x) = \lambda e^{-\lambda x}, \quad 0 \leqslant x < \infty$$

$$= 0 \text{ otherwise.} \tag{5.1}$$

We have $f(x) > 0$ and

$$\int_0^\infty f(x)\, dx = \int_0^\infty \lambda e^{-\lambda x}\, dx = [-e^{-\lambda x}]_0^\infty = 1.$$

Further,

$$\Pr(0 \leqslant X \leqslant x_0) = \int_0^{x_0} \lambda e^{-\lambda x}\, dx = [-e^{-\lambda x}]_0^{x_0} = 1 - e^{-\lambda x_0}$$

from which we immediately obtain

$$\Pr(0 \leqslant x_0 \leqslant X) = 1 - \Pr(0 \leqslant X \leqslant x_0) = e^{-\lambda x_0},$$

and when $0 < x_1 < x_2$,

$$\Pr(x_1 \leqslant X \leqslant x_2) = (1 - e^{-\lambda x_2}) - (1 - e^{-\lambda x_1}) = e^{-\lambda x_1} - e^{-\lambda x_2}.$$

This distribution has a striking property. Suppose we require

$$\Pr(X \leqslant x_2 \mid X \geqslant x_1),$$

where $0 < x_1 < x_2$. Then this conditional probability can be evaluated as

$$\frac{\Pr(X \leqslant x_2 \text{ and } X \geqslant x_1)}{\Pr(X \geqslant x_1)} = \frac{\Pr(x_1 \leqslant X \leqslant x_2)}{\Pr(X \geqslant x_1)}$$

which is

$$\frac{e^{-\lambda x_1} - e^{-\lambda x_2}}{e^{-\lambda x_1}} = 1 - e^{-\lambda(x_1 - x_1)}$$

and this probability *only depends on the difference* $x_2 - x_1$. That is, if $x_2 - x_1$ is a positive constant, k, $\Pr(X \leqslant x_1 + k \,|\, X \geqslant x_1) = 1 - e^{-\lambda k}$, *whatever* x_1. It is this property that is useful in connection with *waiting time problems*. A possible model of the time to failure of a new spare part is that it is exponentially distributed. If the spare part has already been working for x_1 hours, then the probability that it will last a *further k* hours does not depend on x_1. Thus this model does not take wear and tear into account.

Example 2

For bulbs of a certain type, the time to failure is thought to have an exponential distribution with parameter λ. A number of the bulbs are put on test and it is found that 5% have failed by 100 hours. Estimate the probability

(a)   that a new bulb of this type will last more than 200 hours;
(b)   that one of the survivors of the test will complete 200 hours.

Since an exponential distribution is determined by one parameter, our task is first to estimate λ. This estimate must be based on the 5% which failed before 100 hours. It may be regretted that the actual times at which these bulbs failed appears to have been mislaid, for the extra information might have provided, in some sense, a 'better estimator'. Indeed the mean time to failure is $1/\lambda$, as we shall show, and a not unreasonable estimate may be based on the average of the recorded failure times. For the exponential distribution,

$$\Pr(X \leqslant x) = 1 - e^{-\lambda x}.$$

If we estimate $\Pr(X \leqslant 100)$ as 0.05

$$1 - e^{-100\lambda} = 0.05, \quad e^{-100\lambda} = 0.95$$

and from tables of the exponential function, $100\lambda = 0.051$, $\lambda = 0.00051$.
Now,

$$\Pr(X \geqslant x) = 1 - \Pr(X \leqslant x) = e^{-\lambda x}$$

$$\Pr(X \geqslant 200) = e^{-200\lambda}$$

$$= (e^{-100\lambda})^2 = (0.95)^2 = 0.9025.$$

This for a new bulb. For one of the survivors of the original test, the question is: Will it last more than another 100 hours? But we have shown that this conditional probability is the same as the unconditional probability that a new bulb lasts 100 hours, that is $1 - 0.05 = 0.95$.

Problem 4. The working of a machine is entirely dependent on the functioning of two similar components. The times to failure of the components are independent events and each has an exponential distribution with parameter λ. However, the components are 'in series' in the sense that the machine fails as soon as either component fails. Find the p.d.f of the time to failure of the machine.

Problem 5. The probability that a light bulb lasts longer than t hours is $e^{-t/u}$. find the probability density function for the lifetime of a bulb.

Show that the mean lifetime is u (assume this result). If the mean lifetime is 1500 hours, how unlikely is it that a bulb will last more than 3000 hours?

If the manufacturer wants to ensure that less than 1 in 1000 bulbs fail before 5 hours, what is the lowest mean lifetime that he can allow his bulbs to have?

Oxford & Cambridge, A.L., 1967.

Problem 6. The probability density function $p(t)$ of the length of life, t hours of a certain component is given by

$$p(t)\, dt = k\, e^{-kt}\, dt \quad (0 \leqslant t < \infty)$$

where k is a positive constant.

An apparatus contains three components of this type and the failure of one may be assumed independent of the failure of the others.

Find the probability that

(a)   None will have failed at t_0 hours.
(b)   Exactly one will fail in the first t_0 hours, another in the next t_0 hours and the third after more than $2t_0$ hours.

Oxford & Cambridge (S.P.), 1968

5.3 RANDOM STREAM OF EVENTS

We are now in a position to investigate a process in which the exponential and Poisson distributions are closely related. Suppose events are happening in time according to the following rules. Regardless of the number of events that have already happened in the interval $(0, t)$.

(a) The probability of just one more event in the intergal $(t, t + \delta t)$ is $\lambda \delta t + o(\delta t)$.

(b) The probability of no event in the interval $(t, t + \delta t)$ is $1 - \lambda \delta t + o(\delta t)$.

(c) The probability of more than one event in the interval $(t, t + \delta t)$ is $o(\delta t)$.

Then we say that the process is a random stream of events. In this notation, $o(\delta t)$ means some function of δt such that its ratio to δt tends to zero as δt tends to zero, that is, a quantity of smaller order of magnitude than δt itself. Roughly speaking these requirements mean that for a small interval of fixed length, there is a constant probability of just one event in this interval and a negligible probability of more than one event. From the above rules, we can calculate $P_n(t)$, the probability that there will be just n events in the interval $(0, t)$. We obtain $P_n(t)$ indirectly by considering what happens in a further interval $(t, t + \delta t)$. Provided $n \geqslant 1$, there will be just n events in $(0, t + \delta t)$ if either

(i)   there were already n events in $(0, t)$ and *no* further event took place in $(t, t + \delta t)$;

(ii)   there were already $n - 1$ events in $(0, t)$ and *one* further event took place in $(t, t + \delta t)$.

We do not have to specify the details for any other possibility, since its total probability is negligible in comparison with δt. Thus we have

$$P_n(t + \delta t) = P_n(t)[1 - \lambda \delta t + o(\delta t)]$$

$$+ P_{n-1}(t)[\lambda \delta t + o(\delta t)] + o(\delta t),$$

$$\frac{P_n(t + \delta t) - P_n(t)}{\delta t} = -\lambda P_n(t) + \lambda P_{n-1}(t)$$

$$+ \frac{o(\delta t)[P_n(t) + P_{n-1}(t) + 1]}{\delta t}.$$

Now let δt tend to zero. We now have

$$\frac{dP_n(t)}{dt} = \lambda P_{n-1}(t) - \lambda P_n(t), \quad n \geqslant 1. \tag{5.2}$$

The case when $n = 0$ is special, for then the only possibility is no event in $(0, t)$ *and* no further event in $(t, t + \delta t)$, that is

$$P_0(t + \delta t) = P_0(t)[1 - \lambda \delta t + o(\delta t)] + o(\delta t),$$

$$\frac{P_0(t + \delta t) - P_0(t)}{\delta t} = -\lambda P_0(t) + \frac{o(\delta t)[P_0(t) + 1]}{\delta t}.$$

Again, taking the limit as $\delta t \to 0$

$$\frac{dP_0(t)}{dt} = -\lambda P_0(t). \tag{5.3}$$

As it happens, these differential equations can be solved one at a time, commencing with equation (5.3),

$$\frac{dP_0(t)}{dt} = -\lambda P_0(t)$$

$$\frac{1}{P_0(t)} \frac{dP_0(t)}{dt} = -\lambda.$$

$$\frac{d}{dt}[\log_e P_0(t)] = -\lambda,$$

integrating with respect to time, and

$$\log_e P_0(t) = -\lambda t + k_0,$$

where k_0 is a constant. To evaluate this constant we must consider the initial conditions of the process. By time $t = 0$, it is certain that no event has occurred, that is

$$P_n(0) = 0, \quad n \neq 0;$$

but $P_0(0) = 1$. Hence, $k_0 = 0$ and we obtain

$$\log_e P_0(t) = -\lambda t \quad \text{or} \quad P_0(t) = e^{-\lambda t}.$$

Before solving the rest of the equations, we make an immediate deduction based on $P_0(t)$. It is related to an exponential distribution with parameter λ. Indeed it is the probability that a random variable with such a distribution exceeds t. Thus the random variable T, which is the time to the first event starting from 0 has the exponential distribution with parameter λ. Now as soon as an event occurs, we start from scratch again and the time to the *next* event also has an exponential distribution with parameter λ. In general, the waiting time T_n to the nth event is the sum of n independent waiting times to single events, that is the sum of n independent random variables each having an exponential distribution with parameter λ. The p.d.f. of T_n can be found by general methods but also emerges from the evaluation of $P_n(t)$.

Now that we have obtained $P_0(t)$, we return to the differential equations in equation (5.2) and put $n = 1$.

$$\frac{dP_1 t}{dt} = -\lambda P_1(t) + \lambda P_0(t) = -\lambda P_1(t) + \lambda e^{-\lambda t}$$

hence

$$\frac{dP_1(t)}{dt} + \lambda P_1(t) = \lambda e^{-\lambda t}.$$

Multiplying both sides by $e^{\lambda t}$

$$e^{\lambda t} \frac{dP_1(t)}{dt} + \lambda e^{\lambda t} P_1(t) = \lambda,$$

$$\frac{d}{dt}[e^{\lambda t} P_1(t)] = \lambda$$

integrating with respect to time

$$e^{\lambda t} P_1(t) = \lambda t + k_1.$$

But when $t = 0$, $P_1(0) = 0$, and thus $k_1 = 0$

$$P_1(t) = (\lambda t) e^{-\lambda t}.$$

Similarly, by substituting $n = 2$ in equation (5.2)

$$P_2(t) = \frac{(\lambda t)^2 e^{-\lambda t}}{2!}.$$

It is now easy to show by mathematical induction that

$$P_n(t) = \frac{(\lambda t)^n e^{-\lambda t}}{n!}.$$

For suppose

$$P_{n-1}(t) = \frac{(\lambda t)^{n-1} e^{-\lambda t}}{(n-1)!},$$

then from equation (5.2)

$$\frac{dP_n(t)}{dt} + \lambda P_n(t) = \frac{\lambda^n t^{n-1} e^{-\lambda t}}{(n-1)!}.$$

Multiply both sides by $e^{\lambda t}$.

$$\frac{d}{dt}[e^{\lambda t} P_n(t)] = \frac{\lambda^n t^{n-1}}{(n-1)!},$$

$$e^{\lambda t} P_n(t) = \frac{\lambda^n t^n}{n!} + k_n.$$

But $P_n(0) = 0$, hence $k_n = 0$, or

$$P_n(t) = \frac{(\lambda t)^n e^{-\lambda t}}{n!}.$$

Thus the number of events in any fixed interval has the Poisson distribution with parameter λt. The fixed origin is not material to the argument. Hence if t_1, t_2 are fixed times ($t_1 < t_2$) with respect to this same origin, then we may take t_1 as a new origin and the number of events in the interval (t_1, t_2) will have a Poisson distribution with parameter $\lambda(t_2 - t_1)$. This is consistent with the number of events in $(0, t_1)$, $(0, t_2)$ having Poisson distributions with parameters λt_1, λt_2, respectively. (See Problem 9, Chapter 4.)

The model just discussed has been applied in practice to, among other things, the flow of calls to a telephone exchange, the emission of particles from radioactive material, and accident rates. How then is λ to be understood in such contexts? Let k be a positive integer and the interval of time $(0, k)$ be divided into k intervals of unit duration. In each of these, the number of events has a Poisson distribution with parameter λ. Suppose, for any particular realization, there are f_x such intervals in which x events occur. Then the average number of events per interval is $\Sigma x f_x/k$. Now as k increases, we have $f_x/k \to \lambda^x e^{-\lambda}/x!$ and hence

$$\Sigma x f_x/k \to \sum_{x=0}^{\infty} x\lambda^x e^{-\lambda}/x!,$$

$$= \lambda \sum_{x=1}^{\infty} \lambda^{x-1} e^{-\lambda}/(x-1)! = \lambda.$$

That is, the average number of events per unit interval approaches λ.

Problem 7. Telephone calls arrive at an exchange in a random stream for which $\lambda = 1/5$ when t is measured in minutes. Starting from a fixed time, what is the probability that

(a)   there is no incoming call for $1/4$ hour;
(b)   the next call arrives in the course of $1/4$ hour;
(c)   the next but one call does not arrive before $1/4$ hour.

Problem 8. Assuming the model for a random stream of events, show that if two successive intervals of duration t_1 and t_2 contain events with the same λ, then the total number of events in the combined interval has a Poisson distribution with parameter $\lambda(t_1 + t_2)$. If it is known that s events occurred in the combined interval, what is the distribution of the number of events that fell in the interval of duration t_1?   ∎

We now go back and find the p.d.f. $f_n(t)$, of the time to the nth event in a random stream. We exploit again an observation recorded in Chapter 4. The nth event will occur before t if and only if there are at least n events in the interval $(0, t)$. So, in the notation of section 5.3,

$$\int_0^t f_n(x)\,\mathrm{d}x = \sum_{k=n}^{\infty} P_k(t).$$

Differentiating with respect to t,

$$f_n(t) = \sum_{k=n}^{\infty} P_k'(t)$$

$$= \sum_{k=n}^{\infty} [\lambda P_{k-1}(t) - \lambda P_k(t)], \quad \text{from equation (5.2)}$$

$$= \lambda P_{n-1}(t),$$

$$= \frac{\lambda(\lambda t)^{n-1} e^{-\lambda t}}{(n-1)!}, \quad t > 0.$$

This p.d.f. is of a type which is known as the *gamma distribution*, a class of distributions which is discussed in the next section. Note that, as would be expected, when $n = 1$, $f_1(t) = \lambda e^{-\lambda t}$, which is the p.d.f. of an exponential distribution.

5.4 THE GAMMA DISTRIBUTION

A continuous random variable X is said to have the **gamma distribution** with parameters $\lambda > 0, \alpha > 0$ if its p.d.f. $f(x)$ satisfies

$$f(x) = \frac{\lambda(\lambda x)^{\alpha-1} e^{-\lambda x}}{\Gamma(\alpha)}; \quad 0 \leqslant x < \infty$$

$$= 0, \text{ otherwise,}$$

where $\Gamma(\alpha)$ is defined as

$$\int_0^\infty x^{\alpha-1} e^{-x} \, dx,$$

and is known as the gamma function. X is also said to have the $\Gamma(\alpha, \lambda)$ distribution.

Example 3
If X has the $\Gamma(2, 1/2)$ distribution.

$$\Gamma(2) = \int_0^\infty x \, e^{-x} \, dx = 1 \text{ and hence}$$

$$f(x) = [(x/2) e^{-x/2}]/2, \quad 0 \leqslant x < \infty.$$

The reader should sketch a graph of $f(x)$ and verify that it has a maximum at $x = 2$. We can calculate, say, $\Pr(X \leqslant 8)$ as

$$\int_0^8 (1/2)(x/2) e^{-x/2} \, dx = 1 - 5 e^{-4}.$$

When $\alpha > 1$, $f(x)$ has a single maximum and tends to zero as x tends to infinity. For $\alpha < 1$, $f(x)$ tends to infinity as x tends to zero. When $\alpha = 1$, the $\Gamma(1, \lambda)$ distribution is also an exponential distribution with parameter λ. In this text we are mainly interested in the case when α is a positive integer, when $\Gamma(\alpha) = (\alpha - 1)!$.

Problem 9. When n is a positive integer, show that $\Gamma(n) = (n-1)!$.

Problem 10. Using a change of variable $y = \lambda x$, verify that for the gamma distribution,

$$\int_0^\infty f(x)\,dx = 1.$$

By considering the waiting times between events in a random stream of events, show that the sum of n independent random variables, each distributed with an exponential distribution with parameter λ, has a $\Gamma(n, \lambda)$ distribution.

Problem 11. If X_1 has the $\Gamma(n_1, \lambda)$ distribution and X_2 has the $\Gamma(n_2, \lambda)$ distribution, independently of X_1, find the distribution of $X_1 + X_2$.

5.5 THE NORMAL DISTRIBUTION

The continuous random variable X is said to have the **normal distribution** with parameters μ, $\sigma > 0$ if its probability density function $f(x)$ satisfies

$$f(x) = \frac{1}{\sqrt{(2\pi)}\sigma} \exp\left[-\frac{1}{2}\left(\frac{x-\mu}{\sigma}\right)^2\right], \quad -\infty < x < \infty. \qquad (5.4)$$

Such a random variable is said to be distributed $N(\mu, \sigma^2)$. Since $e^x > 0, f(x) > 0$, but it is not easy to verify that

$$\int_{-\infty}^{+\infty} f(x)\,dx = 1.$$

The graph of $f(x)$ is symmetrical about $x = \mu$, since

$$(x - \mu)^2 = (\mu - x)^2.$$

Also, $f(x)$ tends to zero as x tends to plus or minus infinity, since for positive x, $e^x > x$ hence $e^{-x^2/2} < 2/x^2$.

$$f'(x) = -\frac{(x - \mu)}{\sigma^2} f(x).$$

Thus $f'(\mu) = 0$ and consideration of the sign of $f'(x)$ in the neighbourhood of $x = \mu$, shows that this stationary value is a maximum.

The second derivative is

$$\frac{d}{dx}\left[-\frac{(x-\mu)}{\sigma^2}f(x)\right] = -\frac{f(x)}{\sigma^2} - \frac{(x-\mu)f'(x)}{\sigma^2}$$

$$= -\frac{f(x)}{\sigma^2} + \frac{(x-\mu)^2 f(x)}{\sigma^4}$$

$$= \frac{f(x)}{\sigma^4}[(x-\mu)^2 - \sigma^2]$$

and this is zero at the points $x = \mu \pm \sigma$. Consideration of the sign of $f''(x)$ shows that these are points of inflection.

The $\Pr(X \leqslant x_0)$ is of course the area under the curve of $f(x)$ from minus infinity to x_0. Since we cannot integrate $f(x)$ in a simple closed form, these probabilities have to be evaluated by approximate methods for selected values of x_0. At first sight it appears that a vast number of tables are required to cover suitable values of the parameters. The following consideration simplifies matters.

We require

$$\Pr(X \leqslant x_0) = \int_{-\infty}^{x_0} f(x)\,dx.$$

Now, in this integral, change the variable, putting $y = (x - \mu)/\sigma$, when $x = x_0$, $y = (x_0 - \mu)/\sigma$. Hence

$$\int_{-\infty}^{x_0} f(x)\,dx = \sigma \int_{-\infty}^{(x_0-\mu)/\sigma} f(\sigma y + \mu)\,dy$$

$$= \frac{1}{\sqrt{(2\pi)}} \int_{-\infty}^{(x_0-\mu)/\sigma} e^{-y^2/2}\,dy. \tag{5.5}$$

Thus the required area is also the area from minus infinity to $(x_0 - \mu)/\sigma$ under the curve

$$\frac{1}{\sqrt{(2\pi)}} e^{-x^2/2}.$$

This important function has been extensively tabulated. We can squeeze some more information from the above result, for

$$\Pr(X \leqslant x_0) = \Pr(X - \mu \leqslant x_0 - \mu) = \Pr[(X-\mu)/\sigma \leqslant (x_0 - \mu)/\sigma]$$

since σ is positive. Thus if the random variable $Y = (X-\mu)/\sigma$ and $y_0 = (x_0 - \mu)/\sigma$, then

$$\Pr(Y \leqslant y_0) = \frac{1}{\sqrt{(2\pi)}} \int_{-\infty}^{y_0} e^{-y^2/2}\,dy.$$

But

$$\frac{1}{\sqrt{(2\pi)}} \, e^{-y^2/2}$$

is the probability density function of a normal distribution with parameters $\mu = 0, \sigma = 1$. Hence, Y is distributed $N(0, 1)$, the unit normal distribution.

Thus the essential thing is to master the table for the unit normal distribution. Since the curve is symmetrical, only the area under half the curve is usually tabulated. Thus, in Table 5.1 is tabulated

$$\Phi(x) = \int_{-\infty}^{x} \frac{e^{-t^2/2} \, dt}{\sqrt{(2\pi)}}$$

for $x \geqslant 0$, that is to say $\Pr(X \leqslant x)$ for positive x. The values of x increase by 0.01 up to $x = 4$.

In fact, $\Pr(-3 \leqslant X \leqslant +3)$ is nearly one. For $x = 2.10$ we find $\Phi(x) = 0.98214$. Thus if X is distributed $N(0, 1)$, $\Pr(X \leqslant 2.10) = 0.98214$. Hence

$$\Pr(X \geqslant 2.10) = 1 - \Pr(X \leqslant 2.10) = 1 - 0.98214 = 0.01786$$

Moreover, by symmetry,

$$\Pr(X \leqslant -2.10) = \Pr(X \geqslant +2.10) = 0.01786;$$

and

$$\Pr(X \geqslant -2.10) = \Pr(X \leqslant +2.10) = 0.98214.$$

Taking these statements together,

$$\Pr(|X| \geqslant 2.10) = \Pr(X \leqslant -2.10 \text{ or } X \geqslant 2.10)$$

$$= 2(0.01786) = 0.03572.$$

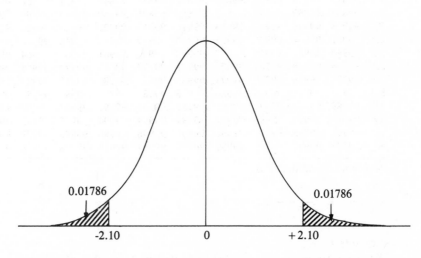

Fig. 5.2

Table 5.1. Table of the normal distribution function

x	.00	.01	.02	.03	.04	.05	.06	.07	.08	.09
0	.5	.50399	.50798	.51197	.51595	.51994	.52392	.5279	.53188	.53586
.1	.53983	.5438	.54776	.55172	.55567	.55962	.56356	.56749	.57142	.57535
.2	.57926	.58317	.58706	.59095	.59483	.59871	.60257	.60642	.61026	.61409
.3	.61791	.62172	.62552	.6293	.63307	.63683	.64058	.64431	.64803	.65173
.4	.65542	.6591	.66276	.6664	.67003	.67364	.67724	.68082	.68439	.68793
.5	.69146	.69497	.69847	.70194	.7054	.70884	.71226	.71566	.71904	.7224
.6	.72575	.72907	.73237	.73565	.73891	.74215	.74537	.74857	.75175	.7549
.7	.75804	.76115	.76424	.7673	.77035	.77337	.77637	.77935	.7823	.78524
.8	.78814	.79103	.79389	.79673	.79955	.80234	.80511	.80785	.81057	.81327
.9	.81594	.81859	.82121	.82381	.82639	.82894	.83147	.83398	.83546	.83891
1	.84134	.84375	.84614	.84849	.85083	.85314	.85543	.85769	.85993	.86214
1.1	.86433	.8665	.86864	.87076	.87286	.87493	.87698	.879	.881	.88298
1.2	.88493	.88686	.88877	.89065	.89251	.89435	.89617	.89796	.89973	.90147
1.3	.9032	.9049	.90658	.90824	.90988	.91149	.91309	.91466	.91621	.91774
1.4	.91924	.92073	.9222	.92364	.92507	.92647	.92785	.92922	.93056	.93189
1.5	.93319	.93448	.93574	.93699	.93822	.93943	.94062	.94179	.94295	.94408
1.6	.9452	.9463	.94738	.94845	.9495	.95053	.95154	.95254	.95352	.95449
1.7	.95543	.95637	.95728	.95818	.95907	.95994	.9608	.96164	.96246	.96327
1.8	.96407	.96485	.96562	.96637	.96712	.96784	.96856	.96926	.96995	.97062
1.9	.97128	.97193	.97257	.9732	.97381	.97441	.975	.97558	.97615	.9767
2	.97725	.97778	.97831	.97882	.97932	.97982	.9803	.98077	.98124	.98169
2.1	.98214	.98257	.983	.98341	.98382	.98422	.98461	.985	.98537	.98574
2.2	.9861	.98645	.98679	.98713	.98745	.98778	.98809	.9884	.9887	.98899
2.3	.98928	.98956	.98983	.9901	.99036	.99061	.99086	.99111	.99134	.99158
2.4	.9918	.99202	.99224	.99245	.99266	.99286	.99305	.99324	.99343	.99361
2.5	.99379	.99396	.99413	.9943	.99446	.99461	.99477	.98491	.99506	.9952
2.6	.99534	.99547	.9956	.99573	.99585	.99598	.99609	.99621	.99632	.99643
2.7	.99653	.99664	.99674	.99683	.99693	.99702	.99711	.9872	.99728	.99736
2.8	.99744	.99752	.9976	.99767	.99774	.99781	.99788	.99795	.99801	.99807
2.9	.99813	.99819	.99825	.99831	.99836	.99841	.99846	.99851	.99856	.99861
3	.99865	.99869	.99874	.99878	.99882	.99886	.99889	.99893	.99896	.999
3.1	.99903	.99906	.9991	.99913	.99916	.99918	.99921	.99924	.99926	.99929
3.2	.99931	.99934	.99936	.99938	.9994	.99942	.99944	.99946	.99948	.9995
3.3	.99952	.99953	.99955	.99957	.99958	.9996	.99961	.99962	.99964	.99965
3.4	.99966	.99968	.99969	.9997	.99971	.99972	.99973	.99374	.99975	.99976
3.5	.99977	.99978	.99978	.99979	.9998	.99981	.99981	.99982	.99983	.99983
3.6	.99984	.99985	.99985	.99986	.99986	.99987	.99987	.99988	.99988	.99989
3.7	.99989	.9999	.9999	.99991	.99991	.99991	.99991	.99992	.99992	.99992
3.8	.99993	.99993	.99993	.99994	.99994	.99994	.99994	.99995	.99995	.99995
3.9	.99995	.99995	.99996	.99996	.99996	.99996	.99996	.99996	.99997	.99997
4	.99997	.99997	.99997	.99997	.99997	.99997	.99998	.99998	.99998	.99998

The function tabulated is

$$\Phi(x) = \int_{-\infty}^{x} \frac{e^{-t^2/2}}{\sqrt{(2\pi)}} \, dt.$$

Example 4

X has a distribution $N(3, 16)$. Calculate $\Pr(X \leqslant 9)$ and $\Pr(X \leqslant 9 | X \geqslant 5)$.

Since X is distributed $N(3, 16)$, $Z = (X - 3)/4$ is distributed $N(0, 1)$,

$$\Pr(X \geqslant 9) = \Pr\left(\frac{X-3}{4} \geqslant \frac{9-3}{4}\right) = \Pr(Z \geqslant 1.5)$$

but from Table 5.1, $\Pr(Z \leqslant 1.5) = 0.9332$, hence $\Pr(Z \geqslant 1.5) = 1 - 0.9332 = 0.0668$. Also

$$\Pr(X \geqslant 9 | X \geqslant 5) = \frac{\Pr(X \geqslant 9 \text{ and } X \geqslant 5)}{\Pr(X \geqslant 5)}$$

$$= \frac{\Pr(X \geqslant 9)}{\Pr(X \geqslant 5)} .$$

$$\Pr(X \geqslant 5) = \Pr\left(\frac{X-3}{4} \geqslant \frac{5-3}{4}\right) = \Pr(Z \geqslant 0.5).$$

From Table 5.1, $\Pr(Z \leqslant 0.5) = 0.6915$, hence $\Pr(Z \geqslant 0.5) = 1 - 0.6915 = 0.3085$. Finally,

$$\Pr(X \geqslant 9 | X \geqslant 5) = 0.0668/0.3085 = 0.217.$$

Example 5
For a certain normal distribution with parameters μ, σ, 10% of the distribution lies below 40 and 20% is greater than 80. Find μ, σ.
    Suppose X is distributed $N(\mu, \sigma^2)$, $Z = (X - \mu)/\sigma$ is distributed $N(0, 1)$. The upper 10% point of the $N(0, 1)$ distribution is 1.2816, hence the lower 10% point is -1.2816.
    Hence, if $\Pr(Z \leqslant -1.2816) = 0.1$,

$$\Pr\left(\frac{X-\mu}{\sigma} \leqslant -1.2816\right) = 0.1 \quad \text{or} \quad \Pr(X \leqslant \mu - 1.2816\sigma) = 0.1.$$

But $\Pr(X \leqslant 40) = 0.1$, hence $\mu - 1.2816\sigma = 40$. Similarly, since the upper 20% of the unit normal distribution is 0.8416,

$$\Pr\left(\frac{X-\mu}{\sigma} \geqslant 0.8416\right) = 0.2 \quad \text{or} \quad \mu + 0.8416\sigma = 80.$$

The two simultaneous equations may now be solved for μ and σ.
    Most tables of the normal distribution display the upper percentage points separately (see Table 5.2). Frequently used values are the upper and lower 2½% points, which are ±1.96, and the upper and lower 5% points, which are ±1.64.

Problem 12. The lifetime in hours of a certain electrical part has the normal distribution with parameters $\mu = 100$ and $\sigma = 5$. What is the probability that a new part lasts at least 105 hours? If a part has already run for 90 hours, what is the probability that it will last at least another 15 hours?

Problem 13. A machine cuts bolts whose length in inches is approximately normally distributed with parameters $\mu = 8$, $\sigma = 0.1$. If only bolts between 7.8 inches and 8.3 inches are acceptable, what proportion of the machine's output is rejected?

Problem 14. The average proportion of p of insects killed by administration of x units of insecticide is given by

$$p = \int_{-\infty}^{(x-\mu)/\sigma} (2\pi)^{-1/2} e^{-t^2/2} \, dt$$

where μ and σ are constants.

When $x = 10$, $p = 0.400$, and when $x = 15$, $p = 0.900$; what dose will be lethal to 50% of the insect population, on average?

If a dose of 17.5 units is administered to each of 100 insects, how many will be expected to die?

Oxford & Cambridge (S.P.), 1968.

Table 5.2. Percentage points of normal distribution

P	0.000	0.001	0.002	0.003	0.004	0.005	0.006	0.007	0.008	0.009
0	∞	3.0902	2.8782	2.7478	2.6521	2.5758	2.5121	2.4573	2.4089	2.3656
.01	2.3263	2.2904	2.2571	2.2262	2.1973	2.1701	2.1444	2.1201	2.0969	2.0749
.02	2.0538	2.0335	1.0141	1.9954	1.9774	1.96	1.9431	1.9268	1.911	1.8957
.03	1.8808	1.8663	1.8522	1.8384	1.825	1.8119	1.7991	1.7866	1.7744	1.7624
.04	1.7507	1.7392	1.7279	1.7169	1.706	1.6954	1.6849	1.6747	1.6646	1.6546
.05	1.8449	1.6352	1.6258	1.6164	1.6072	1.5982	1.5893	1.5805	1.5718	1.5632
.06	1.5548	1.5484	1.5382	1.5301	1.522	1.5141	1.5063	1.4985	1.4909	1.4833
.07	1.4758	1.4684	1.4611	1.4538	1.4466	1.4395	1.4325	1.4355	1.4187	1.4118
.08	1.4051	1.3984	1.3917	1.3852	1.3787	1.3722	1.3658	1.3595	1.3532	1.3469
.09	1.3408	1.3346	1.3285	1.3225	1.3165	1.3106	1.3047	1.2988	1.293	1.2873

	0.00	0.01	0.02	0.03	0.04	0.05	0.06	0.07	0.08	0.09
.1	1.2816	1.2265	1.175	1.1264	1.0803	1.0364	.9945	.9542	.9154	.8779
.2	.8416	.8064	.7722	.7388	.7063	.6745	.6433	.6128	.5828	.5534
.3	.5244	.4959	.4677	.4399	.4125	.3853	.3585	.3319	.3055	.2793
.4	.2533	.2275	.2019	.1764	.151	.1257	.1004	.0753	.0502	.0251
.5	0	−.0251	−.0502	−.0753	−.1004	−.1257	−.151	−.1764	−.2019	−.2275

The table gives values of x_p such that

$$p = \int_{x_p}^{\infty} \frac{e^{-t^2/2}}{\sqrt{(2\pi)}} \, dt$$

Problem 15. Show that if X is distributed $N(\mu, \sigma^2)$ then $(X-a)/b$ is distributed

$$N\left(\frac{\mu-a}{b}, \frac{\sigma^2}{b^2}\right).$$ ■

* To show that the p.d.f. of the distribution $N(\mu, \sigma^2)$ integrates to 1 is equivalent to proving that

$$\frac{1}{\sqrt{(2\pi)}} \int_{-\infty}^{+\infty} e^{-x^2/2} \, dx = 1 \quad \text{or} \quad \frac{1}{\sqrt{(2\pi)}} \int_{0}^{\infty} e^{-x^2/2} \, dx = \frac{1}{2},$$

from symmetry. We now sketch a non-rigorous proof of this result. Suppose

$$I(x_0) = \int_{0}^{x_0} e^{-x^2/2} \, dx$$

then

$$I^2(x_0) = \left(\int_{0}^{x_0} e^{-x^2/2} \, dx\right) \left(\int_{0}^{x_0} e^{-x^2/2} \, dx\right)$$

$$= \left(\int_{0}^{x_0} e^{-x^2/2} \, dx\right) \left(\int_{0}^{x_0} e^{-y^2/2} \, dy\right)$$

$$= \int_{0}^{x_0} \int_{0}^{x_0} e^{-(x^2+y^2)/2} \, dx \, dy.$$

Now, changing variables to $x = r \cos\theta$, $y = r \sin\theta$, then $x^2 + y^2 = r^2$ and

$$I^2(x_0) > \int_{0}^{x_0} \int_{0}^{\pi/2} e^{-r^2/2} \, r \, dr \, d\theta = \frac{\pi}{2} [-e^{-r^2/2}]_{0}^{x_0}$$

$$= \frac{\pi}{2} (1 - e^{-x_0^2/2}).$$

The inequality arises because we are now integrating over a quadrant of a circle rather than a square. We omit showing that in fact the difference tends to zero as x_0 tends to infinity. The limit of

$$\frac{\pi}{2} (1 - e^{-x_0^2/2})$$

is $\pi/2$ or

$$\lim_{x_0 \to \infty} I(x_0) = \sqrt{\frac{\pi}{2}}.$$

.

BRIEF SOLUTIONS AND COMMENTS ON THE PROBLEMS

Problem 1. A sketch shows that $AB > a$ when its midpoint is closer to the centre than $(a^2 - a^2/4)^{1/2} = a\sqrt{3}/2$. Hence required probability is $\sqrt{3}/2$. This result should be compared with Example 1.

Problem 2. The line $x + y = 1.5$ meets the square suggested in the hint at the points $(1/2, 1)$, $(1, 1/2)$. The triangle with co-ordinates $(1/2, 1)$, $(1, 1/2)$, $(1, 1)$ has area $1/8$. Hence from the uniformity of the distributions, the required probability is $1 - 1/8 = 7/8$.

Problem 3.

$$\Pr[k/10 \leqslant \sqrt{X} < (k + 1)/10]$$
$$= \Pr[k^2/100 \leqslant X < (k + 1)^2/100]$$
$$= [(k + 1)^2 - k^2]/100 = (2k + 1)/100.$$

Problem 4. The machine fails before t_0 if either component fails before t_0. The probability of this event is

$$(1 - e^{-\lambda t_0}) + (1 - e^{-\lambda t_0}) - (1 - e^{-\lambda t_0})(1 - e^{-\lambda t_0})$$
$$= 1 - e^{-2\lambda t_0}.$$

From section 5.2, p.d.f. must be exponential with parameter 2λ.

Problem 5. If $u = 1500$, $\Pr(X > 3000) = e^{-2} = 0.135$. Last condition requires $1 - e^{-5/u} = 1/1000$.

Problem 6. e^{-3kt_0}.

$3!(1 - e^{-kt_0})(e^{-kt_0} - e^{-2kt_0})e^{-2kt_0}$.

Problem 7. The number of calls in 15 minutes has a Poisson distribution with parameter $15 \times 1/5 = 3$.

(a) $\Pr(\text{no call in 15 minutes}) = e^{-3} = 0.0498$.
(b)   Next new call arrives during 15 minutes if there is at least one new call, with probability $1 - \text{probability (no new call)} = 1 - 0.0498 = 0.950$.
(c)   There is at most one call in the next 15 minutes, with probability $e^{-3} + 3e^{-3} = 0.199$. The general significance of the parameter in a Poisson distribution (here λt) will be appreciated when the reader comes to the chapter on expectations.

Problem 8. The first part follows immediately from Problem 5 of Chapter 4. For the second part: let r be the number of events in the interval of duration t_1, then

$$\Pr(r|s) = \Pr(r \& s)/\Pr(s)$$
$$= \Pr(r \text{ in } t_1) \Pr(s - r \text{ in } t_2)/\Pr(s),$$

since events in disjoint intervals are independent.

$$\Pr(s) = [\lambda(t_1 + t_2)]^s \, e^{-\lambda t_1 - \lambda t_2}/s!$$
$$\Pr(r \text{ in } t_1) = (\lambda t_1)^r \, e^{-\lambda t_1}/r!$$
$$\Pr(s - r \text{ in } t_2) = (\lambda t_2)^{s-r} \, e^{-\lambda t_2}/(s - r)!$$

After substituting these values, and cancelling,

$$\Pr(r|s) = \binom{s}{r} \left(\frac{t_1}{t_1 + t_2}\right)^r \left(\frac{t_2}{t_1 + t_2}\right)^{s-r}.$$

Evidently the distribution of R given $S = s$ is binomial with parameters s, $t_1/(t_1 + t_2)$.

Problem 9. Integrate by parts, $n \geq 2$,

$$\int_0^\infty x^{n-1} \, e^{-x} \, dx = (-e^{-x} x^{n-1})_0^\infty + \int_0^\infty (n-1)x^{n-1} \, e^{-x} \, dx$$
$$= (n-1) \, \Gamma(n-1).$$

Now

$$\Gamma(1) = \int_0^\infty e^{-x} \, dx = 1,$$

hence, by repeated application, $\Gamma(n) = (n-1)!$.

Problem 10. Since $\lambda > 0$,

$$\int_0^\infty \lambda(\lambda x)^{\alpha-1} \, e^{-\lambda x} \, dx = \int_0^\infty y^{\alpha-1} \, e^{-y} \, dy = \Gamma(\alpha)$$

and hence result.

Problem 11. From section 5.3, $X_1 + X_2$ is the sum of $n_1 + n_2$ exponential distributions with common parameter λ. Hence, $X_1 + X_2$ has a gamma distribution with parameters λ, $n_1 + n_2$.

Problem 12.

$$\Pr(X > 105) = \Pr[(X - 100)/5 > (105 - 100)/5]$$
$$= \Pr(Z > 1) = 1 - \Pr(Z < 1) = 1 - 0.8413$$
$$= 0.1587, \text{ since } Z \text{ is distributed } N(0, 1).$$

$$\begin{aligned}
\Pr(X > 105 \,|\, X > 90) &= \Pr(X \geqslant 105 \text{ and } X \geqslant 90)/\Pr(X \geqslant 90)\\
&= \Pr(X \geqslant 105)/\Pr(X \geqslant 90)\\
&= 0.1587/0.977.
\end{aligned}$$

Problem 13.

$$(8.3 - 8)/0.1 = 3. \quad \Pr(Z > 3) = 0.00135$$
$$(7.8 - 8)/0.1 = -2. \quad \Pr(Z < -2) = 0.02275.$$

Hence proportion ~ 0.024 rejected.

Problem 14. $(10 - \mu)/\sigma = -0.2533$ and $(15 - \mu)/\sigma = 1.2816$, from the table of the $N(0, 1)$ distribution. Thence $\sigma = 3.26$, $\mu = 10.83$. 50% point corresponds to $\mu = 10.83$. About 98 would be expected to die.

Problem 15. Suppose $b > 0$, then

$$\Pr(X \leqslant t) = \Pr[(X - a)/b \leqslant (t - a)/b]$$

$$= \int_{-\infty}^{t} \frac{1}{\sqrt{2\pi}\sigma} \exp\left[-\frac{(x - \mu)^2/\sigma^2}{2}\right] dx.$$

Change the variable to $y = (x - a)/b$,

$$\Pr(X \leqslant t) = \int_{-\infty}^{(t-a)/b} \frac{b}{\sqrt{2\pi}\sigma} \exp\left[-\frac{(yb - \mu + a)^2/\sigma^2}{2}\right] dy.$$

But the integrand is the p.d.f. of a $N[(\mu - a)/b, \sigma^2/b^2]$ distribution. Changes in the inequality and limits of the integral produce the required result for $b < 0$.

6

Distribution Function

6.1 THE MODE

The simplest descriptive measures of the 'centre' of a probability density function are provided by the mode and the median. **A mode** of a distribution is a value of the random variable at which there is a maximum of the probability density function. Confidence in this measure is undermined by the possibility of several modes existing. A distribution which has only one mode is said to be unimodal.

6.2 THE MEDIAN

The **median** of a distribution is a number x_0 such that

$$\Pr(X \leqslant x_0) \geqslant \frac{1}{2} \quad \text{and} \quad \Pr(X \geqslant x_0) \geqslant \frac{1}{2}. \tag{6.1}$$

Roughly speaking, the median is a value above and below which half the probability lies. For a continuous distribution there is always in principle a single value x_0 such that $\Pr(X \leqslant x_0) = 1/2$, though it may be difficult to recover its numerical value. For a discrete distribution there may be an ambiguity. Thus if the discrete random variable X takes the values 0, 1, 2 with probabilities 1/4, 1/4, 1/2, then any x_0 such that $1 < x_0 < 2$ satisfies the definition and is a median. The matter is usually settled by nominating the midpoint of such an interval as the median, in this case $1\frac{1}{2}$.

6.3 CUMULATIVE DISTRIBUTION FUNCTION

A useful picture of the spread of probability in a distribution is provided by the **cumulative distribution function** $F(x)$, which for either a continuous or a discrete distribution, is defined as

$$F(x) = \Pr(X \leqslant x).$$

From its definition, $0 \leqslant F(x) \leqslant 1$,

$$F(-\infty) = 0,$$

$$F(+\infty) = 1,$$

$$\Pr(a < X \leqslant b) = F(b) - F(a) \geqslant 0.$$

Hence, if $a < b$, $F(a) \leqslant F(b)$, and $F(x)$ is a non-decreasing function.

The cumulative distribution function is also known merely as the distribution function. The usual abbreviations are c.d.f. and d.f.

For a continuous random variable,

$$F(x) = \int_{-\infty}^{x} f(t)\,dt \tag{6.2}$$

and its graph is continuous. Its slope need not be everywhere continuous, but where this is so, it is equal to the p.d.f. That is to say, $F'(x) = f(x)$.

Example 1

$$f(x) = 1/k, \ 0 < x < k,$$

$$F(x) = \int_{0}^{x} \frac{1}{k}\,dt = x/k$$

and is sketched in Fig. 6.1.

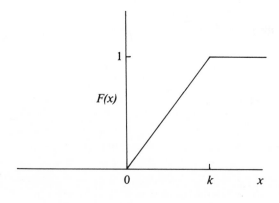

Fig. 6.1

For a discrete random variable,

$$F(x) = \sum_{t \leqslant x} f(t) \tag{6.3}$$

and its graph is a step-function.

Example 2
If

$$f(x) = 1/4, \quad x = 0, 1, 2, 3, \text{ then}$$
$$F(x) = 0, \quad x < 0$$
$$F(x) = 1/4, \quad 0 \leqslant x < 1$$
$$F(x) = 1/2, \quad 1 \leqslant x < 2$$
$$F(x) = 3/4, \quad 2 \leqslant x < 3$$

and

$$F(x) = 1, \quad 3 \leqslant x$$

with graph as shown in Fig. 6.2.

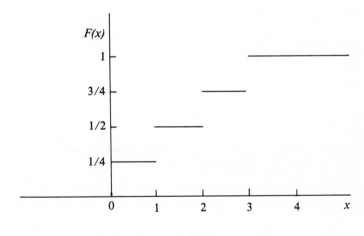

Fig. 6.2

Example 3
The continuous random variable X has p.d.f.

$$f(x) = 2x\, e^{-x^2}, \quad 0 \leqslant x < \infty$$
$$= 0 \text{ otherwise.}$$

Find the mode, distribution function, and median of X. Now,

$$f'(x) = 2\,e^{-x^2} - 4x^2\,e^{-x^2} = 2\,e^{-x^2}(1 - 2x^2),$$

and by considering the sign of this derivative, or otherwise, there is a maximum at $x = 1/\sqrt{2}$, which is the mode.

$$F(x) = \int_0^x 2t\,e^{-t^2}\,dt$$

$$= [-e^{-t^2}]_0^x$$

$$= 1 - e^{-x^2}$$

We observe that $F(0) = 0$, and

$$\lim_{x \to \infty} F(x) = 1.$$

The graph of $F(x)$ should be sketched as an exercise. Note that the mode of $f(x)$ corresponds to a point of inflection in $F(x)$. The median x_0 satisfies $F(x_0) = 1/2$, or $1 - e^{-x_0^2} = 1/2$; that is, $x_0 = \sqrt{(\log_e 2)}$.

Problem 1. Find the distribution function, median, and mode for the following probability density functions.

(a) $f(x) = \dfrac{1}{2\sqrt{x}}$, $\quad 0 \leqslant x < 1$

(b) $f(x) = \frac{1}{4}(\frac{3}{4})^{x-1}$, $\quad x = 1, 2, \ldots$

(c) $f(x) = 1 - |1 - x|$, $\quad 0 \leqslant x \leqslant 2$

(d) $f(x) = 6x(1 - x)$, $\quad 0 \leqslant x \leqslant 1$.

Problem 2. For the following probability density functions find the distribution function. In each case verify the result by differentiation,

(a) $f(x) = \frac{1}{2}x^2\,e^{-x}$, $\quad 0 \leqslant x < \infty$

$\quad = 0$ otherwise

(b) $f(x) = \dfrac{6x}{(1 + x)^4}$, $\quad 0 \leqslant x < \infty$

$\quad = 0$ otherwise.

Problem 3. If

$$\phi_{n,m} = \int_0^1 \frac{(n + m - 1)!}{(n - 1)!(m - 1)!}\, x^{n-1}(1 - x)^{m-1}\,dx,$$

show that $\phi_{n,m} = \phi_{n-1,m+1}$, and hence or otherwise that $\phi_{n,m} = 1$.

6.4 SAMPLING A DISTRIBUTION

Rather surprisingly, every random variable with a continuous p.d.f. can be transformed into another variable which has a rectangular distribution. To be specific, if X has p.d.f. $f(x)$ and d.f.

$$F(x) = \int_{-\infty}^{x} f(t)\,dt$$

then the random variable $F(X)$ has the rectangular distribution over the interval $(0, 1)$. For we have already seen that F is a non-decreasing function of x and, if f is continuous, it is strictly increasing, hence

$$\Pr(X \leqslant x) = \Pr[F(X) \leqslant F(x)].$$

But, $\Pr(X \leqslant x) = F(x)$, in virtue of the definition of $F(x)$. Hence,

$$\Pr[F(X) \leqslant F(x)] = F(x).$$

If we write $Y = F(X), y = F(x)$,

$$\Pr(Y \leqslant y) = y, \quad 0 \leqslant y \leqslant 1.$$

This is the distribution function of the stated rectangular distribution.

This little result has a useful practical application. For suppose we require a random sample from a continuous distribution with distribution function $F(x)$. We draw instead a random sample from the rectangular distribution over $(0, 1)$, let these values be $y_1, y_2, \ldots, y_n$. Then the values x_i such that $y_i = F(x_i)$ are a random sample from the distribution of X. To be sure, not every such transformation can be inverted explicitly or prove suitable for available tables. In such cases, approximate solutions may be used.

Example 4
X has the p.d.f. $f(x) = 2x\,e^{-x^2}, 0 < x < \infty$.

$$F(x) = \int_{0}^{x} 2t\,e^{-t^2}\,dt = 1 - e^{-x^2},$$

hence, if $y_i = F(x_i)$,

$$x_i = \sqrt{[-\log_e(1 - y_i)]}.$$

Problem 4. Explain how to draw a random sample from the distributions with p.d.f.

(a) $f(x) = 1/[\pi(1 + x^2)], \quad -\infty < x < \infty$
(b) $f(x) = ra^r/x^{r+1}, \quad x \geqslant a.$ ∎

The random sample from the rectangular distribution over $(0, 1)$ may be obtained from a table of random numbers. The original presenters of such tables had of course to find a source. For example, Fisher and Yates took 15 000 digits from tables of logarithms and Tippett took 10 400 digits from entries in population census tables. More recently, machines have been used to generate random digits. Such sets have been tested to ensure that they enjoy certain obvious properties, in particular that:

(a)   each digit occurs approximately equally often; similarly for each pair of digits;
(b)   digits in adjacent places are independent.

We start by choosing a page and a row and column on that page at random. We then read off in any direction either one at a time or in pairs, etc. Thus if we need a random sample of ten values from the rectangular distribution over (0, 1) to three decimal places, we draw ten successive sets of three random digits and affix a decimal point.

There are many other applications of such tables. Suppose we wish to compare three drugs on thirty patients. To reduce the probability of bias, we may make a random selection of ten patients from the available thirty to receive treatment A. Then from the remaining twenty patients a further random selection of ten to receive treatment B, the remainder to receive C. Now we could handle this by assigning to each patient a different card from a set of thirty cards numbered 1–30. After shuffling, we draw ten cards without replacement. The patients whose numbers are on these cards are then given drug A, and so on. Such a method becomes cumbersome and time-consuming if there are many cards and it may prove difficult to ensure proper shuffling.

To use the table of random numbers, we first renumber the patients 00, 01, 01, . . . 29 and then draw ten different pairs of random digits, rejecting pairs that have already been noted or which cannot be used. Wastage can be reduced by subtracting 30 from every pair from 30 to 59 inclusive and 60 from every pair from 60 to 89 inclusive. Thus the pair 81 counts as patient number 21. In this way, only the numbers 90 to 99 are wasted.

6.5 EMPIRICAL DISTRIBUTION FUNCTION

We recall that the distribution function, $F(x)$, is the probability that a random value from the distribution does not exceed x. Indeed, for a number of such independent values, the observed proportion of values which do not exceed x ought to be close to $F(x)$. This suggests that we should examine the difference between this observed proportion and the distribution function throughout the range of values of the variable.

Example 5
Suppose four independent values are drawn from a continuous distribution are found to be 0.84, 0.42, 0.53, 0.91, in order of appearance. We wish to compute $S_4(x)$, the proportion of values less than or equal to x. It pays to put the values

in increasing order of magnitude; 0.42, 0.53, 0.84, 0.91. There are no values less than 0.42, so that $S_4(x) = 0$ for $x < 0.42$. At $x = 0.42$ the proportion jumps to 1/4 and remains at this value until $x = 0.53$. We find

$$
\begin{aligned}
S_4(x) &= 0, & x < 0.42; \\
&= 1/4, & 0.42 \leqslant x < 0.53; \\
&= 1/2, & 0.53 \leqslant x < 0.84; \\
&= 3/4, & 0.84 \leqslant x < 0.91; \\
&= 1, & 0.91 \leqslant x.
\end{aligned}
$$

The graph of $S_4(x)$ is a step function, with jumps where the obtained values are located.

Now suppose it is claimed that the four values have been drawn from a distribution with p.d.f. $f(x) = 2x$, $0 < x < 1$. The corresponding c.d.f. is $F(x) = x^2$, $0 < x < 1$. In Fig. 6.3 we have drawn the graphs of $F(x)$ and $S_4(x)$

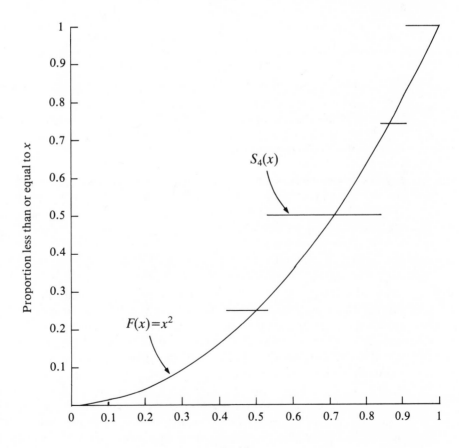

Fig. 6.3

and the reader will find this helpful in computing $F(x) - S_4(x)$. It will be seen that sometimes $S_4(x)$ is below $F(x)$ and sometimes above, so that the difference is sometimes positive and at other times negative. Since large deviations of either sign are damaging to the claim, we consider $|F(x) - S_4(x)|$. Kolmogorov has suggested that we look at the largest value of $|F(x) - S_4(x)|$ and consider whether this is acceptable. This is equivalent to asking whether $S_4(x)$ is ever too far from $F(x)$. Bearing in mind that $F(x)$ is increasing, we must look at the points where the graph of $F(x)$ intersects with that of $S_4(x)$. The reader should verify that

$$\max|F(x) - S_4(x)| = (0.42)^2 - 0 = 0.1764, 0 < x < 0.42;$$

$$= 0.25 - (0.42)^2 = 0.0736, 0.42 \leqslant x < 0.5;$$

$$= (0.53)^2 - 0.25 = 0.0309, 0.5 \leqslant x < 0.53;$$

$$= 0.5 - (0.53)^2 = 0.2191, 0.53 \leqslant x < \sqrt{0.5};$$

$$= (0.84)^2 - 0.5 = 0.2056, \sqrt{0.5} \leqslant x < 0.84;$$

$$= 0.75 - (0.84)^2 = 0.0444, 0.84 \leqslant x < \sqrt{0.75};$$

$$= (0.91)^2 - 0.75 = 0.0781, \sqrt{0.75} \leqslant x < 0.91;$$

$$= 1 - (0.91)^2 = 0.1719, 0.91 \leqslant x < 1.$$

(The reader should recall that $F(x)$ is sometimes less than $S_4(x)$ and sometimes greater, and this determines the order of subtraction.)

Thus $|F(x) - S_4(x)|$ never exceeds 0.2191 for these four values, and we would then need to question whether it was reasonable in a sample of 4 to obtain a deviation of 0.2191.

Problem 5. Five independent values are drawn from a continuous distribution and are found to be 5.1, 3.2, 4.7, 0.5, 6.8. Calculate $S_5(x)$. It is claimed that the sample has been drawn from an exponential distribution with parameter one. The distribution function is accordingly $F(x) = 1 - e^{-x}$. Calculate $\max|F(x) - S_5(x)|$.   ∎

The original question facing the statistician may have been 'is it reasonable to believe that *these* sample values could have been drawn at random from this particular continuous distribution? We may call $S_n(x)$, the proportion of sample values not exceeding x, the **empirical distribution function**. Then another formulation of the question would be 'Does the empirical distribution function appear to be compatible with this cumulative distribution function?' Neither version appears to be decisive in form and, certainly, we cannot prove that a sample *must* have been drawn from a particular distribution. Indeed there may be many distributions with which a sample is compatible (and possibly some with which it is definitely not). What we must try to do is to set a level at which we would be prepared to accept the compatibility and that level would depend on the importance to us of the decision being taken. We will not here discuss how to proceed further with the decision, except to say that we can alleviate some of the difficulties by making an application of the result derived in section 6.4, namely that if X is continuous then $F(X)$ has a rectangular distribution over $(0, 1)$. The cumulative distribution function of this rectangular distribution is just x. We transform the obtained sample values $x_1, x_2, \ldots, x_n$ to $F(x_1), F(x_2), \ldots, F(x_n)$ and compare the empirical distribution function of these transformed values with x.

Example 6

For Example 5, $F(x) = x^2$, $F(0.42) = 0.1764$, $F(0.53) = 0.2809$, $F(0.84) = 0.7056$, $F(0.91) = 0.8281$.   ■

The comparison now to be made is between the step function of the transformed results and the straight line, which is the graph of the distribution function of the rectangular distribution, viz. x. The reader may easily check that the maximum distance between the graphs remains unchanged.

The transformation suggested has the merit of standardizing the comparison to be made. Another potential advantage stemming from the suggested transformation is that it facilitates the study of the sample properties. We have already noted that for n independent values from a rectangular distribution, the number which do not exceed x has a binomial distribution with parameters n, x. That is, the probability that just k values do not exceed x is

$$\binom{n}{k} x^k (1 - x)^{n-k}, \quad 0 < x < 1.$$

When computing the empirical distribution function we were obliged to put the sample values in order of magnitude. We are now in a position to find the probability density function of, say, the rth value in order of magnitude. For this value will not exceed x if there are at *least* r sample values which do not exceed x and this has probability

$$\sum_{k=r}^{n} \binom{n}{k} x^k (1 - x)^{n-k}.$$

The probability density function can be found by differentiating with respect to x, yielding

$$\sum_{k=r}^{n} \binom{n}{k} k x^{k-1} (1 - x)^{n-k} - \sum_{k=r}^{n} \binom{n}{k} (n-k) x^k (1 - x)^{n-k-1}$$

$$= n \left[\sum_{r}^{n} \binom{n-1}{k-1} x^{k-1} (1 - x)^{n-k} \right.$$

$$\left. - \sum_{r}^{n-1} \binom{n-1}{k} x^k (1 - x)^{n-k-1} \right]$$

$$= n\binom{n-1}{r-1} x^{r-1}(1-x)^{n-r}$$

$$= \frac{n!}{(r-1)!(n-r)!}\, x^{r-1}(1-x)^{n-r}, \quad 0<x<1. \tag{6.4}$$

The equation is a particular case of a class of probability densities

$$f(x) = \frac{(m+n-1)!}{(m-1)!(n-1)!}\, x^{m-1}(1-x)^{n-1}, \quad 0<x<1. \tag{6.5}$$

A continuous random variable with such a density is said to have a **beta distribution** with (integral) parameters m, n. When $m = n = 1$, we have a rectangular distribution.

Problem 6. Verify that

$$\int_0^1 f(x)\,dx = 1$$

where f is defined as in equation (6.4).   ∎

Returning to equation (6.4) we note two special cases. When $r = 1$ we have the smallest value in the sample and when $r = n$ the largest value in a sample. The statistician sometimes suspects that one of the extreme values in his sample is untrustworthy. He can use the corresponding probability density function to explore whether it is unacceptably deviant.

Problem 7. For n independent values drawn from the rectangular distribution over (0, 1). state the p.d.f. of the largest value and *hence* calculate the probability that it exceeds x_0 $(0 < x_0 < 1)$. Confirm your answer with a direct argument. If the probability that the smallest value exceeds x_0 is $\phi_n(x_0)$, find

$$\lim_{n\to\infty}\,[\phi_n(x_0/n)].$$

BRIEF SOLUTIONS AND COMMENTS ON THE PROBLEMS

Problem 1

(a) $\displaystyle\int_0^x \frac{1}{2\sqrt{t}}\,dt = \sqrt{x}$, $1/4, 0$.

(b) $\displaystyle F(x) = \frac{1}{4}\sum_1^x \left(\frac{3}{4}\right)^{t-1} = 1 - \left(\frac{3}{4}\right)^x$, $3\frac{1}{2}, 1$.

(c)   Since $f(x) = x\,(0 \leqslant x \leqslant 1)$ but $f(x) = 2 - x\,(1 \leqslant x \leqslant 2)$,

$$F(x) = \frac{x^2}{2}\,(0 \leqslant x \leqslant 1) \text{ and } F(x) = \frac{1}{2} + \int_1^x (2 - t)\,dt$$

$$= 2x - \frac{x^2}{2} - 1\,(1 \leqslant x \leqslant 2). \text{ Median} = 1, \text{ Mode} = 1.$$

(d) $3x^2 - 2x^3, 1/2, 1/2.$

Problem 2

(a) $F(x) = 1 - (1 + x + x^2/2)\,e^{-x}$

(b) $F(x) = \int_0^x \frac{6t}{(1 + t)^4}\,dt$

$$= \int_0^x \frac{6(1 + t)}{(1 + t)^4}\,dt - 6 \int_0^x \frac{1}{(1 + t)^4}\,dt$$

$$= 1 - (1 + 3x)/(1 + x)^3.$$

Problem 3. Integrate by parts

$$\phi_{n,m} = -\frac{(n + m - 1)!}{(n - 1)!m!}\,[y^{n-1}(1 - y)^m]_0^1$$

$$+ \frac{(n + m - 1)!}{(n - 2)!m!} \int_0^1 y^{n-2}(1 - y)^m\,dy$$

$$= \phi_{n-1,m+1} = \phi_{1,n+m-1}, \text{ by repeated application.}$$

This last integral is easily evaluated.

Problem 4

(a) $F(x) = \int_{-\infty}^x \frac{1}{\pi(1 + t^2)}\,dt.$ Substitute $t = \tan y$,

$$F(x) = \int_{-\pi/2}^{\tan^{-1}x} \frac{dy}{\pi} = \frac{1}{2} + \frac{1}{\pi}\tan^{-1} x. \text{ Choose a random value } u$$

between 0 and 1 and set $x = \tan[\pi(2u - 1)/2].$

(b) $F(x) = \int_a^x (ra^r/t^{r+1})\,dt = 1 - (a/x)^r.$ Choose a random value y between 0

and 1 and set $x = a/(1 - y)^{1/r}.$

Problem 5

$$S_5(x) = 0, \qquad x < 0.5;$$
$$= 1/5, \quad 0.5 \leqslant x < 3.2;$$
$$= 2.5, \quad 3.2 \leqslant x < 4.7;$$
$$= 3/5, \quad 4.7 \leqslant x < 5.1;$$
$$= 4/5, \quad 5.1 \leqslant x < 6.8;$$
$$= 1, \qquad 6.8 \leqslant x.$$

From tables of the exponential function,

$F(0.5) = 0.393, F(3.2) = 0.959, F(4.7) = 0.991,$
$F(5.1) = 0.994, F(6.8) = 0.999.$

$$\max|F(x) - S_5(x)| = 0.393, \qquad 0 \leqslant x < 0.5;$$
$$= 0.759, \qquad 0.5 \leqslant x < 3.2;$$
$$= 0.591, \qquad 3.2 \leqslant x < 4.7;$$
$$= 0.394, \qquad 4.7 \leqslant x < 5.1;$$
$$= 0.199, \qquad 5.1 \leqslant x < 6.8;$$
$$= 0.001, \qquad 6.8 \leqslant x.$$

Hence $|F(x) - S_5(x)|$ never exceeds 0.759.

Problem 6. See the solution to Problem 3.

Problem 7. In equation (6.4) set $r = n$, giving the p.d.f. $nx^{n-1} (0 < x < 1)$. Hence probability of exceeding x_0 is

$$\int_{x_0}^{1} nx^{n-1} \, dx = 1 - x_0^n.$$

More directly, notice that the greatest value exceeds x_0 if not all the values are less than x_0. But the probability that any value is less than x_0 is x_0.

The smallest value exceeds x_0 if and only if all the values exceed x_0. Hence $\phi_n(x_0) = (1 - x_0)^n$ and $\phi_n(x_0/n) = (1 - x_0/n)^n$, which has limit e^{-x_0}.

Functions of Random Variables

7.1 INTRODUCTION

Suppose X is a random variable with p.d.f. $f(x)$ and we wish to find the p.d.f. of the random variable $Y = g(X)$, where g is a function. We have evaded the question as to whether $g(X)$ is, in fact, bound to be a random variable. We shall assume that any functions used are to be sufficiently well behaved so that $g(X)$ is a random variable. However, we consider briefly how the matter may be investigated. Now, $Y = g(X)$ will be a random variable if for every y_0, we can evaluate $\Pr(Y \leqslant y_0)$. Let $\delta(y_0)$ be the set of all the real numbers x such that $g(x) \leqslant y_0$. Then $Y \leqslant y_0$ if and only if X assumes a value in $\delta(y_0)$. Thus we can calculate $\Pr(Y \leqslant y_0)$ if we can evaluate $\Pr[X$ falls in $\delta(y_0)]$ for every y_0. That is, if $\delta(y_0)$ is among the sets to which we can assign a probability. Thus, we must restrict our attention to those functions $g(x)$ for which this is the case. For example, if the $\delta(y_0)$ are composed of a number of disjoint intervals, then $g(X)$ is a random variable.

Example 1
$Y = \sin X$. By sketching the curve $y = \sin x$ (Fig. 7.1), it is seen that for each y_0, the values of x such that $\sin x \leqslant y_0$ fall in regularly spaced intervals of equal length. To each of these intervals we can assign a probability. In particular, if $y_0 = 1/\sqrt{2}$, whenever x lies in the interval $2n\pi + (3\pi/4)$, $2n\pi + 2\pi + (\pi/4)$, $n = 0, \pm1, \pm2, \ldots$, then $\sin x \leqslant 1/\sqrt{2}$. We find the probability that X falls in each interval and after summing, we have $\Pr[Y \leqslant (1/\sqrt{2})]$. It may happen of

course that $f(x)$ is zero for most such intervals. Suppose $f(x) = 1/\pi, 0 < x < \pi$, and is zero otherwise, then

$$\Pr\left(Y \leqslant \frac{1}{\sqrt{2}}\right) = \Pr\left(X \leqslant \frac{\pi}{4}\right) + \Pr\left(X \geqslant \frac{3\pi}{4}\right)$$

$$= \frac{\pi/4}{\pi} + 1 - \frac{3\pi/4}{\pi} = \frac{1}{2},$$

as is otherwise obvious.

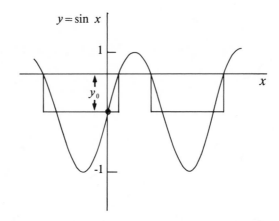

Fig. 7.1

7.2 FUNCTIONS OF DISCRETE RANDOM VARIABLES

If X is a discrete random variable and g is a function, then $Y = g(X)$ is a random variable. For since X is discrete, there are a countable number of points $x_1, x_2, \ldots$ such that $\Pr(X = x_r)$ is positive and at all other values of x, $\Pr(X = x) = 0$. Since g is a function, the numbers $y_r = g(x_r), r = 1. 2, \ldots$ are also countable, though not necessarily different. Then $\Pr(Y = y)$ is found by adding the probabilities for all those x_r such that $g(x_r) = y$. That is, if $h(y)$ is the p.d.f. of Y and $f(x)$ the p.d.f. of X,

$$h(y) = \sum f(x),$$

where the summation is over all those values of x such that $g(x) = y$. If there are no such values of x or if for all such x, $f(x) = 0$, then $h(y)$ is also zero.

Example 2

X is a discrete random variable with p.d.f.

$$f(x) = \frac{1}{n}, \quad x = 1, 2, 3, \ldots, n$$

$$= 0 \text{ otherwise.}$$

If $Y = X^2$, then as x takes the values $1, 2, 3, \ldots, n$, y takes the values $1, 4, 9, \ldots, n^2$. Thus

$$\Pr(Y = r^2) = \Pr(X = r) = \frac{1}{n}, \quad r = 1, 2, \ldots, n$$

$$h(y) = \frac{1}{n}, \quad y = 1, 4, \ldots, n^2$$

$$= 0 \text{ otherwise.}$$

In this case, the values of $g(x)$ are distinct for different values of x.

Example 3
X is a discrete random variable with p.d.f.

$$f(x) = \frac{1}{2n}, \quad x = \pm 1, \pm 2, \ldots, \pm n.$$

If $Y = X^2$, then

$$\Pr(Y = r^2) = \Pr(X = r \text{ or } X = -r)$$
$$= \Pr(X = r) + \Pr(X = -r)$$
$$= \frac{1}{2n} + \frac{1}{2n} = \frac{1}{n}, \quad r = 1, 2, \ldots, n.$$

Thus

$$h(y) = \frac{1}{n}, \quad y = 1, 4, 9, \ldots, n^2$$

$$= 0 \text{ otherwise.}$$

Example 4
X is a discrete random variable with p.d.f.

$$f(x) = \left(\frac{1}{2}\right)^x, \quad x = 1, 2, \ldots$$

$$= 0 \text{ otherwise.}$$

Suppose g is the function such that $g(x) = 0$ if x is even, $g(x) = 1$ if x is odd. The distribution of X has an infinite but countable number of values with positive probability and each of the values of y with positive probability arises from a countable number of values of x.

$$\Pr(Y=0) = \Pr(X \text{ is even}) = \sum_{x \text{ even}} \left(\frac{1}{2}\right)^x = \sum_{x=1}^{\infty} \left(\frac{1}{2}\right)^{2x}$$

$$= \sum_{x=1}^{\infty} \left(\frac{1}{4}\right)^x = \frac{1}{3}.$$

While $\Pr(Y=1) = 1 - \Pr(Y=0) = 2/3$.

Problem 1. X has the Poisson distribution with parameter λ and $g(x) = 0$ if x is even, $g(x) = 1$ if x is odd. Find the p.d.f. of $Y = g(X)$.

7.3 FUNCTIONS OF CONTINUOUS RANDOM VARIABLES

There are two cases to consider. The first of these arises when X is continuous but $g(X)$ is discrete.

Example 5
X has the exponential distribution with parameter λ,

$$g(x) = 0 \text{ if } x \geqslant \frac{1}{\lambda}, \quad g(x) = 1 \text{ if } 0 \leqslant x < \frac{1}{\lambda}.$$

Thus the whole of the distribution of X is mapped into two points. Since $f(x) = \lambda e^{-\lambda x}$, $x > 0$, $\Pr(X \geqslant x) = e^{-\lambda x}$. Hence, $\Pr(Y=0) = \Pr(X \geqslant 1/\lambda) = e^{-1}$. $\Pr(Y=1) = 1 - \Pr(Y=0) = 1 - e^{-1}$. Hence, $h(y)$, the p.d.f. of $Y = g(X)$ satisfies $h(0) = e^{-1}$, $h(1) = 1 - e^{-1}$, $h(y) = 0$, otherwise.

The second, and more important, case is when X is a continuous random variable and $g(X)$ is also a continuous random variable. This will be the case if $g(x)$ is a continuous function of x. The following theorem deals with a class of functions g which leads to a simple formula for $h(y)$.

Theorem 7.1. If X is a continuous random variable with p.d.f. $f(x)$ and $g'(x)$ is positive, then the random variable $Y = g(X)$ is continuous and

$$h(y) = \left[f(x) \frac{dx}{dy} \right]_y = \left[\frac{f(x)}{g'(x)} \right]_y \tag{7.1}$$

where the right-hand side is to be evaluated as a function of y. Since $g'(x)$ exists, then $g(x)$ is continuous. Since $g'(x) > 0$, g is a strictly increasing function of x. Hence, if $y_0 = g(x_0)$, $y \leqslant y_0$ if and only if $x \leqslant x_0$. Thus

$$\Pr(Y \leqslant y_0) = \Pr(X \leqslant x_0) = \int_{-\infty}^{x_0} f(x)\, dx.$$

Now, changing the variable in this integral, put $y = g(x)$, $dy/dx = g'(x)$, when $x = x_0$, $y = g(x_0) = y_0$. Suppose as $x \to -\infty$, $g(x) \to l_1$ where l_1 may be $-\infty$.

$$\int_{-\infty}^{x_0} f(x)\,dx = \int_{l_1}^{y_0} \frac{f(x)}{g'(x)}\,dy = \Pr(Y \leqslant y_0) = \int_{l_1}^{y_0} h(y)\,dy.$$

Hence,

$$h(y) = \left[\frac{f(x)}{g'(x)}\right]_y, \quad l_1 \leqslant y,$$

$$= 0 \text{ otherwise.}$$

We make some comments. Since $y = g(x)$ is strictly increasing, we can always, in principle, solve for x in the form $x = g^{-1}(y)$. Secondly, if l_1 is *not* $-\infty$, then if $y_1 < l_1$, there is *no* x such that $y = g(x)$, and this explains why $h(y) = 0$ if $y < l_1$.

Continuous random variables

Example 6

If X is a continuous random variable with p.d.f.

$$f(x) = \frac{1}{\pi(1 + x^2)}, \quad -\infty < x < +\infty$$

find the p.d.f. of $Y = \tan^{-1} X$.

Since $y = \tan^{-1} x$, $x = \tan y$, $dx/dy = \sec^2 y > 0$, and the conditions of the theorem apply. Further, $1 + x^2 = 1 + \tan^2 y = \sec^2 y$. Hence,

$$h(y) = \left[\frac{f(x)}{g'(x)}\right]_y = \left[f(x)\frac{dx}{dy}\right]_y = \left[\frac{\sec^2 y}{\pi(1 + x^2)}\right]_y = \frac{1}{\pi}.$$

As $x \to +\infty$, $\tan^{-1} x \to +\pi/2$ and as $x \to -\infty$, $\tan^{-1} x \to -\pi/2$, then

$$h(y) = \frac{1}{\pi}, \quad -\frac{\pi}{2} \leqslant y \leqslant +\frac{\pi}{2}$$

$$= 0 \text{ otherwise.}$$

The distribution of Y is rectangular. X is said to have a **Cauchy** distribution.

Only a small amendment is required to the above theorem to deal with the case when $g'(x) < 0$. For, from a sketch (Fig. 7.2) we see

$$\Pr(Y \leqslant y_0) = \Pr(X \geqslant x_0) = \int_{x_0}^{\infty} f(x)\,dx$$

where $y_0 = g(x_0)$. If as $x \to +\infty$, $g(x) \to l_2$, on changing the variable $y = g(x)$.

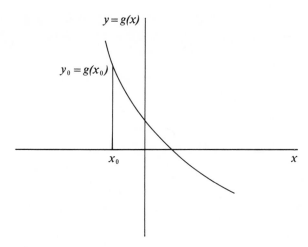

Fig. 7.2

$$\int_{x_0}^{\infty} f(x)\,dx = \int_{y_0}^{l_2} \frac{f(x)}{g'(x)}\,dy = -\int_{l_2}^{y_0} \frac{f(x)}{g'(x)}\,dy = \int_{l_0}^{y_0} h(y)\,dy$$

and in this case,

$$h(y) = \left[\frac{-f(x)}{g'(x)}\right]_y, \quad l_2 \leqslant y$$

$$= 0 \text{ otherwise.} \tag{7.2}$$

Example 7

If X has the rectangular distribution over $(0, 1)$, find the p.d.f. of $Y = -\log_e X$.

In this case, when $y = -\log_e x$, $dy/dx = -(1/x)$, which is <0 for $0 \leqslant x \leqslant 1$. But

$$f(x) = 1, \quad 0 < x < 1, \quad \text{hence } h(y) = \left[-\frac{f(x)}{g'(x)}\right]_y = (+x)_y = e^{-y}.$$

When $x \to 0, y \to +\infty$, $x = 1, y = 0$.

Thus, $h(y) = e^{-y}, 0 < y < \infty$, $h(y) = 0$ otherwise and, Y has an exponential distribution.

Both versions of the theorem are combined in the statement

$$h(y) = [f(x)/|g'(x)|]_y,$$

where $|g'(x)|$ means the absolute value of $g'(x)$.

7.4 NON-MONOTONE FUNCTION

We have excluded the case when $g'(x) = 0$ for one or more values of x. In general, when $g'(x) = 0$, x may be a turning value of g. This means that for some values of y, there is more than one value of x such that $y = g(x)$. Hence, as seen in Fig. 7.3, there may be several intervals of x for which $g(x) < y_0$. The probabilities for these intervals must be added to obtain $\Pr(Y < y_0)$.

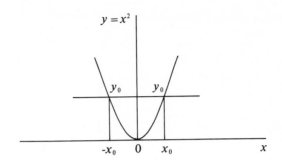

Fig. 7.3

We shall not state a theorem to cover this case, but suggest that each problem be given individual treatment.

Example 8

$Y = X^2$. If $y = x^2$, $dy/dx = 2x$ and is zero when $x = 0$, which corresponds to a minimum of $y = x^2$. Hence, if X is a continuous random variable which assumes both positive and negative values we must take account of the remarks in the last paragraph. If $y_0 = x_0^2$ and $x_0 > 0$, it is clear that $y < y_0$, whenever $-x_0 < x < +x_0$.

$$\Pr(Y \leqslant y_0) = \Pr(-x_0 \leqslant X \leqslant x_0) = \int_{-x_0}^{x_0} f(x)\, dx$$

$$= \int_{-x_0}^{0} f(x)\, dx + \int_{0}^{x_0} f(x)\, dx.$$

We put $y = x^2$ in both integrals since y is monotone (though in opposite senses in each). $dy/dx = 2x$, which is $2\sqrt{y}$ for $0 \leqslant x \leqslant x_0$ and $-2\sqrt{y}$ for $-x_0 \leqslant x \leqslant 0$.

$$\Pr(Y \leqslant y_0) = \int_{y_0}^{0} \frac{f(x)}{g'(x)}\, dy + \int_{0}^{y_0} \frac{f(x)}{g'(x)}\, dy$$

$$\int_{y_0}^{0} \frac{f(-\sqrt{y})}{-2\sqrt{y}}\, dy + \int_{0}^{y_0} \frac{f(\sqrt{y})}{+2\sqrt{y}}\, dy$$

$$= \int_{0}^{y_0} \frac{f(\sqrt{y}) + f(-\sqrt{y})}{2\sqrt{y}}\, dy.$$

Hence,

$$h(y) = \frac{f(\sqrt{y}) + f(-\sqrt{y})}{2\sqrt{y}}.$$ (7.3)

If the p.d.f. of X, is symmetrical about $x = 0$, then $f(-\sqrt{y}) = f(+\sqrt{y})$ and the last formular reduces to $f(\sqrt{y})/\sqrt{y}$.

Example 9
X has the distribution $N(0, 1)$, find the p.d.f. of $Y = X^2$. In this case

$$f(x) = \frac{1}{\sqrt{(2\pi)}} e^{-x^2/2}$$

$-\infty \leqslant x \leqslant +\infty$, and $f(x) = f(-x)$. Hence, by equation (7.3),

$$h(y) = \frac{f(\sqrt{y})}{\sqrt{y}}, \quad 0 \leqslant y$$

$$= \frac{1}{\sqrt{(2\pi)}} y^{-1/2} e^{-(\sqrt{y})^2/2}$$

$$= \frac{y^{-1/2} e^{-y/2}}{\sqrt{(2\pi)}}.$$

Y has a gamma distribution with parameters $(1/2, 1/2)$.

The difficulty we have been discussing will not arise if $g'(x) = 0$ only at the endpoints of an interval in which all the probability of X is concentrated. For example, if $f(x) > 0$ for $a \leqslant x \leqslant b$ and is zero otherwise, while $g'(x) \neq 0$, $a < x < b$ but possibly $g'(a)$ or $g'(b) = 0$.

Example 10
X has the rectangular distribution over $(0, 1)$; find the p.d.f. of $Y = X^2$. Here $g'(x) = 2x > 0$, except at $x = 0$, but $f(x)$ is zero for $x < 0$.

$$h(y) = \left[\frac{f(x)}{g'(x)} \right]_y = \left(\frac{1}{2x} \right)_y = \frac{1}{2\sqrt{y}}, \quad 0 < y.$$

Problem 2. X is a continuous random variable with p.d.f.

$$f(x) = \frac{(m + n - 1)!}{(m - 1)!(n - 1)!} x^{m-1} (1 - x)^{n-1}, \quad 0 \leqslant x \leqslant 1$$

$$= 0 \text{ otherwise } (m, n \text{ positive integers}).$$

Such a random variable is said to have the **beta distribution** with parameters m, n. Show that $Y = 1 - X$ also has a beta distribution.

Problem 3. If X has the normal distribution with parameters μ, σ show that $Y = aX + b$ has the normal distribution with parameters $a\mu + b$, $|a|\sigma$. For what values of a, b is the distribution of Y unit normal?

Problem 4. If X has the rectangular distribution over $-\pi/2$, $+\pi/2$, show that $Y = \tan X$ has the Cauchy distribution.

Problem 5. If X has the rectangular distribution over $(0, 1)$, show that $Y = e^X$ has p.d.f. $h(y) = 1/y$, $1 \leqslant y \leqslant e$.

Problem 6. If X is a continuous random variable with p.d.f. $f(x) = 2xe^{-x^2}$, $x > 0$, show that $Y = X^2$ has p.d.f. $h(y) = e^{-y}$, $y > 0$.

Problem 7. If X has the rectangular distribution over $(-\pi/2, +\pi/2)$, show that $Y = \sin X$ has p.d.f.

$$h(y) = \frac{1}{\pi(1 - y^2)^{1/2}}, \quad -1 < y < 1.$$

Problem 8. If X is a continuous random variable, show that $Y = F(X)$, where $F(x)$ is the distribution function of X, has the rectangular distribution over $0, 1$).

BRIEF SOLUTIONS AND COMMENTS ON THE PROBLEMS

Problem 1

$$\Pr(g(X) = 0) = \Pr(X \text{ is even})$$

$$= \sum_{n=0}^{\infty} \Pr(X = 2n)$$

$$= \sum_{0}^{\infty} \frac{\lambda^{2n}}{2n!} e^{-\lambda}$$

$$= \left(\frac{e^\lambda + e^{-\lambda}}{2}\right) e^{-\lambda} = \frac{1}{2} + \frac{1}{2} e^{-2\lambda},$$

and clearly $\Pr[g(X) = 1] = 1 - \Pr[g(X) = 0]$.

Problem 2

$$\left|\frac{dx}{dy}\right| = 1, \quad \text{and } f(1 - y) = \frac{(m + n - 1)!}{(m - 1)!(n - 1)!} (1 - y)^{m-1} y^{n-1},$$

$$0 < y < 1$$

which is, from the definition, the p.d.f. of a beta distribution with parameter n, m. (m, n have changed places!.)

Problem 3. We require

$$\left[f(x) \left| \frac{dx}{dy} \right| \right]_y .$$

Now

$$\left| \frac{dx}{dy} \right| = \frac{1}{|a|}$$

and

$$[f(x)]_y = f[(y-b)/a]$$

$$= \frac{1}{\sqrt{(2\pi)}\sigma} \exp \left[-\frac{1}{2} \{(y-b)/a - \mu\}^2 / \sigma^2 \right],$$

$$= \frac{1}{\sqrt{(2\pi)}\sigma} \exp \left[-\frac{1}{2} \{y - (a\mu + b)\}^2 / a^2 \sigma^2 \right],$$

and the result follows.

Problem 4. $f(x) = 1/\pi, -\pi/2 \leqslant x \leqslant \pi/2$.
If $y = \tan x$, then

$$\frac{dy}{dx} = \sec^2 x = 1 + \tan^2 x = 1 + y^2$$

or

$$\frac{dx}{dy} = \frac{1}{1+y^2} > 0.$$

Hence p.d.f. of Y is $1/[\pi(1+y^2)], -\infty < y < +\infty$.

Problem 5. $f(x) = 1, 0 \leqslant x \leqslant 1$. If $y = e^x$, then

$$\frac{dy}{dx} = e^x = y,$$

and

$$\frac{dx}{dy} = \frac{1}{y} > 0.$$

Hence the p.d.f. of Y is $1/y, 1 \leqslant y \leqslant e$. (Check that the function integrates to 1.)

Problem 6. $f(x) = 2x\,e^{-x^2}, 0 \leqslant x < \infty$. If $y = x^2$, then

$$\frac{dy}{dx} = 2x \quad \text{and} \quad \frac{dx}{dy} = \frac{1}{2x} > 0.$$

Hence the p.d.f. of Y is $(2x\,e^{-x^2}/2x)_y = (e^{-x^2})_y = e^{-y}, 0 \leqslant y < \infty$.

Problem 7. $f(x) = 1/\pi, -\pi/2 \leqslant x < \pi/2$. If $y = \sin x$ then

$$\frac{dy}{dx} = \cos x = (1 - \sin^2 x)^{1/2} = (1 - y^2)^{1/2}, \; -1 \leqslant y \leqslant 1.$$

Hence result.

Problem 8. If $y = F(x)$, then

$$\frac{dy}{dx} = f(x) \quad \text{and} \quad \frac{dx}{dy} = \frac{1}{f(x)} > 0.$$

Hence the p.d.f. of Y is $[f(x)/f(x)]_y = 1$.

8

Bivariate Distributions

8.1 DISCRETE BIVARIATE DISTRIBUTIONS

So far we have concentrated on single random variables. Consider now a bag containing four discs numbered one to four. From this population, two discs are drawn, one at a time, without replacement, and we write X_1 as the number on the first disc and X_2 as the number of the second disc. Both X_1 and X_2 are random variables on the sample space generated by the experiment. It is also convenient to think of X_1, X_2 as the components of a two-dimensional random variable on the same sample space. That is, a random vector X which has ordered components X_1, X_2. For each ordered pair of real numbers x_1, x_2 we can find $\Pr(X_1 = x_1 \text{ and } X_2 = x_2)$. In fact

$$\Pr(X_1 = x_1 \text{ and } X_2 = x_2) = \frac{1}{12}, \quad x_1 \neq x_2, \quad x_1 = 1, 2, 3, 4,$$

$$x_2 = 1, 2, 3, 4.$$

This probability is the joint probability density function, $f(x_1, x_2)$ of the bivariate distribution of the random variables X_1 and X_2. Starting from the joint p.d.f. we may recover the p.d.f. of X_1, denoted by $f_1(x_1)$, as follows:

$$f_1(x_1) = \Pr(X_1 = x_1) = \sum_{x_2} \Pr(X_1 = x_1 \text{ and } X_2 = x_2)$$

$$= \sum_{x_2} f(x_1, x_2).$$

In the present case,

$$\sum_{x_2} f(x_1, x_2) = \sum_{x_2 \neq x_1} \frac{1}{12} = \frac{3}{12} = \frac{1}{4} = f_1(x_1),$$

which of course agrees with considering X_1 alone and ignoring X_2. $f_1(x_1)$ is known as the marginal probability density function of X_1. Similarly, the marginal p.d.f. of X_2, is $f_2(x_2) = 1/4$.

Definition. If the random variables X_1, X_2 take a countable number of pairs of values (x_1, x_2) and there is a function $f(x_1, x_2)$ such that

$$f(x_1, x_2) \geqslant 0 \tag{8.1}$$

$$\sum_{x_1} \sum_{x_2} f(x_1, x_2) = 1 \tag{8.2}$$

$$\Pr(a \leqslant X_1 \leqslant b \text{ and } c \leqslant X_2 \leqslant d) = \sum_{x_1 = a}^{b} \sum_{x_2 = c}^{d} f(x_1, x_2), \tag{8.3}$$

then X_1, X_2 are said to have a joint discrete distribution with (joint) probability density function $f(x_1, x_2)$. The marginal p.d.f. of X_1 is defined as

$$f_1(x_1) = \sum_{x_2} f(x_1, x_2)$$

and, similarly, the marginal p.d.f. of X_2 is

$$f_2(x_2) = \sum_{x_1} f(x_1, x_2),$$

where $f_1(x_1)$, $f_2(x_2)$ are none other than the separate p.d.f.s of X_1 and X_2. In agreement with our earlier work on independence, we shall say that X_1 and X_2 are independent if and only if for every (a, c), (b, d),

$$\Pr(a \leqslant X_1 \leqslant b \text{ and } c \leqslant X_2 \leqslant d) = \Pr(a \leqslant X_1 \leqslant b) \Pr(c \leqslant X_2 \leqslant d)$$

which for a discrete distribution implies

$$\sum_{x_1 = a}^{b} \sum_{x_2 = c}^{d} f(x_1, x_2) = \sum_{x_1 = a}^{b} f_1(x_1) \sum_{x_2 = c}^{d} f_2(x_2). \tag{8.4}$$

We need a criterion for determining whether or not two random variables are independent without having to verify the above relationship for all (a, c), (b, d):

Theorem 8.1. A necessary and sufficient condition for X_1 and X_2 to be independent is

$$f(x_1, x_2) = f_1(x_1) f_2(x_2). \tag{8.5}$$

The condition is *sufficient* for if it holds

$$\sum_{x_1=a}^{b} \sum_{x_2=c}^{d} f(x_1, x_2) = \sum_{x_1=a}^{b} \sum_{x_2=c}^{d} f_1(x_1) f_2(x_2)$$

$$= \sum_{x_1=a}^{b} f_1(x_1) \sum_{x_2=c}^{d} f_2(x_2).$$

It is also necessary, for in (8.4) put $a = b = x_1$ and $c = d = x_2$ then

$$f(x_1, x_2) = f_1(x_1) f_2(x_2).$$

This theorem tempts us to assert something further, namely that it is enough if $f(x_1, x_2)$ factorizes into a function of x_1 and another of x_2. We should then be spared the labour of evaluating $f_1(x_1)$ and $f_2(x_2)$. Unfortunately, the result needs qualifying. The following is true.

Theorem 8.2. If $f(x_1, x_2) = g(x_1)h(x_2)$ then X_1, X_2 are independent if and only if both

(a) $g(x_1) > 0$ when $f_1(x_1) > 0$ and is zero otherwise;
(b) $h(x_2) > 0$ when $f_2(x_2) > 0$ and is zero otherwise.

With these conditions it can be shown that

$$g(x_1) = c_1 f_1(x_1), \quad h(x_2) = c_2 f_2(x_2),$$

where c_1, c_2 are positive constraints and $c_1 c_2 = 1$. The commonest situation leading to failure to satisfy conditions (a) and (b) is when the region in which $f(x_1, x_2)$ is positive is restricted by a relation between x_1 and x_2.
Whether or not X_1 and X_2 are independent, if $\Pr(X_2 = x_2) > 0$

$$\Pr(X_1 = x_1 | X_2 = x_2) = \frac{\Pr(X_1 = x_1 \text{ and } X_2 = x_2)}{\Pr(X_2 = x_2)},$$

which we write

$$f(x_1 | x_2) = \frac{f(x_1, x_2)}{f_2(x_2)}$$

where $f(x_1|x_2)$ is called the conditional probability density function of X_1 given $X_2 = x_2$. Similarly, if $f_1(x_1) > 0$ we have

$$f(x_2 | x_1) = \frac{f(x_1, x_2)}{f_1(x_1)}.$$

Both the above conditional p.d.f.s satisfy the usual conditions for p.d.f.s. For example,

$$f(x_1 | x_2) \geqslant 0$$

and

$$\sum_{x_1} f(x_1 | x_2) = \frac{\sum\limits_{x_1} f(x_1, x_2)}{f_2(x_2)} = \frac{f_2(x_2)}{f_2(x_2)} = 1.$$

8.2 THE TRINOMIAL DISTRIBUTION

Definition. The discrete random variables X_1, X_2 are said to have the trinomial distribution with positive parameters n, p_1, p_2 such that n is an integer and $0 < p_1 + p_2 < 1$, if the joint p.d.f. of X_1, X_2 satisfies

$$f(x_1, x_2) = \frac{n!}{x_1! x_2! (n - x_1 - x_2)!} p_1^{x_1} p_2^{x_2} (1 - p_1 - p_2)^{n - x_1 - x_2},$$

$$x_1, x_2 \text{ non-negative integers such that } x_1 + x_2 \leqslant n$$

$$= 0 \text{ otherwise.} \tag{8.6}$$

This joint distribution would be appropriate if in n independent trials

(a)   on each trial there is a probability p_1 of an outcome of type O_1;
(b)   on each trial there is a probability p_2 of an outcome of type O_2;
(c) O_1 and O_2 are mutually exclusive.

Then any particular sequences of outcomes of which x_1 are of O_1, x_2 are of O_2 and $n - x_1 - x_2$ are neither O_1 nor O_2 has probability

$$p_1^{x_1} p_2^{x_2} (1 - p_1 - p_2)^{n - x_1 - x_2}.$$

But there are

$$\binom{n}{x_1} \binom{n - x_1}{x_2} = \frac{n!}{x_1! x_2! (n - x_1 - x_2)!}$$

distinguishable arrangements for such a set of outcomes.

We now find the marginal distributions of X_1 and X_2.

$$f_1(x_1) = \sum_{x_2 = 0}^{n - x_1} \frac{n!}{x_1! x_2! (n - x_1 - x_2)!} p_1^{x_1} p_2^{x_2} (1 - p_1 - p_2)^{n - x_1 - x_2}$$

$$= \frac{n!}{x_1! (n - x_1)!} p_1^{x_1} \sum_{x_2 = 0}^{n - x_1} \frac{(n - x_1)!}{x_2! (n - x_1 - x_2)!}$$

$$p_2^{x_2} (1 - p_1 - p_2)^{n - x_1 - x_2}$$

$$= \frac{n!}{x_1! (n - x_1)!} p_1^{x_1} (1 - p_1 - p_2 + p_2)^{n - x_1}$$

$$= \frac{n!}{x_1! (n - x_1)!} p_1^{x_1} (1 - p_1)^{n - x_1}.$$

Thus the marginal distribution of X_1 is the binomial distribution with parameters n and p_1. This will occasion no surprise since we may think of O_1 as 'success' and all other outcomes *including* O_2 as 'failure'. Similarly,

$$f_2(x_2) = \frac{n!}{x_2!(n-x_2)!} p_2^{x_2} (1-p_2)^{n-x_2}.$$

Since $f(x_1, x_2) \neq f_1(x_1) f_2(x_2)$, X_1 and X_2 are not independent, agreeing with the fact that when $X_1 \neq 0$, the range of X_2 is restricted. Furthermore

$$f(x_1 | x_2) = \frac{n!/[x_1! x_2!(n-x_1-x_2)!] \, p_1^{x_1} \, p_2^{x_2} \, (1-p_1-p_2)^{n-x_1-x_2}}{n!/[x_2!(n-x_2)!] \, p_2^{x_2} \, (1-p_2)^{n-x_2}}$$

$$= \frac{(n-x_2)!}{x_1!(n-x_1-x_2)!} \cdot \frac{p_1^{x_1} \, (1-p_1-p_2)^{n-x_1-x_2}}{(1-p_2)^{n-x_2}}$$

$$= \binom{n-x_2}{x_1} \left(\frac{p_1}{1-p_2}\right)^{x_1} \left(\frac{1-p_2-p_1}{1-p_2}\right)^{n-x_1-x_2},$$

the p.d.f. of a binomial distribution with parameters $n - x_2$ and $p_1/(1-p_2)$. Since $X_2 = x_2$, we have indeed only $n - x_2$ trials left to consider. The value of the other parameter needs a little explanation. The probability that any one of the remaining trials has outcome O_1 is no longer p_1, for we know it cannot be O_2.

$$\Pr(O_1 | \text{not } O_2) = \frac{\Pr(O_1 \text{ and not } O_2)}{\Pr(\text{not } O_2)} = \frac{\Pr(O_1)}{\Pr(\text{not } O_2)} = \frac{p_1}{1-p_2}.$$

The trinomial distribution can be generalized in an obvious way to the multinomial distribution with parameters $n, p_1, p_2, \ldots, p_k$.

Example 1
If a bag contains three white, two black, and four red balls, and four balls are drawn at random with replacement, calculate the probabilities that

(a) the sample contains just one white ball;
(b) the sample contains just one white ball given that it contains just one red ball;
(c) the sample contains at least one ball of each colour.

Since there are nine balls, on each draw,

$$\Pr(\text{white}) = \frac{3}{9} = \frac{1}{3}, \quad \Pr(\text{black}) = \frac{2}{9}, \quad \Pr(\text{red}) = \frac{4}{9}.$$

(a) The marginal distribution of the number of whites in samples of four is binomial with parameters $n = 4, p = 1/3$.

$$\Pr(\text{just one white}) = \binom{4}{1}\left(\frac{1}{3}\right)^1\left(\frac{2}{3}\right)^3 = \frac{32}{81}.$$

(b) The conditional distribution of the number of whites given that the sample contains just one red is again binomial but this time with parameters.

$$n = 4 - 1 = 3, \quad p = \frac{1/3}{1 - (4/9)} = \frac{3}{5}.$$

Hence

$$\Pr(\text{just one white} \mid \text{just one red}) = \binom{3}{1}\left(\frac{3}{5}\right)^1\left(\frac{2}{5}\right)^2 = \frac{36}{125}.$$

We do not really need a formula, since one red ball is already counted and we now have to draw three more with replacement from the remaining five which are not red. This conditional probability may of course also be obtained from first principles,

$$\Pr(\text{just one white} \mid \text{just one red}) = \frac{\Pr(\text{just one white and just one red})}{\text{pr(just one red)}}$$

$$= \frac{\Pr(\text{one white and two blacks and one red})}{\Pr(\text{just one red})}.$$

The joint distribution of the number of whites and blacks is multinomial with parameters, $n = 4, p_1 = 1/3, p_2 = 2/9$. Hence the numerator is

$$\frac{4!}{1!\,2!\,1!}\left(\frac{1}{3}\right)^1\left(\frac{2}{9}\right)^2\left(\frac{4}{9}\right)^1$$

while the denominator is

$$\binom{4}{1}\left(\frac{4}{9}\right)\left(\frac{5}{9}\right)^3$$

and the ratio of these is again 36/125.

(c) Pr(one white and one black and two reds)

$$= \frac{4!}{2!\,1!\,1!}\left(\frac{1}{3}\right)\left(\frac{2}{9}\right)\left(\frac{4}{9}\right)^2 = \frac{12 \times 32}{2187},$$

Pr(two whites and one black and one red)

$$= \frac{4!}{2!\,1!\,1!}\left(\frac{1}{3}\right)^2\left(\frac{2}{9}\right)\left(\frac{4}{9}\right) = \frac{12 \times 24}{2187},$$

Pr(one white and two blacks and one red)

$$= \frac{4!}{2!1!1!} \left(\frac{1}{3}\right)\left(\frac{2}{9}\right)^2\left(\frac{4}{9}\right) = \frac{12 \times 16}{2187}.$$

The sum of these probabilities is 32/81. As an exercise the reader should now check that the corresponding answer for (c) when the sample is drawn without replacement is 4/7.

Problem 1. If $3m$ fair dice are each rolled $2n$ times, find the probability that scores $1, 2, 3, 4, 5, 6$ each appear mn times.

Problem 2. Six cards are drawn at random with replacement from a normal pack of playing cards. Calculate the probabilities that the sample contains:

(a)   three cards in one suit and one card in each of the others;
(b)   just three spades;
(c)   just three cards in at least one unit.

Problem 3. A sample of n is drawn with replacement from a population and each element is reported as having or not having each of two characteristics A and B. An element has character A with probability p_1 and characteristic B with probability p_2. If these characteristics are independent, write down the probability for the following description: r_1 elements with both A and B, r_2 elements with A but not B, r_3 elements with B but not A, r_4 elements with neither A nor B, where $r_1 + r_2 + r_3 + r_4 = n$.

Problem 4. A line of items passes an inspection point. Let p_1 be the probability that an item is inspected and p_2 the probability that an item is defective. If the nth item is the first defective found, calculate the probability that k defective items have been missed.

Example 2
The joint distribution of the discrete random variables X_1, X_2 has p.d.f.

$$f(x_1, x_2) = \frac{\lambda^{x_1+x_2}\, e^{-2\lambda}}{x_1!x_2!}, \quad x_1 = 0, 1, \ldots, \quad x_2 = 0, 1, \ldots .$$

Find the marginal distributions of X_1, X_2 and say whether they are independent

$$f_1(x_1) = \sum_{x_2=0}^{\infty} \frac{\lambda^{x_1+x_2}\, e^{-2\lambda}}{x_1!x_2!}$$

$$= \frac{\lambda^{x_1}\, e^{-2\lambda}}{x_1!} \sum_{x_2=0}^{\infty} \frac{\lambda^{x_2}}{x_2!}$$

$$= \frac{\lambda^{x_1}}{x_1!} e^{-2\lambda} e^{+\lambda} = \frac{\lambda^{x_1} e^{-\lambda}}{x_1!}.$$

But this is the p.d.f. of a Poisson distribution with parameter λ. Similarly X_2 has a Poisson distribution with parameter λ. Since $f(x_1, x_2) = f_1(x_1)f_2(x_2)$, X_1 and X_2 are independent. Alternatively, $f(x_1, x_2)$ may be factorized into $(\lambda^{x_1} e^{-\lambda}/x_1!)(\lambda^{x_2} e^{-\lambda}/x_2!)$ and the result follows. This easy example should be compared carefully with Example 3.

Example 3

The discrete random variables X_1, X_2 have joint p.d.f.

$$f(x_1, x_2) = \frac{\lambda^{x_2} e^{-2\lambda}}{x_1!(x_2 - x_1)!}, \quad x_1 = 0, 1, 2, \ldots, x_2,$$

$$x_2 = 0, 1, 2, \ldots$$

$$= 0 \text{ otherwise.}$$

Find the marginal distributions of X_1, X_2 and the conditional distribution of X_1 given $X_2 = x_2$.

$$f_1(x_1) = \sum_{x_2 = x_1}^{\infty} \frac{\lambda^{x_2} e^{-2\lambda}}{x_1!(x_2 - x_1)!}$$

$$= \frac{e^{-2\lambda} \lambda^{x_1}}{x_1!} \sum_{x_2 = x_1}^{\infty} \frac{\lambda^{x_2 - x_1}}{(x_2 - x_1)!}$$

$$= \frac{e^{-2\lambda} \lambda^{x_1}}{x_1!} e^{+\lambda} = \frac{\lambda^{x_1} e^{-\lambda}}{x_1!}, \quad x_1 = 0, 1, \ldots .$$

Thus X_1 has a Poisson distribution with parameter λ.

$$f_2(x_2) = \sum_{x_1 = 0}^{x_2} \frac{\lambda^{x_2} e^{-2\lambda}}{x_1!(x_2 - x_1)!}$$

$$= \frac{\lambda^{x_2} e^{-2\lambda}}{x_2!} \sum_{x_1 = 0}^{x_2} \frac{x_2!}{x_1!(x_2 - x_1)!}.$$

Now,

$$\sum_{x_1 = 0}^{x_2} \binom{x_2}{x_1} = (1 + 1)^{x_2} = 2^{x_2},$$

thus

$$f_2(x_2) = \frac{(2\lambda)^{x_2} e^{-2\lambda}}{x_2!}, \quad x_2 = 0, 1, 2, \ldots .$$

Thus X_2 has a Poisson distribution with parameter 2λ. Since, however, $f(x_1, x_2) \neq f_1(x_1) f_2(x_2), X_1, X_2$ are, in this instance, not independent, and

$$f(x_1 | x_2) = \frac{f(x_1, x_2)}{f_2(x_2)}$$

$$= \frac{\lambda^{x_2} e^{-2\lambda} / [x_1! (x_2 - x_1)!]}{(2\lambda)^{x_2} e^{-2\lambda} / x_2!}$$

$$= \binom{x_2}{x_1} \left(\frac{1}{2}\right)^{x_2}, \quad x_1 \leqslant x_2.$$

But this is the p.d.f. of the binomial distribution with parameters x_2 and $1/2$. Because of the dependence, we must return to the joint p.d.f. should we require to evaluate probabilities relating to both random variables. Thus,

$$\Pr(X_1 = X_2) = \sum_{x_1 = x_2} f(x_1, x_2) = \sum_{x_1} f(x_1, x_1)$$

$$= \sum_{x_1 = 0}^{\infty} \frac{\lambda^{x_1} e^{-2\lambda}}{x_1! 0!} = e^{-\lambda}.$$

If we require $\Pr(0 \leqslant X_1 \leqslant 1$ and $0 \leqslant X_2 \leqslant 1)$, then this is

$$f(0, 0) + f(0, 1) + f(1, 1),$$

since $\Pr(X_1 = 1$ and $X_2 = 0)$ is zero.

Problem 5. If the discrete random variables X_1, X_2 have the following joint p.d.f.s, find in each case the marginal distributions of X_1, X_2 and the conditional distribution of X_1 given X_2.

(a) $f(x_1, x_2) = q^2 p^{x_2 - 2}, \quad x_1 = 1, 2, 3, \ldots, x_2 - 1, \quad x_2 = 2, 3, 4, \ldots$.

$\quad\quad = 0$ otherwise.

(b) $f(x_1, x_2) = \binom{n_1}{x_1} \binom{n_2}{x_2 - x_1} p^{x_2} q^{n_1 + n_2 - x_2}, \quad x_1 \leqslant x_2 \leqslant n_2 + x_1,$

$$0 \leqslant x_1 \leqslant n_1$$

$\quad\quad = 0$ otherwise.

Problem 6. r different balls are placed at random into n different cells. Let X_1 be the number of balls in cell number one and X_2 be the number of balls in cell number two. Find the p.d.f. of X_2, of X_1 given x_2, and hence derive the joint p.d.f. of X_1 and X_2.

8.3 CONTINUOUS BIVARIATE DISTRIBUTIONS

We now consider pairs of random variables which have a joint continuous distribution. This case is at the opposite extreme to a joint discrete distribution in the sense that $\Pr(X_1 = x_1$, and $X_2 = x_2)$ is zero for every pair of points (x_1, x_2). On the other hand, there are some regions R in the plane such that the probability that (X_1, X_2) falls in R is not zero. In continuous distributions there will be a non-negative function $f(x_1, x_2)$ such that the approximate probability of (X_1, X_2) falling in a small rectangle centred on (x_1, x_2) and with sides of length $\delta x_1, \delta x_2$ is given by $f(x_1, x_2)\delta x_1 \delta x_2$. Actually, we can tolerate a little less than this, namely, that such an approximation holds except perhaps at certain points of the plane. These points should lie so that the plane can be divided into separate zones for which the approximation holds at interior points but not on the boundaries. The distribution of probability is now represented by a volume in three dimensions. At most points (x_1, x_2) we can erect a perpendicular of height $f(x_1, x_2)$, and the probability for the above small rectangle is approximately $f(x_1, x_2)\delta x_1 \delta x_2$. To find the probability that (X_1, X_2) falls in a region R of the plane, we require the volume between R and the surface $f(x_1, x_2)$ for (x_1, x_2) in R. This volume is found by integration. Bivariate distributions frequently arise in practice as a result of making measure- on two different variables on the same element from a population. It is often convenient to suppose that the joint distribution is continuous. Thus, suppose we measure the height X_1 and weight X_2 of each person in a large population, then, although there are only a finite number of persons, we may wish to construct a continuous model for the joint distribution of X_1, X_2, since conceptually X_1 and X_2 may assume any value in certain intervals.

Definition. The random variables X_1, X_2 are said to have a joint continous distribution with probability density function $f(x_1, x_2)$ if

(a) $f(x_1, x_2) \geqslant 0$,

(b) $f(x_1, x_2)$ is continuous except perhaps along a countable number of curves;

(c) $\displaystyle\int_{-\infty}^{+\infty} \int_{-\infty}^{+\infty} f(x_1, x_2)\mathrm{d}x_1\,\mathrm{d}x_2 = 1;$

(d) $\Pr(a \leqslant X_1 \leqslant b \text{ and } c \leqslant X_2 \leqslant d) = \displaystyle\int_a^b \int_c^d f(x_1, x_2)\mathrm{d}x_1\,\mathrm{d}x_2.$

The effect of condition (b) is that any definite integrals required may be evaluated by the familiar process of dividing into sections.

The marginal p.d.f. of X_1 is defined as:

$$f_1(x_1) = \int_{-\infty}^{+\infty} f(x_1, x_2)\mathrm{d}x_2. \qquad (8.7)$$

The marginal p.d.f. of X_2 is similarly

$$f_2(x_2) = \int_{-\infty}^{+\infty} f(x_1, x_2) \, dx_1 . \tag{8.8}$$

At points where $f_2(x_2) > 0$, the conditional p.d.f. of X_1 given $X_2 = x_2$ is

$$f(x_1 | x_2) = \frac{f(x_1, x_2)}{f_2(x_2)} \tag{8.9}$$

and, similarly, where $f_1(x_1) > 0$, the conditional p.d.f. of X_2 given $X_1 = x_1$ is

$$f(x_2 | x_1) = \frac{f(x_1, x_2)}{f_1(x_1)}. \tag{8.10}$$

As with the discrete case, X_1 and X_2 are independent if and only if

$$f(x_1, x_2) = f_1(x_1) f_2(x_2).$$

Example 4

$$f(x_1, x_2) = 4x_1 x_2, \quad 0 \leqslant x_1 \leqslant 1, \quad 0 \leqslant x_2 \leqslant 1$$
$$= 0 \text{ otherwise.}$$

Here,

$$f_1(x_1) = \int_0^1 4x_1 x_2 \, dx_2 = 2x_1 [x_2^2]_0^1 = 2x_1 \quad \text{and} \quad f_2(x_2) = 2x_2 .$$

Since $f(x_1, x_2) = f_1(x_1) f_2(x_2)$, X_1, X_2 are independent.

Example 5

$$f(x_1, x_2) = 8x_1 x_2, \quad 0 \leqslant x_1 \leqslant x_2 \leqslant 1.$$

In this case,

$$f_2(x_2) = \int_0^{x_2} 8x_1 x_2 \, dx_1 = 4x_2 [x_1^2]_0^{x_2} = 4x_2^3$$

$$f_1(x_1) = \int_{x_1}^1 8x_1 x_2 \, dx_2 = 4x_1 [x_2^2]_{x_1}^1 = 4x_1(1 - x_1^2)$$

and now, $f_1(x_1) f_2(x_2) \neq f(x_1, x_2)$.

This brings out again that the 'obvious' factorization of

$$f(x_1, x_2) = g(x_1) h(x_2)$$

fails here, since $g(x_1)$ is not always positive when $f_1(x_1)$ is positive, since $f(x_1, x_2)$ is zero when $x_1 > x_2$.

Example 6

X_1, X_2 are continuous random variables with joint p.d.f.

$$f(x_1, x_2) = e^{-(x_1 + x_2)}, \quad x_1 \geqslant 0, \quad x_2 \geqslant 0$$

$$= 0 \text{ otherwise.}$$

Calculate

(a) $\Pr(X_1 > 2 \text{ and } X_2 > 1)$;
(b) $\Pr(X_1 > 2 | X_2 > 1)$;
(c) $\Pr(X_1 > X_2 | 2X_1 > X_2)$.

Now, the joint p.d.f. factorizes into $e^{-x_1} e^{-x_2}$ and as the ranges of x_1, x_2 are not related, X_1, X_2 are independent. By inspection the marginal distributions have p.d.f.s $f_1(x_1) = e^{-x_1}$, $f_2(x_2) = e^{-x_2}$.

(a) $\Pr(X_1 > 2 \text{ and } X_2 > 1) = \Pr(X_1 > 2) \Pr(X_2 > 1)$

$$= \left(\int_2^\infty e^{-x_1} \, dx_1 \right) \left(\int_1^\infty e^{-x_2} \, dx_2 \right)$$

$$= (e^{-2})(e^{-1}) = e^{-3}.$$

(b) $\Pr(X_1 > 2 | X_2 > 1) = \Pr(X_1 > 2)$, since X_1, X_2 are independent, $= e^{-2}$.

(c) $\Pr(X_1 > X_2 | 2X_1 > X_2) = \dfrac{\Pr(X_1 > X_2 \text{ and } 2X_1 > X_2)}{\Pr(2X_1 > X_2)}$

$$= \frac{\Pr(X_1 > X_2)}{\Pr(2X_1 > X_2)}.$$

Now,

$$\Pr(X_1 > X_2) = \int_0^\infty \int_0^{x_1} f(x_1, x_2) \, dx_1 \, dx_2 = \int_0^\infty \int_0^{x_1} e^{-x_1 - x_2} \, dx_1 \, dx_2$$

$$= \int_0^\infty e^{-x_1} \left[\int_0^{x_1} e^{-x_2} \, dx_2 \right] dx_1 = \int_0^\infty e^{-x_1} (1 - e^{-x_1}) \, dx_1$$

$$= \int_0^\infty (e^{-x_1} - e^{-2x_1}) \, dx_1 = \frac{1}{2},$$

which is entirely reasonable, for since X_1, X_2 are independent, and have the same distribution, $\Pr(X_1 > X_2) = \Pr(X_2 > X_2) = 1/2$. Also,

$$\Pr(2X_1 > X_2) = \int_0^\infty \int_0^{2x_1} f(x_1, x_2) \, dx_1 \, dx_2 = \frac{2}{3},$$

which should be verified as an exercise, and the required conditional probability is $(1/2)/(2/3) = 3/4$.

Problem 7. The continuous random variables X_1, X_2 have joint p.d.f.

$$f(x_1, x_2) = n(n-1)(x_2 - x_1)^{n-2}, \quad 0 < x_1 < x_2 < 1.$$

Calculate $\Pr(X_1 < X_2/2)$.

Problem 8. The joint distribution of the continuous random variables X_1, X_2 has p.d.f.

$$f(x_1, x_2) = n(n-1)\lambda^2 \, e^{-\lambda(x_1 + x_2)} (e^{-\lambda x_1} - e^{-\lambda x_2})^{n-2},$$

$$0 \leqslant x_1 \leqslant x_2 < \infty$$

find the p.d.f.s of the marginal distributions of X_1, X_2 and evaluate

$$\Pr(X_2 \geqslant 1/\lambda), \Pr(X_1 \leqslant 1/\lambda).$$

Problem 9. The continuous random variables R, Θ, have joint p.d.f.

$$f(r, \theta) = \frac{1}{2\pi} r e^{-r^2/2}, \quad 0 \leqslant r < \infty, \quad 0 \leqslant \theta \leqslant 2\pi.$$

Find the p.d.f.s of the marginal distributions of R and Θ. Evaluate $\Pr(R \leqslant r$ and $\Theta \leqslant \pi)$.

Problem 10. The continuous random variables X, Y have joint p.d.f.

$$f(x, y) = 3/(2x^3 y^2), \quad 1/x \leqslant y \leqslant x, \quad 1 \leqslant x < \infty$$

$$= 0 \text{ otherwise.}$$

Find the p.d.f.s of the marginal distributions of X, Y and calculate $\Pr(X > 2)$, $\Pr(Y > 2)$, $\Pr(X > 2 \text{ and } Y > 2)$.

Problem 11. The continuous random variables X, Y have joint p.d.f.

$$f(x, y) = \frac{(m + n + 2)!}{m! n!} (1 - x)^n y^m, \quad 0 \leqslant y \leqslant x < 1$$

$$= 0 \text{ otherwise, } m, n \text{ positive integers.}$$

Find the p.d.f.s of the marginal distributions of X, Y and calculate

$$\Pr(Y \leqslant y_0 | X = x_0).$$

8.4 THE BIVARIATE NORMAL DISTRIBUTION

The continuous random variables X_1, X_2 are said to have the bivariate normal distribution, with parameters $\mu_1, \mu_2, \sigma_1, \sigma_2$ and ρ, if their joint p.d.f. satisfies

$$f(x_1, x_2) = \frac{1}{2\pi\sigma_1\sigma_2\sqrt{(1-\rho^2)}} \exp\left\{-\frac{1}{2(1-\rho^2)}\left[\left(\frac{x_1-\mu_1}{\sigma_1}\right)^2\right.\right.$$

$$\left.\left. -2\rho\left(\frac{x_1-\mu_1}{\sigma_1}\right)\left(\frac{x_2-\mu_2}{\sigma_2}\right) + \left(\frac{x_2-\mu_2}{\sigma_2}\right)^2\right]\right\}$$

$$-\infty < x_1 < +\infty, \quad -\infty < x_2 < +\infty, \qquad (8.11)$$

where the parameters satisfy, $\sigma_1 > 0, \sigma_2 > 0$ and $-1 < \rho < +1$.

It is convenient to write

$$f(x_1, x_2) = \frac{e^{-Q/2}}{2\pi\sigma_1\sigma_2\sqrt{(1-\rho^2)}},$$

where Q is a quadratic form in x_1, x_2. Some properties of $f(x_1, x_2)$ can be found by completing the square on $(x_1 - \mu_1)/\sigma_1$. Thus

$$(1-\rho^2)Q = \frac{(x_1-\mu_1)^2}{\sigma_1^2} - 2\rho\frac{(x_1-\mu_1)(x_2-\mu_2)}{\sigma_1\sigma_2}$$

$$+ \rho^2\frac{(x_2-\mu_2)^2}{\sigma_2^2} + \frac{(x_2-\mu_2)^2}{\sigma_2^2}(1-\rho_2)$$

$$= \left[\frac{x_1-\mu_1-\rho\sigma_1(x_2-\mu_2)/\sigma_2}{\sigma_1}\right]^2$$

$$+ \left(\frac{x_2-\mu_2}{\sigma_2}\right)^2(1-\rho^2). \qquad (8.12)$$

Since $\rho^2 < 1$, we have $Q(x_1, x_2) \geqslant 0$ for all x_1 and x_2. In fact, when $x_1 = \mu_1$, $x_2 = \mu_2$, Q is zero. Thus when $x_1 \neq \mu_1, x_2 \neq \mu_2, Q > 0$, and hence $f(x_1, x_2)$ has a maximum at $x_1 = \mu_1, x_2 = \mu_2$.

This way of expressing Q, allows the marginal distributions of X_1, X_2 to be obtained.

$$f(x_1, x_2) = \frac{1}{\sqrt{(2\pi)}\sigma_2}\exp\left[-\frac{1}{2}\left(\frac{x_2-\mu_2}{\sigma_2}\right)^2\right]$$

$$\times \frac{1}{\sqrt{(2\pi)}\sigma_1\sqrt{(1-\rho^2)}}$$

$$\times \exp\left\{-\frac{1}{2}\left[\frac{x_1-\mu_1-\rho\sigma_1(x_2-\mu_2)/\sigma_2}{\sigma_1\sqrt{(1-\rho^2)}}\right]^2\right\}.$$

$$f_2(x_2) = \int_{-\infty}^{+\infty} f(x_1, x_2)dx_1. \qquad (8.13)$$

But the second of the two factors making up $f(x_1, x_2)$, is the p.d.f. of a normal distribution with parameters $\mu = \mu_1 + \rho\sigma_1(x_2 - \mu_2)/\sigma_2$ and $\sigma = \sigma_1\sqrt{(1 - \rho^2)}$. Hence, if x_2 is fixed and we integrate this term from minus infinity to plus infinity with respect to x_1, we obtain unity. Thus,

$$f_2(x_2) = \frac{1}{\sqrt{(2\pi)}\sigma_2} \exp\left[-\frac{1}{2}\left(\frac{x_2 - \mu_2}{\sigma_2}\right)^2\right],$$

$$-\infty < x_2 < +\infty. \tag{8.14}$$

But this is the p.d.f. of a normal distribution with parameters, μ_2, σ_2. Similarly,

$$f_1(x_1) = \frac{1}{\sqrt{(2\pi)}\sigma_1} \exp\left[-\frac{1}{2}\left(\frac{x_1 - \mu_1}{\sigma_1}\right)^2\right],$$

$$-\infty < x_1 < +\infty,$$

Now X_1, X_2 will be independent, if and only if $f_1(x_1)f_2(x_2) = f(x_1, x_2)$ for all x_1, x_2 in this case, if and only if

$$\frac{1}{\sqrt{(2\pi)}\sigma_1\sqrt{(1 - \rho^2)}} \exp\left\{-\frac{1}{2}\left[\frac{x_1 - \mu_1 - \rho\sigma_1(x_2 - \mu_2)/\sigma_2}{\sigma_1\sqrt{(1 - \rho^2)}}\right]^2\right\}$$

$$= \frac{1}{\sqrt{(2\pi)}\sigma_1} \exp\left[-\frac{1}{2}\left(\frac{x_1 - \mu_1}{\sigma_1}\right)^2\right].$$

Now choose $x_1 = \mu_1, x_2 = \mu_2$, and we require $1/\sqrt{(1 - \rho^2)} = 1$ or $\rho = 0$.

Thus $\rho = 0$ is necessary, and by putting $\rho = 0$ it is easily seen that this is sufficient for independence. The wider significance of this parameter ρ will appear at a later stage.

It now appears 'almost self-evident' that if the random variables X_1, X_2 have marginal distributions which are normal, then the joint distribution of X_1, X_2 must be bivariate normal. This is *not* necessarily the case as is shown in the counter-example in Problem 13.

It is also regrettable that probabilities for the bivariate normal distribution cannot be obtained by a judicious use of the ordinary normal distribution. Bivariate tables are available in Pearson's *Tables for Statisticians*.

Now we turn our attention to the conditional distributions. Since

$$f(x_1|x_2) = \frac{f(x_1, x_2)}{f_2(x_2)},$$

by equations (8.13) and (8.14),

$$f(x_1|x_2) = \frac{1}{\sqrt{(2\pi)}\sigma_1\sqrt{(1 - \rho^2)}}$$

$$X \exp\left\{-\frac{1}{2}\left[\frac{x_1 - \mu_1 - \rho\sigma_1(x_2 - \mu_2)/\sigma_2}{\sigma_1\sqrt{(1-\rho^2)}}\right]^2\right\}.$$

But this is again the p.d.f. of a normal distribution. Hence the conditional distribution of X_1 given x_2 is $N[\mu_1 + \rho\sigma_1(x_2 - \mu_2)/\sigma_2, \sigma_1^2(1-\rho^2)]$.

It will be noticed that the second of these parameters does not depend on x_2, while the first is a linear function of x_2. Similarly the conditional distribution of X_2 given x_1 is $N[\mu_2 + \rho\sigma_2(x_1 - \mu_1)/\sigma_1, \sigma_2^2(1-\rho^2)]$.

We can squeeze a little more information from the result for the conditional distribution of X_2 given X_1, which will have important consequences. For it follows immediately that the conditional distribution of

$$\frac{(X_2 - \mu_2) - \rho\sigma_2(x_1 - \mu_1)/\sigma_1}{\sigma_2\sqrt{(1-\rho^2)}},$$

given $X_1 = x_1$, is $N(0, 1)$.

But this distribution does not depend on x_1, hence the unconditional distribution of

$$\frac{(X_2 - \mu_2) - \rho\sigma_2(X_1 - \mu_1)/\sigma_1}{\sigma_2\sqrt{(1-\rho^2)}}$$

is $N(0, 1)$ *and is independent* of X_1. In consequence, $X_2 - \rho\sigma_2 X_1/\sigma_1$, X_1 are independent random variables.

Example 7

If X_1, X_2 have the bivariate normal distribution with parameters $\mu_1 = \mu_2 = 1$, $\sigma_1 = \sigma_2 = 2$, $\rho = 3/5$, calculate $\Pr(X_1 > 4)$ and $\Pr(X_1 > 4 | X_2 = 3)$. Now X_1 is distributed $N(1, 2)$, hence $(X_1 - 1)/2 = Y_1$ is distributed $N(0, 1)$.

$$\Pr(X_1 > 4) = \Pr\left(\frac{X_1 - 1}{2} > \frac{4 - 1}{2}\right) = \Pr(Y_1 > 1.5)$$

$$= 1 - \Pr(Y_1 < 1.5) = 1 - 0.9332 = 0.0668.$$

For the conditional distribution,

$$\mu = \mu_1 + \frac{\rho\sigma_1(x_2 - \mu_2)}{\sigma_2}$$

$$= 1 + \frac{3}{5} \cdot \frac{2}{2}(3 - 1) = 2.2$$

and $\sigma = \sigma_1\sqrt{(1-\rho^2)} = 2\sqrt{[1 - (9/25)]} = 1.6$, hence the distribution of X_1 given $X_2 = 3$ is $N(2.2, 2.56)$.

$$\Pr(X_1 > 4 | X_2 = 3) = \Pr\left(\frac{X_1 - 2.2}{1.6} > \frac{4 - 2.2}{1.6} \,\Big|\, X_2 = 3\right)$$

$$= \Pr\left(\frac{X_1 - 2.2}{1.6} > 1.125 \,\Big|\, X_2 = 3\right)$$

$$= 1 - \Pr\left(\frac{X_1 - 2.2}{1.6} \leqslant 1.125 \,\Big|\, X_2 = 3\right)$$

$$\approx 1 - 0.8697 = 0.1303.$$

It is precisely the factorization in equation (8.13) which leads one to hope that the bivariate distribution can be handled, using tables of the normal distribution. All that this yields, however, is that the random variables $(X_2 - \mu_2)/\sigma_2$ and

$$\frac{X_1 - \mu_1 - \rho\sigma_1(X_2 - \mu_2)/\sigma_2}{\sigma_1\sqrt{(1 - \rho^2)}}$$

are independent, and X_2 appears in the second of these.

Problem 12. The random variables X_1, X_2 have the bivariate normal distribution with parameters $\mu_1 = 1, \mu_2 = 2, \sigma_1 = 4, \sigma_2 = 5, \rho = 12/13$. Calculate $\Pr(X_1 > 2)$ and $\Pr(X_1 > 2 | X_2 = 2)$.

Problem 13. The joint p.d.f. of X_1, X_2 is

$$f(x_1, x_2) = \frac{1}{2\pi\sqrt{(1 - \rho^2)}} \left\{ \exp\left[-\frac{1}{2}\left(\frac{x_1^2 - 2\rho x_1 x_2 + x_2^2}{1 - \rho^2}\right)\right] \right.$$

$$\left. + \exp\left[-\frac{1}{2}\left(\frac{x_1^2 + 2\rho x_1 x_2 + x_2^2}{1 - \rho^2}\right)\right] \right\}$$

for $x_1 > 0$ and $x_2 > 0$ or $x_1 < 0$ and $x_2 < 0$. $f(x_1, x_2) = 0$ otherwise. Show that the marginal distributions of X_1, X_2 are normal.

8.5 RANDOM PARAMETERS

When the distribution of a random variable X depended on some parameter, θ, then we have regarded the parameter as fixed. It may happen that the parameter also has a distribution. That is, θ is the value of a random variable Θ. Then the probability density function of X is a **conditional** p.d.f. given θ. It is legitimate to compute the unconditional distribution of X, after removing any variation in θ.

Example 8

There are k urns in the ith of which the proportion of white balls is p_i. An urn is selected at random from which a ball is selected at random. We seek the probability that this ball is white.

$$\text{Pr(white ball)} = \sum_{i=1}^{k} \text{Pr(white ball from urn } i)$$

$$= \sum_{1}^{k} \text{Pr(white ball} \mid \text{urn } i) \text{ Pr(urn } i \text{ selected)}$$

$$= \sum_{1}^{k} p_i/k,$$

since any urn is selected with probability $1/k$. If we obtain a white ball, Bayes's theorem readily supplies the probability that it in fact came from urn i.

$$\text{Pr(urn } i \text{ selected} \mid \text{white ball)} = \frac{\text{Pr(white ball} \mid \text{urn } i) \text{ Pr(urn } i)}{\text{Pr(white ball)}}$$

$$= \frac{p_i/k}{\Sigma p_i/k} = \frac{p_i}{\Sigma p_i}. \qquad\blacksquare$$

In Example 8 the method for selecting an urn is open to our inspection and the resulting probability distribution non-controversial.

Example 9

Items are made in a factory on two machines, one of them is new and produces items for each of which there is a probability p_1 that it is defective, the other is old and produces a defective with constant probability p_2. The new machine produces a constant proportion θ of the factory output. Find the probability that a randomly chosen item from the output is defective.

Pr(item is defective)

= pr(defective and from new machine or defective and from old machine)

= Pr(defective and from new machine) +
+ Pr(defective and from old machine)

= Pr(item is defective | new machine)
Pr(item is from new machine)

+ Pr(item is defective | old machine)
Pr(item is from old machine)

$$= p_1\theta + p_2(1-\theta).$$

Hence if a random sample of n is drawn without replacement from the (large) factory output, the number of defectives has a binomial distribution with parameters n and $p_1\theta + p_2(1-\theta)$.

In Example 9, the key assumption about the proportion of output from the new machine is buried in the production records.

In general X and Θ may be thought of as having a joint p.d.f. $f(x, \theta)$ which we may factorize as

$$f(x, \theta) = f(x \mid \theta) f_2(\theta).$$

Provided that we know $f_2(\theta)$, the marginal p.d.f. of X can be obtained as

$$f_1(x) = \sum_\theta f(x \mid \theta) f_2(\theta) \quad \text{or} \quad \int f(x \mid \theta) f_2(\theta) \, d\theta$$

according as the distribution of Θ is discrete or continuous. The marginal distribution of Θ is known as the **prior distribution** for Θ and we shall frequently use $g(.)$ for its probability density function.

Example 10

The discrete random variable X has the conditional probability density function

$$\Pr(X = 1 \mid p) = p, \quad \Pr(X = 0 \mid p) = 1 - p.$$

[Thus for *fixed* p, X has a Bernoulli distribution.] The prior distribution of P is continuous and has p.d.f. $g(p) = 2p$, $0 < p < 1$. It is required to find the unconditional distribution of X.

$$\Pr[X = 1] = \int_0^1 \Pr[X = 1 \mid p] \, g(p) \, dp = \int_0^1 p \cdot 2p \, dp$$

$$= \int_0^1 2p^2 \, dp = 2/3.$$

$$\Pr[X = 0] = \int_0^1 \Pr[X = 0 \mid p] \, g(p) \, dp = \int_0^1 (1 - p) \, 2p \, dp = 1/3.$$

The prior density, $g(p) = 2p (0 < p < 1)$, used in Example 10 has been 'plucked from the air'. Such a prior distribution entails acceptance of any conclusions which can then be legitimately drawn. For instance,

$$\Pr(0 < P < 1/2) = \int_0^{1/2} 2p \, dp = 1/4.$$

We are obliged to conclude that the value of P obtaining is three times as likely to be greater than $1/2$ as it is to be less than $1/2$. Suppose a value of p is selected and *then* the distribution of X is sampled. Let us say $x = 1$. We are now in a position to revise our opinion as to $\Pr(0 < P < 1/2)$, for a more appropriate measure would be $\Pr(0 < P < 1/2|X = 1)$. We must obtain the corresponding conditional probability density function using

$$g(p|x) = \frac{f(x, p)}{f_1(x)} = \frac{f(x|p)\,g(p)}{f_1(x)}$$

which here translates to

$$g(p|x = 1) = \frac{\Pr(X = 1|p)\,2p}{2/3} = 3p^2, \quad 0 < p < 1.$$

Thus

$$\Pr(0 < P < 1/2|X = 1) = \int_0^{1/2} 3p^2\,dp = 1/8.$$

This somewhat diminished probability corresponds to intuition. For $\Pr(X = 1|p)$ increases with p, hence, if $X = 1$ is in fact observed, it suggests that a larger value of p had probably been selected. The conditional distribution of a random parameter Θ given the obtained value, x, is called the **posterior distribution** of Θ.

Problem 14. If $f(x|\theta) = \theta^{x-1}(1 - \theta)$, $x = 1, 2, \ldots$, and the prior distribution of Θ has density $g(\theta) = m\theta^{m-1}\,(0 < \theta < 1)$, calculate the marginal distribution of X and the posterior distribution of Θ.

Problem 15. If $f(x|\theta) = \theta\,e^{-\theta x}$ and the prior distribution of Θ has density $g(\theta) = \lambda\,e^{-\lambda\theta}$, find the marginal distribution of X and the posterior distribution of θ. Calculate $\Pr(X > x_0)$,

(i)   from $f_1(x)$,

(ii) from $\int [\Pr(X > x_0)|\theta]\,g(\theta)\,d\theta$.

Problem 16. A loom stops from time to time and the number, X, of stops in unit running time may be assumed to have a Poisson distribution with parameter μ. For each stop, there is a probability θ that a fault will be produced in the fabric being woven. Occurrences associated with different stops may be assumed independent. Let Y be the number of fabric faults so produced in unit running time. By first finding the conditional distribution of Y, given that $X = x$, or otherwise, prove that Y has a Poisson distribution with parameter $\theta\mu$.

U.L. B.Sc. General, Pt II, 1962. ∎

If the interval in which the distribution of X given θ is positive has end points which are also determined by θ, then this must be taken into account when recovering the marginal distribution of X.

Example 11

The continuous random variable X has the rectangular distribution over the interval $[0, \theta]$, given θ. The distribution of Θ is rectangular over the interval $[0, l]$ where l is a known constant. It is required to find the marginal distribution of X. We have,

$$f(x|\theta) = 1/\theta, \quad 0 \leqslant x \leqslant \theta; \quad g(\theta) = 1/l, \quad 0 \leqslant \theta \leqslant l$$

Hence the p.d.f. of the joint distribution of X, Θ is zero when $\theta < x$. Thus

$$f(x, \theta) = f(x|\theta) g(\theta) = \left(\frac{1}{\theta}\right)\left(\frac{1}{l}\right), \quad 0 \leqslant x \leqslant \theta \leqslant l.$$

Hence

$$f_1(x) = \int_x^l \frac{1}{l\theta} \, d\theta = \frac{1}{l} \left[\log_e (l/x)\right], \quad 0 \leqslant x \leqslant l.$$

Problem 17. For Example 11, verify that

$$\int_0^l f_1(x) \, dx = 1.$$

Find the p.d.f. of the posterior distribution of Θ given x.

Problem 18. The distribution of X given θ is rectangular over $(0, \theta)$. The prior distribution of Θ has probability density function $g(\theta) = \theta_0/\theta^2$, $\theta > \theta_0$. Find the posterior distribution of Θ given $X = x$.

Problem 19. The continuous random variable X has p.d.f. $f(x|\theta) = e^{\theta-x}, x \geqslant \theta$, and Θ has the rectangular distribution over $[0, 1]$. Calculate the marginal distribution of X. ∎

In Example 9, the source of the prior information was in the production figures. In more complicated processes an appropriate prior distribution may be suggested by earlier studies of similar processes and would ordinarily only be obvious to experts in that particular field. The next problems illustrate how prior distributions have been used in practice.

Problem 20. Greenwood and Yule studied the number of accidents experienced by women working on the manufacture of shells [1]. In a long period, the number of shells handled by a particular woman would be large and naturally the probability of an accident in the couse of any particular transaction ought to have been small. Therefore the number of accidents experienced by a particular woman might be taken as having a Poisson distribution. From the data it appeared that the parameter, θ, of this distribution varied from woman to woman. Greenwood and Yule suggested that Θ had a $\Gamma(\alpha, \lambda)$ distribution. On this basis calculate the probability that a woman chosen at random from a large workforce will experience x accidents.

Problem 21. Chatfield and Goodhardt employed a prior distribution in connection with a model for consumer purchasing behaviour [2]. The main probabilistic features of the model were:

(1)   The probability, p, that a given customer will buy at least one unit of a brand in a particular week is constant and independent of previous purchases. Therefore in a time period of n weeks, the number of weeks in which the consumer buys at least one unit will follow a binomial distribution with parameters n and p.
(2)   The probability, p, varies from consumer to consumer and has a beta distribution in the whole population.

Assuming that the beta distribution has parameters α, β, i.e. that the prior distribution for P has density

$$g(p) = \frac{\Gamma(\alpha + \beta)}{\Gamma(\alpha)\,\Gamma(\beta)}\, p^{\alpha-1}(1-p)^{\beta-1}, \qquad 0<p<1,$$

calculate the proportion of people who buy at least one unit on exactly r out of n weeks.   ∎

The assignment of a particular prior distribution to a parameter has substantial implications for the statistician. One of his major preoccupations is to judge the probable whereabouts of unknown parameters. A prior distribution allows a calculation of the probability that Θ lies in any prescribed interval. Suppose Θ has a continuous distribution with p.d.f. $g(\theta)$ consider

$$\Pr(\theta \leqslant \Theta \leqslant \theta + l) = p(\theta)$$

where l is a fixed constant (see Fig. 8.1). Then it is worth searching for the interval $(\theta, \theta + l)$ which encloses the most probability. Now if $G(.)$ is the c.d.f. of Θ then

$$p(\theta) = G(l + \theta) - G(\theta),$$

$$\Rightarrow \qquad \frac{dp(\theta)}{d\theta} = g(l + \theta) - g(\theta).$$

If $g(.)$ has a single maximum, then $p(\theta)$ will have a turning value at θ^* such that $g(l + \theta^*) = g(\theta^*)$ (an equal ordinate solution). On the other hand if $g(.)$ is monotone, $p(\theta)$ will have a maximum at θ^* such that $(\theta^*, \theta^* + l)$ is one of the tails of the distribution of Θ (see Fig. 8.2)

Example 12
If $g(\theta) = \lambda e^{-\lambda\theta}$, $\theta > 0$ then $dp(\theta)/d\theta = \lambda\,(e^{-\lambda\theta - \lambda l} - e^{-\lambda\theta}) < 0$. Hence $p(\theta)$ is

decreasing with θ and hence is a maximum when $\theta = 0$. The required interval is $(0, l)$.

Problem 22. If $g(\theta) = 12\theta^2(1 - \theta), 0 < \theta < 1$, find θ_1 such that

$$\Pr(\theta_1 \leqslant \Theta \leqslant \theta_1 + 3/7)$$

is a maximum.

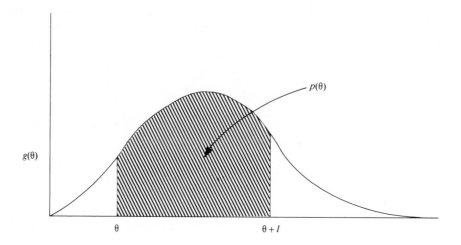

Fig. 8.1

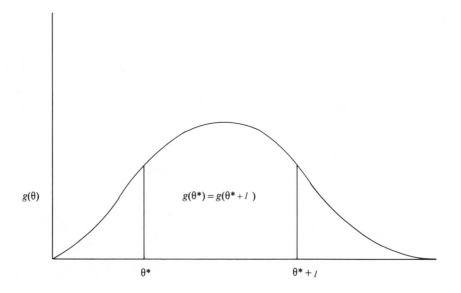

Fig. 8.2

REFERENCES

[1]   M. Greenwood, & G. Yule, *J. R. Stat. Soc.* 1920, **83**.
[2]   C. Chatfield, & G. Goodhardt, *J. R. Stat. Soc. Series C*, 1970, **19**.

BRIEF SOLUTIONS AND COMMENTS ON THE PROBLEMS

Problem 1. Same situation as rolling $6mn$ dice once each. Thus a multinomial distribution with $p_i = 1/6$, $i = 1, 2, \ldots, 6$. Probability of mn scores of each type thus

$$\frac{(6mn)!}{[(mn)!]^6} \left(\frac{1}{6}\right)^{6mn}.$$

Problem 2.
(a)   The suit with three cards can be chosen in four ways. For each, the probability of three in that suit and one card in each of the others is

$$\frac{6!}{3!1!1!1!} \left(\frac{1}{4}\right)^3 \left(\frac{1}{4}\right)^1 \left(\frac{1}{4}\right)^1 \left(\frac{1}{4}\right)^1.$$

The total probability is $15/128$.

(b)   The number of spades has a binomial distribution with parameters $n = 6$ and $p = 1/4$.

$$\text{Pr(just three spades)} = \binom{6}{3} \left(\frac{1}{4}\right)^3 \left(\frac{3}{4}\right)^3 = 135/1024.$$

(c)   We require Pr(3 spades or 3 hearts or 3 diamonds or 3 clubs), where the events are *not* mutually exclusive since we may have 3 spades and 3 hearts, which has probability

$$\frac{6!}{3!3!0!0!} \left(\frac{1}{4}\right)^3 \left(\frac{1}{4}\right)^3 = \frac{5}{1024}.$$

But there are $\binom{4}{2} = 6$ kinds of event involving two triples. Hence, using (b), the required probability is $(4 \times 135 - 6 \times 5)/1024 = 255/512$.

Problem 3.
The descriptions are mutually exclusive and exhaustive and, since the characteristics are independent, have probabilities $p_1 p_2$, $p_1 q_2$, $q_1 p_2$, $q_1 q_2$ for each element, where $q_1 = 1 - p_1$, $q_2 = 1 - p_2$. Hence total probability of r_i of each type $(i = 1, 2, 3, 4)$ is

$$\frac{n!}{r_1! r_2! r_3! r_4!} (p_1 p_2)^{r_1} (p_1 q_2)^{r_2} (q_1 p_2)^{r_3} (q_1 q_2)^{r_4}.$$

Problem 4. The nth item must be both inspected and be defective, with probability $p_1 p_2$. The previous $n - 1$ items must contain just k defectives which have *not* been inspected, with probability

$$\binom{n-1}{k} (q_1 p_2)^k q_2^{n-k-1}.$$

Hence the required probability is

$$\left[\binom{n-1}{k} (q_1 p_2)^k q_2^{n-k-1} \right] p_1 p_2.$$

Problem 5.

(a) $\quad f_1(x_1) = \displaystyle\sum_{x_2 = x_1 + 1}^{\infty} q^2 p^{x_2 - 2} = q^2 p^{x_1 - 1}/(1 - p),$

since the series is a geometric progression,

$$= q p^{x_1 - 1}.$$

$$f_2(x_2) = \sum_{x_1 = 1}^{x_2 - 1} q^2 p^{x_2 - 2} = (x_2 - 1) q^2 p^{x_2 - 2}.$$

Hence, $f(x_1 | x_2) = f(x_1, x_2)/f_2(x_2) = 1/(x_2 - 1),\ 1 \leqslant x_1 \leqslant x_2 - 1.$

(b) For $f_1(x_1)$, the essential computation involves

$$\sum_{x_2 = x_1}^{n_2 + x_1} \binom{n_2}{x_2 - x_1} p^{x_2} q^{n_2 - x_2}$$

$$= p^{x_1} q^{-x_1} \sum_{x_2 - x_1 = 0}^{n_2} \binom{n_2}{x_2 - x_1} p^{x_2 - x_1} q^{n_2 - (x_2 - x_1)}$$

$$= p^{x_1} q^{-x_1} (p + q)^{n_2} = p^{x_1} q^{-x_1}.$$

Hence

$$f_1(x_1) = \binom{n_1}{x_1} q^{n_1} \cdot p^{x_1} q^{-x_1}$$

$$= \binom{n_1}{x_1} p^{x_1} q^{n_1 - x_1}.$$

It is now clear that X_1 has a binomial distribution with parameters n_1, p. On the other hand,

$$f_2(x_2) = \sum_{x_1=0}^{x_2} \binom{n_1}{x_1} \binom{n_2}{x_2-x_1} p^{x_2} q^{n_1+n_2-x_2},$$

$$x_1 \leqslant x_2 \leqslant n_2 + x_1, \qquad 0 \leqslant x_1 \leqslant n_1,$$

$$= \binom{n_1+n_2}{x_2} p^{x_2} q^{n_1+n_2-x_2}, \quad 0 \leqslant x_2 \leqslant n_1 + n_2.$$

X_2 has another binomial distribution with parameters $n_1 + n_2, p$. Finally,

$$f(x_1 | x_2) = \binom{n_1}{x_1} \binom{n_1}{x_2-x_1} \Big/ \binom{n_1+n_2}{x_2}, \quad x_1 \leqslant \min(x_2, n_1).$$

Problem 6. The probability, in r independent placings of r balls, just x_2 finish up in cell number two is

$$\binom{r}{x_2} \left(\frac{1}{n}\right)^{x_2} \left(1 - \frac{1}{n}\right)^{r-x_2}, \quad 0 \leqslant x_2 \leqslant r.$$

Given that there *are* x_2 in cell two, we are placing $r - x_2$ balls in $n - 1$ cells. The conditional probability that just x_1 are in cell number one is (similarly)

$$\binom{r-x_2}{x_1} \left(\frac{1}{n-1}\right)^{x_1} \left(1 - \frac{1}{n-1}\right)^{r-x_2-x_1}, \quad 0 \leqslant x_1 \leqslant r - x_2.$$

But

$$\Pr[X_1 = x_1 \text{ and } X_2 = x_2] = \Pr[X_1 = x_1 | X_2 = x_2] \Pr[X_2 = x_2]$$

$$= \binom{r}{x_2} \binom{r-x_2}{x_1} \frac{(n-2)^{r-x_1-x_2}}{n^r}.$$

Of course this answer can be obtained directly, since having chosen x_1, x_2 balls for the two particular cells in $\binom{r}{x_2} \binom{r-x_2}{x_1}$ ways, the remaining $r - x_1 - x_2$ balls may be placed in $(n-2)$ cells in $(n-2)^{r-x_1-x_2}$ different ways.

Problem 7. First integrate with respect to x_1 from 0 to $x_2/2$.

$$\int_0^{x_2/2} n(n-1)(x_2-x_1)^{n-2} dx_1 = \Big[-n(x_2-x_1)^{n-1}\Big]_0^{x_2/2}$$

$$= nx_2^{n-1} - n(x_2/2)^{n-1}.$$

The result is now integrated with respect to x_2 between 0 and 1 yielding $1 - (1/2)^{n-1}$. The reader should confirm the result by reversing the order of integration noting that $1 > x_2 > 2x_1, 0 < x_1 < 1/2$ are the limits.

Problem 8.

$$f_2(x_2) = \int_0^{x_2} f(x_1, x_2)\, dx_1$$

$$= [-n(e^{-\lambda x_1} - e^{-\lambda x_2})^{n-1}\, \lambda e^{-\lambda x_2}]_0^{x_2}$$

$$= n(1 - e^{-\lambda x_2})^{n-1}\, \lambda e^{-\lambda x_2}.$$

Hence

$$\Pr(X_2 \geqslant 1/\lambda) = \int_{1/\lambda}^{\infty} f_2(x_2)\, dx_2$$

$$= [(1 - e^{-\lambda x_2})^n]_{1/\lambda}^{\infty} = 1 - (1 - e^{-1})^n.$$

Similarly,

$$f_1(x_1) = n\lambda e^{-n\lambda x_1} \quad \text{and} \quad \Pr(X_1 \leqslant 1/\lambda) = 1 - e^{-n}.$$

Problem 9.

$$f_1(\theta) = \int_0^{\infty} f(r, \theta)\, dr = \frac{1}{2\pi}, \quad 0 \leqslant \theta \leqslant 2\pi.$$

$$f_2(r) = \int_0^{2\pi} f(r, \theta)\, d\theta = r\, e^{-r^2/2}, \quad 0 \leqslant r < \infty.$$

R and Θ are clearly independent. Hence

$$\Pr(R \leqslant r \text{ and } \Theta \leqslant \pi) = \Pr(R \leqslant r)\Pr(\Theta \leqslant \pi) = (1 - e^{-r^2/2})/2.$$

Problem 10.

$$f_1(x) = \frac{3}{2x^3} \int_{1/x}^{x} \frac{1}{y^2}\, dy = \frac{3}{2x^3} \left(-\frac{1}{y}\right)_{1/x}^{x}$$

$$= \frac{3}{2x^3} \left(x - \frac{1}{x}\right) = \frac{3}{2x^2} - \frac{3}{2x^4}, \quad 1 \leqslant x < \infty.$$

Hence,

$$\Pr(X > 2) = \frac{3}{2} \int_2^{\infty} \left(\frac{1}{x^2} - \frac{1}{x^4}\right) dx = 11/16.$$

The computation of $f_2(y)$ is more tricky, for $f(x, y) > 0$ for $1/x < y < x$. That is $x > y$ and $x > 1/y$. Thus if $y > 1$, then $x > y$; but if $0 < y < 1$, then $x > 1/y$. Hence, for $y < 1$,

$$f_2(y) = \frac{3}{2y^2} \int_{1/y}^{\infty} \frac{1}{x^3} \, dx = \frac{3}{4}.$$

But, for $y \geqslant 1$,

$$f_2(y) = \frac{3}{2y^2} \int_{y}^{\infty} \frac{1}{x^3} \, dx = \frac{3}{4y^4}.$$

Thus,

$$\Pr(Y > 2) = \int_{2}^{\infty} \frac{3}{4y^4} \, dy = \frac{1}{32}.$$

Finally, since $Y > 2 \Rightarrow X > 2$,

$$\Pr(X > 2 \text{ and } Y > 2) = \Pr(Y > 2) = 1/32.$$

Problem 11. The integrations are straightforward.

$$f_1(x) = \frac{(m + n + 2)!}{(m + 1)! \, n!} \, (1 - x)^n x^{m+1}, \quad 0 < x < 1.$$

$$f_2(y) = \frac{(m + n + 2)!}{m! \, (n + 1)!} \, y^m (1 - y)^{n+1}, \quad 0 < y < 1.$$

$$f(y|x) = (m + 1) \, y^m / x^{m+1}, \quad 0 < y < x.$$

$$\Pr(Y \leqslant y_0 | X = x_0) = \int_{0}^{y_0} f(y|x_0) \, dy = (y_0/x_0)^{m+1}, \quad y_0 < x_0.$$

Problem 12. X_1 is $N(1, 16)$ hence

$$\Pr(X_1 > 2) = \Pr[(X_1 - 1)/4 > (2 - 1)/4]$$
$$= \Pr[(X_1 - 1)4 > 1/4] = 0.40,$$

since $(X_1 - 1)/4$ is distributed $N(0, 1)$. The distribution of X_1 given $X_2 = 2$ is $N(1, 400/169)$. $\Pr(X_1 > 2|X_2 = 2) = 0.26$.

Problem 13. The displayed p.d.f. is not, despite appearances, the sum of two different bivariate normal densities with parameter $\rho, -\rho$, since all the points in the second and fourth quadrants receive zero density. By the same method as outlined in section 8.4,

$$f_1(x_1) = \frac{1}{2} \frac{1}{\sqrt{2\pi}} \exp\left(-\frac{1}{2}x_1^2\right) + \frac{1}{2} \frac{1}{\sqrt{2\pi}} \exp\left(-\frac{1}{2}x_1^2\right),$$

$$x_1 > 0$$

$$= \frac{1}{\sqrt{2\pi}} \exp\left(-\frac{1}{2}x_1^2\right), \quad x_1 > 0$$

and the same result is obtained for $x_1 < 0$. Hence, the distribution of X_1 is $N(0, 1)$. Similarly, that of X_2 is $N(0, 1)$.

Problem 14

$$f_1(x) = \int_0^1 f(x|\theta) g(\theta) d\theta$$

$$= \int_0^1 \theta^{x-1} (1-\theta) m\theta^{m-1} d\theta$$

$$= m \int_0^1 (\theta^{x+m-2} - \theta^{x+m-1}) d\theta = m/[(x+m)(x+m-1)].$$

Hence

$$g(\theta|x) = f(x|\theta) g(\theta)/f_1(x)$$

$$= (x+m)(x+m-1)\theta^{x+m-2}(1-\theta), \quad 0 < \theta < 1.$$

Problem 15

$$f_1(x) = \int_0^\infty \theta e^{-\theta x} \lambda e^{-\lambda\theta} d\theta = \lambda \int_0^\infty \theta e^{-\theta(\lambda+x)} d\theta.$$

Substitute

$$y = \theta(\lambda+x), \quad f_1(x) = \frac{\lambda}{(\lambda+x)^2} \int_0^\infty y e^{-y} dy = \frac{\lambda}{(\lambda+x)^2}.$$

Hence

$$\Pr[X > x_0] = \int_{x_0}^\infty \frac{\lambda}{(\lambda+x)^2} dx = \frac{\lambda}{\lambda+x_0}.$$

On the other hand,

$$\Pr(X > x_0|\theta) = \int_{x_0}^\infty \theta e^{-\theta x} dx = e^{-\theta x_0} \quad \text{and} \quad \int (e^{-\theta x_0}) g(\theta) d\theta$$

$$= \int_0^\infty e^{-\theta x_0} \lambda e^{-\lambda\theta} d\theta$$

$$= \lambda \int_0^\infty e^{-\theta(\lambda+x_0)} d\theta = \frac{\lambda}{\lambda+x_0}.$$

We have $g(\theta \mid x) = \lambda\theta\, e^{-\lambda(\theta+x)}/[\lambda/(\lambda+x)^2] = (\lambda+x)\,[(\lambda+x)\theta]\,e^{-\theta(\lambda+x)}]$.
The distribution of Θ given $X = x$ is $\Gamma[2, (\lambda+x)]$.

Problem 16. There cannot be more fabric faults that there are stops. Hence

$$\Pr(Y = y) = \sum_{x=y}^{\infty} \Pr(Y = y \text{ and } X = x)$$

$$= \sum_{x=y}^{\infty} \Pr(Y = y \mid X = x)\,\Pr(X = x)$$

$$= \sum_{x=y}^{\infty} \binom{x}{y}\,\theta^y(1-\theta)^{x-y}\,\frac{\mu^x e^{-\mu}}{x!},$$

since, given x, the number of faults has the binomial distribution with parameters $x,\,\theta$.

After a little rearranging, we have

$$\Pr(Y = y) = e^{-\mu}\,\frac{(\theta\mu)^y}{y!}\sum_{x=y}^{\infty}\frac{[(1-\theta)\mu]^{x-y}}{(x-y)!}$$

$$= \frac{e^{-\mu}\,(\theta\mu)^y}{y!}\cdot e^{(1-\theta)\mu}$$

$$= \frac{(\theta\mu)^y\, e^{-\theta\mu}}{y!}\,.$$

That is, the unconditional distribution of Y is Poisson with parameter $\theta\mu$.

Problem 17

$$\cdot \int_0^l \log_e x \, dx = [x \log_e x]_0^l - \int_0^l 1 \cdot dx = l \log_e l - l,$$

since

$$\lim_{x\to 0}\,(x \log_e x) = 0.$$

$$g(\theta \mid x) = \frac{1/\theta l}{(1/l)\log_e (l/x)} = \frac{1}{\theta \log_e (l/x)},\qquad x \leqslant \theta \leqslant l.$$

Problem 18. The marginal p.d.f. depends on whether $x < \theta_0$ or $x > \theta_0$. Thus if $x < \theta_0 < \theta$,

$$f(x,\,\theta) = \frac{1}{\theta}\cdot\frac{\theta_0}{\theta^2} \quad\text{and}\quad f_1(x) = \int_{\theta_0}^{\infty} f(x,\,\theta)\,d\theta = 1/2\theta_0.$$

But if $\theta_0 < x < \theta$ we have

$$f_1(x) = \int_x^\infty f(x, \theta)\, d\theta = \theta_0/2x^2 .$$

There is a 'knock-on' effect on the form of $g(\theta|x)$. Thus, if

$$x < \theta_0, \quad g(\theta|x) = 2\theta_0^2/\theta^3 \quad (\theta > \theta_0);$$

but if

$$x > \theta_0, \quad g(\theta|x) = 2x^2/\theta^3 \quad (\theta > x).$$

Problem 19. Everything turns on whether $x < 1$ or $x > 1$. In the first case θ is constrained to lie between 0 and x and

$$f_1(x) = \int_0^x e^{\theta - x}\, d\theta = 1 - e^{-x} .$$

In the second case, $0 < \theta < 1$ and

$$f_1(x) = \int_0^1 e^{\theta - x}\, d\theta = e^{1 - x} - e^{-x} .$$

Problem 20

$$f(x|\theta) = \theta^x e^{-\theta}/x!, \quad x = 0, 1, 2, \ldots .$$
$$g(\theta) = \lambda(\lambda\theta)^{\alpha - 1} e^{-\lambda\theta}/\Gamma(\alpha), \quad \theta > 0.$$

Hence

$$f_1(x) = \frac{\lambda^\alpha}{x!\,\Gamma(\alpha)} \int_0^\infty \theta^{\alpha + x - 1} e^{-\theta(\lambda + 1)}\, d\theta .$$

After putting $y = \theta(\lambda + 1)$,

$$f_1(x) = \frac{\lambda^\alpha}{x!\,\Gamma(\alpha)} \cdot \frac{\Gamma(\alpha + x)}{(\alpha + 1)^{\alpha + x}} .$$

Note if α is an integer, then X has a negative binomial distribution. The posterior distribution of Θ given x is easily recovered and is $\Gamma(\alpha + x, \lambda + 1)$. There remains the little matter of nominating values for α and λ.

Problem 21

$$f(r|p) = \binom{n}{r} p^r (1 - p)^{n - r} .$$

$$f(r, p) = \binom{n}{r} \frac{\Gamma(\alpha + \beta)}{\Gamma(\alpha)\,\Gamma(\beta)} p^{r + \alpha - 1} (1 - p)^{n - r + \beta - 1} .$$

Since

$$\int_0^1 p^{r+\alpha-1} (1-p)^{n-r+\beta-1} \, dp = \Gamma(r+\alpha) \, \Gamma(n-r+\beta)/\Gamma(\alpha+\beta+n),$$

$$f_1(r) = \binom{n}{r} \frac{\Gamma(\alpha+\beta)}{\Gamma(\alpha) \, \Gamma(\beta)} \cdot \frac{\Gamma(r+\alpha) \, \Gamma(n-r+\beta)}{\Gamma(\alpha+\beta+n)}.$$

Note the special case when $\alpha = \beta = 1$, when $f_1(r)$ reduces to $1/(n+1)$.

Problem 22. $g(\theta)$ has a single maximum at $\theta = 2/3$. If $g(\theta_1) = g(\theta_2)$ then $\theta_1^2 - \theta_1^3 = \theta_2^2 - \theta_2^3$ or $(\theta_2^3 - \theta_1^3) - (\theta_2^2 - \theta_1^2) = 0$. Cancelling $\theta_2 - \theta_1 \neq 0$ we have $\theta_2^2 + \theta_1\theta_2 + \theta_1^2 - (\theta_2 + \theta_1) = 0$. Substituting $\theta_2 = \theta_1 + 3/7$, after collecting terms $(7\theta_1 - 3)(21\theta_1 + 4) = 0$. That is $\theta_1 = 3/7$.

9

Expectation of a Random Variable

9.1 INTRODUCTION

When presented with a mass of data, our ability to grasp the information may be suitably strengthened by computing various **statistics**, or functions of the observations. These are chosen to highlight important features of the data, for example, the spread of the numbers. It seems reasonable to suppose that the corresponding measures for the original population from which the data are obtained would play a similar role in summarizing the population. In this section we restrict ourselves to populations which may be described by probability density functions. By 'a random sample of n values from a distribution' we shall mean $X_1, X_2, \ldots, X_n$ are independent random variables each with the same probability density function. We now investigate the sense in which the mean of a sample can be generalized to the mean of a discrete distribution.

Suppose a sample of n values is drawn from a discrete distribution with p.d.f. $f(x)$ which is positive for $x_1, x_2, \ldots, x_m \ldots$ and is zero otherwise. If in the sample the value x_i appears f_i times, we compute the (arithmetic) mean of the sample as

$$\bar{x} = \sum_i \frac{f_i x_i}{n} = \sum_i \left(\frac{f_i}{n}\right) x_i$$

in which f_i/n is the relative frequency of the value x_i. Now, it is our basic assumption that as $n \to \infty, f_i/n \to f(x_i)$, that is the relative frequency approaches the probability that a random value from the distribution assumes the value x_i.

The analogue for the population mean would seem to be based on the behaviour of $\bar{x}$ as $n \to \infty$. Any particular $\bar{x}$ is a number, but since the numbers of which it is composed are subject to random variation, it may be that $\bar{x}$ fails to tend to a limit. In view of the above remarks about the relative frequencies, a suitable definition of the mean of the distribution would be

$$\sum_{i=1}^{\infty} x_i f(x_i),$$

when this quantity exists. We consider the meaning of the word 'exists' in this context. We require the series to converge. Moreover, the manner in which we label the x_i at which $f(x_i) > 0$ is arbitrary. If there are an infinity of negative values and also of positive values x_i at which $f(x_i) > 0$, the value of

$$\sum_{i=1}^{\infty} x_i f(x_i)$$

will, in general, depend on the order in which we sum the terms. This difficulty will not arise if

$$\sum_{i=1}^{\infty} |x_i| f(x_i)$$

converges. For if this is the case,

$$\sum_{i=1}^{\infty} x_i f(x_i)$$

will converge in whatever order the terms are taken, and, moreover, the 'sum' is always the same.

Definition. (a) If X is a discrete random variable with p.d.f. $f(x)$, then the mean or **expectation** of X, written as $E(X)$, is defined as

$$E(X) = \sum_{i=1}^{\infty} x_i f(x_i)$$

provided

$$\sum_{i=1}^{\infty} |x_i| f(x_i)$$

converges. If

$$\sum_{i=1}^{\infty} |x_i| f(x_i)$$

does not converge, then $E(X)$ is said not to exist.

(b) If X is a continuous random variable with p.d.f. $f(x)$, then the mean or **expectation** of X, written as $E(X)$, is defined as

$$E(X) = \int_{-\infty}^{+\infty} xf(x)\,dx$$

provided that

$$\int_{-\infty}^{+\infty} |x|\,f(x)\,dx$$

converges. If

$$\int_{-\infty}^{+\infty} |x|\,f(x)\,dx$$

does not converge, then $E(X)$ is said not to exist.

In either case, $E(X)$ is usually denoted by μ.   ∎

In practice, the requirement of absolute convergence causes few difficulties. For most discrete random variables, of interest, the values of x_i where $f(x_i) > 0$ will all be positive and all we need to know is whether $\sum x_i\,f(x_i)$ is finite. If there is only a finite number of x_i at which $f(x_i) > 0$, then no question of convergence can arise. For continuous distributions it will suffice if both

$$\int_{0}^{\infty} xf(x)\,dx \quad \text{and} \quad \int_{-\infty}^{0} xf(x)\,dx$$

exist separately.

Intuition suggests the following theorem, 'If $X_1, X_2, \ldots, X_n$ is a random sample from a distribution for which $E(X)$ exists, then

$$\bar{X} = \sum_{i=1}^{n} \frac{X_i}{n}$$

converges (in probability) to $E(X)$'.

The phrase 'in probability' needs amplifying. It does not mean converge in the ordinary sense of a mathematical limit, since $\bar{X}$ is not a function of n but is subject to fluctuation. Roughly it means that if n is large enough, then it is almost certain that $\bar{X}$ is not far from $E(X)$. More precisely, for every $\epsilon > 0$, $\delta > 0$ we can find n_0, such that $\Pr[|\bar{X} - E(X)| \leqslant \epsilon] \geqslant 1 - \delta$ when $n \geqslant n_0$.

The above theorem, which is known as the *weak law of large nunbers* (W.L.L.N.) is true, but the proof is not elementary. At a later stage we shall prove a similar result for $\bar{X}$ when slightly more is known than that the expectation of X exists.

The word 'expectation' arose in connection with games of chance. Suppose a fair die is rolled; if it shows up even a gambler wins £2, if odd, he pays £2. In each game he stands to win £2 with probability 1/2 and win −£2 with probability 1/2. His expected win (or his expectation) on each game is $2(1/2) - 2(1/2) = 0$. In view of the W.L.L.N., his average win in a long series of independent games is also zero. Naturally, most games of chance which are organized as a business yield the gambler a negative expectation.

9.2 THE STANDARD DISTRIBUTIONS

The binomial distribution

$$f(x) = \binom{n}{x} p^x (1-p)^{n-x}, \quad x = 0, 1, 2, \ldots, n.$$

$$E(X) = \sum_i x_i f(x_i) = \sum_{x=0}^{n} x \binom{n}{x} p^x (1-p)^{n-x}$$

and since the first term is zero

$$= \sum_{x=1}^{n} x \frac{n!}{x!(n-x)!} p^x (1-p)^{n-x},$$

cancelling x into $x!$,

$$= \sum_{x=1}^{n} \frac{n!}{(x-1)!(n-x)!} p^x (1-p)^{n-x}$$

and this is very nearly another binomial expansion. Taking np outside the summation sign, we have

$$E(X) = \sum_i x_i f(x_i) = np \sum_{x=1}^{n} \frac{(n-1)!}{(x-1)!(n-x)!} p^{x-1} (1-p)^{n-x}.$$

Now put $x - 1 = y$

$$= np \sum_{y=0}^{n-1} \frac{(n-1)!}{y!(n-1-y)!} p^y (1-p)^{n-1-y}$$

$$= np [p + (1-p)]^{n-1}$$

$$= np. \tag{9.1}$$

In view of our previous remarks, we interpret this as follows: if $X_1, X_2, \ldots, X_N$ is a random sample of N values from a binomial distribution with parameters n, p then since $E(X) = np$,

$$\bar{X} = \frac{\sum_{i=1}^{N} X_i}{N}$$

tends in probability to np as N tends to infinity. That is, for large N, $\bar{X}$ is likely to be near np.

The Poisson distribution

$$f(x) = \frac{\lambda^x e^{-\lambda}}{x!}, \quad x = 0, 1, 2, \ldots$$

$$E(X) = \sum_i x_i f(x_i) = \sum_{x=0}^{\infty} \frac{x \cdot \lambda^x e^{-\lambda}}{x!} = \sum_{x=1}^{\infty} \frac{\lambda^x e^{-\lambda}}{(x-1)!}$$

$$= \lambda e^{-\lambda} \sum_{x=1}^{\infty} \frac{\lambda^{x-1}}{(x-1)!}$$

$$= \lambda e^{-\lambda} \cdot e^{+\lambda} = \lambda. \tag{9.2}$$

This result is consistent with that obtained for the binomial distribution which tends to the Poisson with parameter np.

The geometric distribution

$$f(x) = (1-p)^{x-1} p = q^{x-1} p, \quad x = 1, 2, 3, \ldots$$

$$E(X) = \sum_i x_i f(x_i) = \sum_{x=1}^{\infty} x q^{x-1} p = p \sum_{x=1}^{\infty} x q^{x-1}$$

$$= p(1-q)^{-2} = 1/p. \tag{9.3}$$

The exponential distribution

$$f(x) = \lambda e^{-\lambda x}, \quad 0 < x < \infty.$$

$$E(X) = \int x f(x) \, dx = \int_0^{\infty} \lambda x \, e^{-\lambda x} \, dx, \text{ integrating by parts}$$

$$= [-x e^{-\lambda x}]_0^{\infty} + \int_0^{\infty} e^{-\lambda x} \, dx$$

$$= \int_0^{\infty} e^{-\lambda x} \, dx, \text{ since } \lim_{x \to \infty} (x e^{-\lambda x}) = 0$$

$$= \left[-\frac{e^{-\lambda x}}{\lambda} \right]_0^{\infty} = \frac{1}{\lambda}. \tag{9.4}$$

The normal distribution

$$f(x) = \frac{1}{\sqrt{(2\pi)}\sigma} \exp \left[-\frac{1}{2} \left(\frac{x-\mu}{\sigma} \right)^2 \right], \quad -\infty < x < +\infty.$$

$$E(X) = \int_{-\infty}^{+\infty} \frac{1}{\sqrt{(2\pi)}\sigma} \, x \exp\left[-\frac{1}{2}\left(\frac{x-\mu}{\sigma}\right)^2\right] dx.$$

Put $(x - \mu)/\sigma = y$, then this integral becomes

$$\int_{-\infty}^{+\infty} \frac{\sigma}{\sqrt{(2\pi)}\sigma}(\sigma y + \mu) \, e^{-y^2/2} \, dy = \frac{\sigma}{\sqrt{(2\pi)}} \int_{-\infty}^{+\infty} y \, e^{-y^2/2} \, dy$$

$$+ \mu \int_{-\infty}^{+\infty} \frac{e^{-y^2/2}}{\sqrt{(2\pi)}} \, dy.$$

The second of these integrals has value one, the first has value zero. Hence

$$E(X) = \mu. \tag{9.5}$$

The rectangular distribution

$$f(x) = \frac{1}{b-a}, \quad a \leqslant x \leqslant b.$$

$$E(X) = \int xf(x) \, dx = \int_a^b \frac{x}{b-a} \, dx$$

$$= \left[\frac{x^2}{2(b-a)}\right]_a^b = \frac{b^2 - a^2}{2(b-a)} = \frac{b+a}{2}.$$

The Cauchy distribution

$$f(x) = \frac{1}{\pi(1+x^2)}, \quad -\infty < x + \infty.$$

This is an example which shows that the expectation of a random variable need not exist.

$$E(X) = \int xf(x) \, dx = \int_{-\infty}^{+\infty} \frac{x}{\pi(1+x^2)} \, dx.$$

Now we require

$$\int_0^\infty \frac{x}{\pi(1+x^2)} \, dx \quad and \quad \int_{-\infty}^0 \frac{x}{\pi(1+x^2)} \, dx$$

to be finite, but

$$\int_0^t \frac{x}{\pi(1+x^2)} \, dx = \frac{1}{2\pi} \int_0^t \frac{2x}{1+x^2} \, dx = \frac{1}{2\pi} [\log_e(1+x^2)]_0^t$$

$$= \frac{1}{2\pi} \log_e(1+t^2)$$

and as $t \to \infty$, $\log_e(1 + t^2)$ also tends to infinity. Thus $E(X)$ does not exist.
We have excluded the principal value of

$$\int_{-\infty}^{+\infty} \frac{x}{\pi(1 + x^2)} \, dx,$$

which is defined as

$$\lim_{t \to \infty} \frac{1}{2\pi} \int_{-t}^{t} \frac{2x}{1 + x^2} \, dx = \lim_{t \to \infty} \frac{1}{2\pi} [\log_e(1 + x^2)]_{-t}^{+t} = 0.$$

Problem 1. Without quoting previous formulae, find $E(X)$ for the discrete random variables with p.d.f.s

(a) $f(x) = \binom{5}{x} \binom{4}{3-x} / \binom{9}{3},$ $x = 0, 1, 2, 3.$

(b) $f(x) = \binom{3}{x} \left(\frac{5}{9}\right)^x \left(\frac{4}{9}\right)^{3-x},$ $x = 0, 1, 2, 3.$

(c) $f(x) = \frac{1}{3} \left(\frac{2}{3}\right)^{x-1},$ $x = 1, 2, 3, \dots .$

Problem 2. Find $E(X)$ for the continuous random variables with p.d.f.s

(a) $f(x) = 1/(2\sqrt{x}), \quad 0 < x < 1.$

(b) $f(x) = 6x(1 - x), \quad 0 < x < 1.$

(c) $f(x) = (1/2)x^2 \, e^{-x}, \quad 0 < x < \infty.$

(d) $f(x) = 1/x^2, \quad 1 \leqslant x < \infty.$

(e) $f(x) = 1 - |1 - x|, \quad 0 \leqslant x \leqslant 2.$

(f) $f(x) = \lambda(\lambda x)^{\alpha-1} \, e^{-\lambda x} / \Gamma(\alpha), \quad 0 \leqslant x < \infty.$

(g) $f(x) = (n + m - 1)! \, x^{n-1}(1 - x)^{m-1} / [(n - 1)!(m - 1)!], \quad 0 < x < 1.$

Problem 3. Families of two children each are exposed to an infectious disease. Each child has independently a probability p of being infected initially. If only one child in a family is infected initially the second child has probability p of catching the infection from the first. Show that the expectation of the number of children infected is $2p(1 + p - p^2)$ and sketch this as a function of p.

     What would the result be if, at the second stage when only one child is infected initially, the second child has probability 1 (instead of p) of catching the infection from the first?

London, B.Sc. General, Pt. I, 1968.

Problem 4
(a)   A random variable X has distribution function

$$F(u) = \begin{cases} 0 & u < 0 \\[2mm] \dfrac{1 - \cos u}{2}, & 0 \leqslant u < \pi. \\[2mm] 1 & u > \pi \end{cases}$$

(i)   Find the expected value of X;
(ii)   find the median of X;
(iii) find the mode of X;
(iv)   calculate the probability $(\pi/4 \leqslant X \leqslant 3\pi/4)$.

(b)   Articles are drawn at random from a shelf and tested one at a time. If the shelf contains three defectives and five non-defective articles,

(i)   show that the probability that r tests are required to find all the defective articles is given by

$$P(r) = \frac{1}{112}(r - 1)(r - 2), \quad 3 \leqslant r \leqslant 8;$$

(ii)   calculate the expected number of tests required to locate all the defective articles.

University Belfast, Subsidiary, 1968.

Problem 5. A player throws an ordinary die with faces numbered from 1 to 6. If he throws a 1, he has a second throw.

(a)   Find his average total score.
(b)   If, instead, whenever he throws a 1, he has a further throw, show that the probability of obtaining a total score of exactly r is

$$\frac{1}{5}\left[1 - \left(\frac{1}{6}\right)^{r-1}\right] \qquad \text{for } 1 < r \leqslant 6,$$

$$\frac{1}{5}\left[\left(\frac{1}{6}\right)^{r-6} - \left(\frac{1}{6}\right)^{r-1}\right] \qquad \text{for } r > 6,$$

and find the average total score in this case.

Oxford and Cambridge, A.L., 1980.

Problem 6. Tests are to be carried out on samples of blood, one sample having been obtained from each of N persons (where N is large). Each sample can be tested separately (method I), in which case a total of N tests are required, or the samples of k persons can be pooled and analysed together (method II). Under method II, a negative result means that the single test suffices for the k persons; if the result is positive, however, each of the k samples must then be analysed individually (making a total of $k + 1$ tests for this group of k persons). Whatever method is adopted, all tests may be considered independent, each with probability p of giving a positive result. Show that

(i)   the probability of a positive result from a test on a pooled sample of k persons is equal to $1 - q^k$, where $q = 1 - p$;

(ii)   the expected total number of tests required under method II is given by $N(1 - q^k + k^{-1})$. Under what circumstances would method II be the more economical procedure? Assuming that method II is to be used, find an equation for the value of k which minimizes the expectation given in (ii), and show that if p is very small (so that $p^2 \approx 0$ and $\log(1 - p) \approx -p$) then k is close to $p^{-1/2}$.

University College London, 1977.

Problem 7

(a) A simple weather forecasting procedure classifies days as 'wet' or 'dry' and is based on there being a constant probability p that any day is of the same type as the immediately preceding day. If α_n denotes the probability that the nth day of the year is dry, express α_n in terms of p and α_{n-1}. If the probability that day 1 was dry is taken to be θ, use your relationship to show, by induction or otherwise, that

$$\alpha_n = \frac{1}{2} + (2p - 1)^{n-1} \left(\theta - \frac{1}{2} \right).$$

What happens as $n \to \infty$? Comment on this result.

(b) A small car-hire firm has two cars for daily hire; any demand in excess of 2 is therefore refused. The daily demand is a Poisson variable of mean 2. Find in terms of e the expected number of refused demands per day.

University College of Wales, Aberystwyth, 1982.

9.3 EXPECTATION AND CHOICE OF ACTION

Example 1

Suppose a factory has to decide the amount y of lengths of a certain cloth to produce against the following information. The demand, X lengths, is uniformly distributed over the interval (a, b) and for each length sold, a profit of m units is made, while for each length not sold, a loss of n units is suffered. One possible strategy is to find the expected profit and maximize this with respect to y. If the profit is $P(X)$, then

$$E[P(X)] = \int_a^b P(x) f(x) \, dx = \frac{1}{(b - a)} \int_a^b P(x) \, dx.$$

The form of $P(x)$ is related to the value of y. Thus:

if $X \geqslant y$, $P(X) = my$

if $X < y$, $P(X) = mX - n(y - X) = (m + n)X - ny$.

In particular, if $y \leqslant a$, then always $X \geqslant a \geqslant y$ and

$$E[P(X)] = \frac{1}{(b-a)} \int_a^b my\, dx = my,$$

which increases until $y = a$.

If $y \geqslant b$, then always $X \leqslant b \leqslant y$,

$$E[P(X)] = \frac{1}{(b-a)} \int_a^b [(m+n)x - ny]\, dx$$

$$= \left(\frac{m+n}{2}\right)(b+a) - ny,$$

which decreases from $y = b$, while when $a < y < b$,

$$E[P(X)] = \frac{1}{b-a} \left[\int_a^y P(x)\, dx + \int_y^b P(x)\, dx \right]$$

$$= \frac{1}{b-a} \left\{ \int_a^y [(m+n)x - ny]\, dx + \int_y^b my\, dx \right\}$$

$$= \frac{1}{2(a-b)} [(m+n)y^2 - 2(na + mb)y + (m+n)a^2].$$

The reader is asked to verify that this has a maximum when

$$y = \frac{(na + mb)}{(m+n)},$$

and that this choice of y gives the greatest value of $E[P(X)]$.

9.4 EXPECTATION OF A FUNCTION OF A RANDOM VARIABLE

Given that X is a random variable with p.d.f. $f(x)$ and that $Y = g(X)$ is also a random variable, how should we set about finding the expected value of Y? Following the definition, we would *first* find the p.d.f., $h(y)$, of Y and *then* evaluate either

$$\int yh(y)\, dy \quad \text{or} \quad \sum_y yh(y)$$

according as Y is continuous or discrete. However, some of this labour turns out to be unnecessary as the same result may be obtained by evaluating

$$\int g(x) f(x) \, dx \quad \text{or} \quad \sum_x g(x) f(x)$$

according as X is continuous or discrete. This result will be proved for two important cases.

(1) When X is discrete and Y is discrete

We need direct our attention only to those values of x which X assumes with positive probability. For each value y_j there may be several values of x such that $g(x) = y_j$. Suppose, in fact

$$g(x_{ij}) = y_j, \quad i = 1, 2, \ldots, n_j \quad \text{and} \quad f(x_{ij}) \neq 0.$$

Then,

$$E(Y) = \sum_y y h(y) = \sum_j y_j h(y_j)$$

$$= \sum_j y_j \Pr(Y = y_j)$$

$$= \sum_j y_j \left[\sum_{i=1}^{n_j} \Pr(X = x_{ij}) \right]$$

$$= \sum_j \sum_{i=1}^{n_j} [y_j \Pr(X = x_{ij})]$$

$$= \sum_j \sum_{i=1}^{n_j} [g(x_{ij}) \Pr(X = x_{ij})]$$

$$= \sum_x g(x) f(x) = E[g(X)] .$$

Thus the expectation of Y with respect to the p.d.f. of Y is also the expectation of $g(X)$ with respect to the p.d.f. of X.

(2) When X is continuous and $g(x)$ is a differentiable and monotone increasing function of x

Consider

$$\int g(x) f(x) \, dx,$$

in which we change the variable so that $y = g(x)$,

$$\int g(x) f(x) \, dx = \int y f [g^{-1}(y)] \frac{dx}{dy} \, dy$$

but we have already shown that in the case under consideration,

$$\left[f(x) \; \frac{dx}{dy} \right]_y$$

is $h(y)$, the p.d.f. of $Y = g(X)$. Hence the integral is

$$\int y h(y) \, dy.$$

Problem 8. If X has the rectangular distribution over $(0, 1)$ evaluate $E(-\log_e X)$. Find also the p.d.f. of the random variable $Y = -\log_e X$, and *hence* $E(Y)$.

Problem 9. If X has the exponential distribution with parameter λ evaluate $E(e^{tX})$ where $0 < t < \lambda$. Find also the p.d.f. of the random variable $Y = e^{tX}$, and *hence* evaluate $E(Y)$.

9.5 APPLICATIONS

The statistician is often faced with problems of the following type. He is provided with some data and is obliged to enquire whether they can be reasonably supposed to have been derived from some particular source. The first stage is to compare the actual observations with what could be reasonably expected in virtue of the manner in which they are supposed to have been generated.

Example 2
For a variety of peas, Mendel crossed plants with round yellow seeds with other plants having wrinkled green seeds and classified the seeds of plants in the next generation. He wish to check the conjecture that, for such crossings, the characteristics round and yellow were independent, each having probability 3/4. The conjecture implied that, from such a crossing, the probability of a plant having round yellow seeds would be 9/16 and this would be the expected proportion in a large sample of such plants.

Problem 10. Recent research has isolated five distinct red cell acid phosphate patterns in human blood. An argument based on the theory of genetics, however, predicts six patterns as follows. Each of the parents donates independently a gene of type A, B, or C with probability p, q, r respectively where $p + q + r = 1$. Calculate the probability of each of the six possible (unordered) pairs of genes in an offspring in terms of p, q, and r. A recent study of 139 adults was carried out and reported the following distribution:

Type	AA	AB	BB	CA	CB	CC
Frequency	14	64	48	5	8	0

Calculate the observed proportions of genes of types A, B, C. Explain how these proportions can be used to predict expected numbers for each type.　　■

When the data are supposed to have been drawn from a particular distribution, then the statistician attempts to 'fit the distribution' to the sample. In the discrete case the fitted distribution can be used to calculate the probabilities of an observation taking particular values. Sometimes the parameters of the fitted distribution have to be based on the sample and the derived probabilities may have to be grouped to permit comparison with scarce sample values.

Problem 11. Fit a Poisson distribution to the following counts of bacteria found on 500 slides.

Number of bacteria	0	1	2	3	4	5	6
Number of slides	73	142	133	91	41	15	5

■

Comparing a sample with a continuous distribution presents a new difficulty. For the probability that a random variable takes a particular value is now zero. We can, however, compare intervals of values. That is to say, divide the range of variation of the random variable and use the cumulative distribution function to compute the probabilities for the resulting intervals. How should the intervals be chosen? Ideally without being swayed by the run of the data. Obvious choices are (i) intervals of equal length (ii) intervals of equiprobability.

Example 3
Suppose X has distribution function $F(x) = x^2, 0 < x < 1$.

(i)　If $(0, 1)$ is divided into four intervals of equal length, these are $(0, 1/4)$, $(1/4, 1/2)$, $(1/2, 3/4)$, $(3/4, 1)$ with probabilities $1/16, 3/16, 5/16, 7/16$ respectively.

(ii)　If $(0, 1)$ is divided into four intervals of probability 0.25, then the points of subdivision fall at $0.5, \sqrt{0.5}, \sqrt{0.75}$.

Method (ii) enjoys the added attraction of giving equal weight to each interval. We have skirted the question as to the number of intervals to be constructed.

BRIEF SOLUTIONS AND COMMENTS ON THE PROBLEMS

Problem 1

(a)　$E(X) = \left[1 \binom{5}{1}\binom{4}{2} + 2\binom{5}{2}\binom{4}{1} + 3\binom{5}{3}\binom{4}{0} \right] / \binom{9}{3}$

　　　$= 5/3.$

(b) $E(X) = 1\binom{3}{1}\left(\frac{5}{9}\right)\left(\frac{4}{9}\right)^2 + 2\binom{3}{2}\left(\frac{5}{9}\right)^2\left(\frac{4}{9}\right) + 3\binom{3}{3}\left(\frac{5}{9}\right)^3$

$= 5/3.$

(c) $E(X) = \frac{1}{3}\left[1 + 2\left(\frac{2}{3}\right) + 3\left(\frac{2}{3}\right)^2 \dots \right]$

$= \frac{1}{3}\left(1 - \frac{2}{3}\right)^{-2} = 3.$

Problem 2

(a) $E(X) = \frac{1}{2}\int_0^1 \sqrt{x}\, dx = 1/3.$

(b) $E(X) = \int_0^1 (6x^2 - 6x^3)\, dx = 1/2.$

(c) $E(X) = \frac{1}{2}\int_0^\infty x^3\, e^{-x}\, dx$

$= \left(-\frac{x^3}{2}\, e^{-x}\right)_0^\infty + 3\int_0^\infty \frac{x^2}{2}\, e^{-x}\, dx = 3.$

(d) $E(X) = \int_1^\infty \frac{1}{x}\, dx = \lim_{x\to\infty} \log_e x, \quad$ does not exist.

(e) $E(X) = \int_0^1 x^2\, dx + \int_1^2 (2x - x^2)\, dx = 1.$

(f) $E(X) = \frac{\Gamma(\alpha + 1)}{\Gamma(\alpha)\lambda}\int_0^\infty \frac{\lambda(\lambda x)^\alpha\, e^{-\lambda x}}{\Gamma(\alpha + 1)}\, dx$

$= \alpha/\lambda, \quad$ since $\quad \Gamma(\alpha + 1) = \alpha\Gamma(\alpha).$

(g) $E(X) = \frac{n}{n + m}\int_0^1 \frac{(n + m)!\, x^n(1 - x)^{m-1}}{n!\,(m-1)!}\, dx$

$= \frac{n}{n + m}.$

Notice that the computational difficulties in (f), (g) are eased by combining in a form that allows immediate identification of some other integral as having value 1, since the integrands are p.d.f.s.

Problem 3. Neither child infected (stage 1), with probability $(1-p)^2$. One child infected if either is infected at stage 1 and the other not at either stage with probability $2p(1-p)(1-p)$. Both infected if either two at stage 1 or one at stage 1 and the other at the second stage. Probability of both infected is thus $p^2 + 2p(1-p)p$. Expected number infected is hence

$$1.2p(1-p)^2 + 2[p^2 + 2p^2(1-p)]$$
$$= 2p(1+p-p^2).$$

For the second part, one child infected is impossible. Neither infected has probability $(1-p)^2$, probability both $1 - (1-p)^2$. Expected number is $2[1-(1-p)^2] = 2(2p-p^2)$.

Problem 4

(a) $f(u) = F'(u) = \sin(u)/2$. Hence

$$E(U) = \int_0^\pi uf(u)\, du = \pi/2.$$

The median satisfies $(1-\cos u)/2 = 1/2$ whence $u = \pi/2$. Mode is $u = \pi/2$. $\Pr(\pi/4 \leqslant X \leqslant 3\pi/4) = F(3\pi/4) - F(\pi/4) = 1/\sqrt{2}$.

(b)   Articles are drawn without replacement. If the rth test is the third defective, then the previous $(r-1)$ tests contained 2 defectives and $(r-1) - 2 = r - 3$ non-defectives in some order. This assortment can be selected in $\binom{3}{2}\binom{5}{r-3}$ ways and there are $\binom{8}{r-1}$ selections in all. The rth article is defective with conditional probability $1/(9-r)$. Hence the required probability is

$$\left[\binom{3}{2}\binom{5}{r-3} \Big/ \binom{8}{r-1}\right] \Big/ (9-r) = (r-1)(r-2)/112.$$

Finally

$$\sum_{r=3}^8 r(r-1)(r-2)/112 = 6.75.$$

Problem 5

(a)   Ans. 49/12. Hint, the second score is zero if the first is not one.

(b)   If $2 \leqslant r \leqslant 6$, there can be j scores of one in sequence followed by a score of $r - j$. If $r > 6$, the last score cannot exceed 6. Average total score 21/5.

Problem 6. $\Pr(\text{all negative}) = q^k$, hence $\Pr(\text{positive}) = 1 - q^k$. Hence expected number of tests $1.q^k + (k+1)(1-q^k)$, per group of k. There are N/k groups. differentiate with respect to k, $-q^k \log q - 1/k^2 = 0$ and use the suggested approximation, giving $-(1-pk)(-p) \approx 1/k^2$ or $p \approx 1/k^2$.

Problem 7

(a) Day n is dry if either; day $n-1$ is dry, with probability α_{n-1}, and the next is dry with probability p or day $n-1$ is wet, with probability $1-\alpha_{n-1}$, and the next is dry with probability $1-p$.

$$\alpha_n = p\alpha_{n-1} + (1-\alpha_{n-1})(1-p)$$
$$= (2p-1)\alpha_{n-1} + 1 - p,$$

and by repeated application,

$$= (2p-1)^{n-1}\alpha_1 + (1-p)[1 + (2p-1) + (2p-1)^2 + \dots$$
$$+ (2p-1)^{n-2}]$$
$$= (2p-1)^{n-1}\theta + [1 - (2p-1)^{n-1}]/2$$
$$= \frac{1}{2} + (2p-1)^{n-1}\left(\theta - \frac{1}{2}\right) \rightarrow \frac{1}{2} \quad \text{as } n \rightarrow \infty.$$

(b) Expected number of refused demands, D, is

$$\sum_2^\infty (d-2)2^d\, e^{-2}/d!$$

$$= \left[2\,e^{-2} \sum_2^\infty 2^{d-1}/(d-1)!\right] - \left[2\,e^{-2} \sum_2^\infty 2^d/d!\right]$$

$$= 2\,e^{-2}(e^2 - 1) - 2\,e^{-2}(e^2 - 1 - 2)$$

$$= 4\,e^{-2}.$$

Problem 8

$$E(-\log_e X) = - \int_0^1 \log_e x\; dx = -(x\log_e x)_0^1 + \int_0^1 1\; dx = 1.$$

Note that

$$\lim_{x \to 0} [x\log_e x] = 0.$$

Now the p.d.f. of $Y = -\log_e X$ is e^{-y}, $y > 0$. Hence also

$$E(Y) = \int_0^\infty y\,e^{-y}\; dy = 1.$$

Problem 9

$$E(e^{tX}) = \int_0^\infty e^{tx}\,\lambda e^{-\lambda x}\; dx = \left[\frac{\lambda}{t-\lambda}\,e^{(t-\lambda)x}\right]_0^\infty = \lambda/(\lambda - t),$$

since $t < \lambda$. The p.d.f. of $Y = e^{tX}$ is slightly more difficult in this case.

$$h(y) = [f(x)\, dx/dy]_y = (\lambda\, e^{-\lambda x}/ty)_y = \lambda y^{-\lambda/t}/ty, \quad 1 < y < \infty.$$

Hence,

$$E(Y) = \int_1^\infty yh(y)\, dy = [\lambda y^{1-\lambda/t}/(t-\lambda)]_1^\infty = \lambda/(\lambda - t).$$

Problem 10. Note that AB cannot be distinguished from BA etc.

Type	AA	AB	BB	AC	BC	CC
Probability	p^2	$2pq$	q^2	$2pr$	$2qr$	r^2.

Blood of types AA, AB, AC contain an A gene. Hence the total observed number of A genes is $2(14) + 64 + 5 = 97$ and hence the proportion in the 278 genes is 97/278. The proportions of B, C genes are 168/278, 13/278. If these proportions are substituted for p, q, r then, for example, this would suggest an expected number of type AB equal to $139(2pq) = (278)(97/278)(168/278)$. Since the estimate of r is small, it suggests that the probability of CC is negligible and it is not surprising that this sixth pattern was not detected.

Problem 11. The parameter of the Poisson distribution has not been stated. It may be estimated by the average number of bacteria per slide, which is $950/500 = 1.9$. Start with $e^{-1.9} = 0.1496$ and calculate $(1.9)^x\, e^{-1.9}/x!$ for $x = 1, 2, 3, 4, 5$. Since no value is recorded with more than 6 bacteria, group all the remaining values and find $\Pr(\text{more than six})$ by subtraction. These probabilities, when multiplied by 500, provide expected frequencies for a sample of 500 slides.

10

Variance of a Random Variable

10.1 VARIANCE AND PROBABILITY

Generally a random variable is not adequately described by just stating its expected value. In practical terms, we should not know much about a population if only its mean were available. We should also like to have some idea of how the values were dispersed about the mean. One such measure is the **variance** of the random variable, denoted by $V(X)$ and defined as:

$$V(X) = E[X - E(X)]^2. \tag{10.1}$$

The positive square root of $V(X)$ is known as the standard deviation of X, and is denoted by σ.

In virtue of the result in section 9.4, for a continuous random variable,

$$V(X) = \int [x - E(X)]^2 f(x)\, dx$$

$$= \int [x^2 - 2xE(X) + E^2(X)]\, f(x)\, dx$$

$$= \int x^2 f(x)\, dx - 2E(X) \int xf(x)\, dx + E^2(X) \int f(x)\, dx$$

$$= \int x^2 f(x)\, dx - 2.E(X).E(X) + E^2(X)$$

$$= E(X^2) - E^2(X). \tag{10.2}$$

By a precisely similar expansion, for a discrete random variable

$$V(X) = \sum_x [x - E(X)]^2 \, f(x)$$

$$= E(X^2) - E^2(X).$$

In both cases, $V(X)$ is clearly non-negative and it should be stressed that in general $E[g(X)] \neq g[E(X)]$ and in particular $E(X^2) \neq E^2(X)$, unless X can only assume one value with positive probability. We shall later use the relation $E(X^2) = V(X) + E^2(X)$.

For the discrete case, if $V(X)$ is small, then so is every term in the sum defining it, since they are all positive, in which case, large deviations $[x - E(X)]$ must have small probability. That something analogous holds for the continuous case will appear from the next theorem.

The variance has another interesting property,

$$
\begin{aligned}
E(X-C)^2 &= E[X - E(X) + E(X) - C]^2 \\
&= E[X - E(X)]^2 + 2[E(X) - C] \, [E(X) - E(X)] \\
&\quad + [E(X) - C]^2 \\
&= V(X) + [E(X) - C]^2 \geqslant V(X),
\end{aligned}
$$

with equality if and only if $C = E(X)$.

Theorem 10.1. If $g(X)$ is a non-negative random variable, then

$$\Pr[g(X) \geqslant k] \leqslant \frac{E[g(X)]}{k} \quad \text{for } k > 0. \tag{10.3}$$

Proof. Suppose X is a continuous random variable, with p.d.f. $f(x)$, then

$$E[g(X)] = \int g(x) \, f(x) \, dx,$$

but since $g(x) \geqslant 0$,

$$E[g(X)] \geqslant \int_{g(x) \geqslant k} g(x) \, f(x) \, dx,$$

where the integration is over those x such that $g(x) \geqslant k$. Thus

$$E[g(X)] \geqslant \int_{g(x) \geqslant k} k \cdot f(x) \, dx,$$

since $g(x) \geqslant k$, over the stated range of x,

$$= k \int_{g(x) \geqslant k} f(x) \, dx = k \Pr[g(X) \geqslant k].$$

Hence,

$$\Pr[g(X) \geqslant k] \leqslant E[g(X)]/k. \tag{10.4}$$

The reader should prove the same result for a discrete random variable.   ∎

We now specialize the theorem for particular functions g, subject to the existence of $E[g(X)]$. Suppose $g(X) = X$ and X is non-negative, then the inequality states

$$\Pr(X \geqslant k) \leqslant \frac{E(X)}{k}.$$

But, if k is sufficiently large, $E(X)/k$ is as small as we please. If X is not restricted in sign and $E(X)$ exists, then $E(|X|)$ exists. Hence

$$\Pr(|X| \geqslant k) \leqslant E(|X|)/k.$$

We note that

$$\Pr(|X| \geqslant k) = \Pr(X \leqslant -k \quad \text{or} \quad X \geqslant +k).$$

Once again, we have something about extreme deviations, but not anything very sharp.

Example 1
If X has the rectangular distribution over $(0, l)$ then we know that $E(X) = l/2$, in this case we obtain

$$\Pr(X \geqslant k) \leqslant \frac{l}{2k}.$$

But from the exact distribution of X, $\Pr(X \geqslant k) = (l - k)/l$.

Example 2
If X has the exponential distribution with parameter λ, then we have $E(X) = 1/\lambda$. If we choose $k = n/\lambda$.

$$\Pr(X \geqslant n/\lambda) \leqslant \frac{1/\lambda}{n/\lambda} = \frac{1}{n}.$$

Once again, for this distribution we can evaluate the probability exactly.

$$\Pr(X \geqslant n/\lambda) = \int_{n/\lambda}^{\infty} \lambda\, e^{-\lambda x} dx = [-e^{-\lambda x}]_{n/\lambda}^{\infty} = e^{-n}.$$

For theoretical purposes, a fruitful choice of $g(X)$ is $[X - E(X)]^2$, which is certainly non-negative and provided $V(X)$ exists we have,

$$\Pr\{[X - E(X)]^2 \geqslant k^2\} \leqslant \frac{E[X - E(X)]^2}{k^2} = \frac{V(X)}{k^2},$$

or

$$\Pr[X - E(X) \leqslant -k \ \text{ or } \ X - E(X) \geqslant k] \leqslant V(X)/k^2,$$

or

$$\Pr[|X - E(X)| \geqslant k] \leqslant V(X)/k^2, \tag{10.5}$$

in which form the result is known as **Tchebychev's inequality**. Clearly, if k is large enough, $V(X)/k^2$ is as small as we please. Also, if $V(X)$ is small, relative to k, we see that the probability of a deviation far from $E(X)$ is small. Evidently, $V(X)$ is related to the dispersion of probability in the distribution of X.

10.2 THE STANDARD DISTRIBUTIONS

In finding the variances of the standard distributions, we shall use the result $E[ag(X) + bh(X)] = aE[g(X)] + bE[h(X)]$ in one form or another. For the continuous case we show this as follows.

$$E[ag(X) + bh(X)] = \int [ag(x) + bh(x)] \, f(x) \, dx$$

$$= a \int g(x) f(x) \, dx + b \int h(x) f(x) \, dx$$

$$= aE[g(X)] + bE[h(X)].$$

The binomial distribution

$$f(x) = \binom{n}{x} p^x (1-p)^{n-x}, \quad x = 0, 1, 2, \ldots, n$$

The cancellation noted in deriving $E(X)$ suggests that we find $E[X(X-1)]$.

$$E[X(X-1)] = \sum_{x=0}^{n} x(x-1) f(x) = \sum_{x=2}^{n} x(x-1) f(x)$$

since the first two terms are zero.

$$E[X(X-1)] = \sum_{2}^{n} x(x-1) \, \frac{n!}{x!(n-x)!} \, p^x (1-p)^{n-x}$$

$$= \sum_{x=2}^{n} \frac{n!}{(x-2)!(n-x)!} \, p^x (1-p)^{n-x}$$

$$= n(n-1)p^2 \sum_{x=2}^{n} \frac{(n-2)!}{(x-2)!(n-x)!} \, p^{x-2} (1-p)^{n-x}$$

put $x - 2 = y$,

$$= n(n-1)p^2 \sum_{y=0}^{n-2} \binom{n-2}{y} p^y (1-p)^{n-2-y}$$

$$= n(n-1)p^2[p+(1-p)]^{n-2}$$
$$= n^2p^2 - np^2.$$

Now,

$$E[X(X-1)] = E(X^2 - X) = E(X^2) - E(X) = E(X^2) - np.$$

That is,

$$n^2p^2 - np^2 = E(X^2) - np \quad \text{or} \quad E(X^2) = n^2p^2 - np^2 + np.$$

Hence,

$$V(X) = E(X^2) - E^2(X) = n^2p^2 - np^2 + np - n^2p^2 = np - np^2$$
$$= np(1-p) = \underline{npq}.$$
$$V(X) = npq. \tag{10.6}$$

Problem 1. If X has the Poisson distribution, with parameter λ, find $E[X(X-1)]$ and hence show that $V(X) = \lambda$.

The geometric distribution

$$f(x) = q^{x-1}p, \quad x = 1, 2, 3, \ldots.$$

We have already shown that

$$E(X) = \sum_{x=1}^{\infty} xq^{x-1}p = \frac{1}{p}.$$

Now,

$$E[X(X-1)] = pq \sum_{2}^{\infty} x(x-1)q^{x-2}$$
$$= pq[2(1-q)^{-3}]$$
$$= \frac{2q}{p^2}.$$

Hence,

$$V(X) = \frac{2q}{p^2} + \frac{1}{p} - \frac{1}{p^2} = \frac{q}{p^2}. \tag{10.7}$$

Problem 2. If X has the negative binomial distribution,

$$f(x) = \binom{r+x-1}{x} p^r q^x, \quad x = 0, 1, 2, \ldots.$$

Show that $V(X) = rq/p^2$.

The exponential distribution

$$f(x) = \lambda e^{-\lambda x}, \quad x > 0.$$

We have shown that $E(X) = 1/\lambda$.

$$E(X^2) = \int x^2 f(x) \, dx = \int_0^{\infty} x^2 \lambda e^{-\lambda x} \, dx,$$

integrating by parts

$$= [-x^2 e^{-\lambda x}]_0^{\infty} + 2 \int_0^{\infty} x e^{-\lambda x} \, dx.$$

Since $\lim_{x \to \infty} x^2 e^{-\lambda x} = 0$,

$$E(X^2) = 2 \int_0^{\infty} x e^{-\lambda x} \, dx$$

$$= \frac{2}{\lambda} \int_0^{\infty} x \lambda e^{-\lambda x} \, dx = \frac{2}{\lambda} E(X) = \frac{2}{\lambda} \cdot \frac{1}{\lambda} = \frac{2}{\lambda^2}.$$

Thus

$$V(X) = E(X^2) - E^2(X) = \frac{2}{\lambda^2} - \frac{1}{\lambda^2} = \frac{1}{\lambda^2}$$

$$V(X) = 1/\lambda^2. \tag{10.8}$$

The normal distribution

$$f(x) = \frac{1}{\sqrt{(2\pi)}\sigma} e^{-[(x-\mu)/\sigma^2]/2}, \quad -\infty < x < \infty.$$

We have shown that $E(X) = \mu$. In this case it pays to evaluate $E[X - E(X)]^2$ directly.

$$V(X) = \int (x - \mu)^2 f(x) \, dx$$

$$= \int_{-\infty}^{+\infty} \frac{(x-\mu)^2}{\sqrt{(2\pi)}\sigma} \exp\left[-\frac{1}{2}\left(\frac{x-\mu}{\sigma}\right)^2\right] dx.$$

Change the variable, put $y = (x - \mu)/\sigma$,

$$V(X) = \sigma^2 \int_{-\infty}^{+\infty} \frac{y^2}{\sqrt{(2\pi)}} e^{-y^2/2} \, dy$$

$$= \sigma^2 \int_{-\infty}^{+\infty} \frac{y}{\sqrt{(2\pi)}} \, y \, e^{-y^2/2} \, dy,$$

integrating by parts

$$= \sigma^2 \left[\left(-\frac{y}{\sqrt{(2\pi)}} e^{-y^2/2} \right)_{-\infty}^{+\infty} + \int_{-\infty}^{+\infty} \frac{1}{\sqrt{(2\pi)}} e^{-y^2/2} \, dy \right].$$

Both

$$\lim_{y \to \infty} y \, e^{-y^2/2} = 0 \quad \text{and} \quad \lim_{y \to -\infty} y \, e^{-y^2/2} = 0,$$

while

$$\int_{-\infty}^{+\infty} \frac{1}{\sqrt{(2\pi)}} e^{-y^2/2} \, dy = 1.$$

Thus,

$$\underline{V(X) = \sigma^2} \tag{10.9}$$

The hypergeometric distribution

The discrete random variable X is said to have the hypergeometric distribution with positive integral parameters r, N, n with $N \geqslant n$, $N \geqslant r$ if its p.d.f. $f(x)$ satisfies

$$f(x) = \frac{\binom{r}{x}\binom{N-r}{n-x}}{\binom{N}{n}}, \quad x = 0, 1, 2, \ldots, r$$

$$= 0 \text{ otherwise.}$$

This is the appropriate p.d.f. when X is the number of reds in a random sample of n drawn without replacement from a population, containing just r reds. If it happens that $r > n$, then $f(x)$ is zero for $x > n$.

$$E(X) = \sum_x x f(x) = \sum_{x=0}^{r} x \binom{r}{x}\binom{N-r}{n-x} \bigg/ \binom{N}{n}$$

$$= \sum_{x=1}^{r} x \frac{r!}{x!(r-x)!} \binom{N-r}{n-x} \bigg/ \binom{N}{n}$$

$$= r \sum_{x=1}^{r} \frac{(r-1)!}{(x-1)!(r-x)!} \binom{N-r}{n-x} \bigg/ \binom{N}{n}$$

$$= r \sum_{x=1}^{r} \binom{r-1}{x-1} \binom{N-r}{n-x} / \binom{N}{n}$$

$$= r \sum_{y=0}^{r-1} \binom{r-1}{y} \binom{N-1-(r-1)}{n-1-y} / \binom{N}{n}$$

$$= r \binom{N-1}{n-1} / \binom{N}{n}$$

$$= rn/N.$$

$$E[X(X-1)] = \sum_{x=2}^{r} x(x-1) \frac{r!}{x!(r-x)!} \binom{N-r}{n-x} / \binom{N}{n}$$

$$= r(r-1) \binom{N-2}{n-2} / \binom{N}{n}$$

$$= r(r-1) n(n-1)/N(N-1) = E(X^2) - E(X).$$

Hence,

$$E(X^2) = \frac{r(r-1) n(n-1)}{N(N-1)} + \frac{rn}{N}.$$

$$V(X) = E(X^2) - E^2(X) = \frac{r(r-1) n(n-1)}{N(N-1)} + \frac{rn}{N} - \frac{r^2 n^2}{N^2}$$

$$= \frac{nr}{N} \left(1 - \frac{r}{N}\right) \left(\frac{N-n}{N-1}\right).$$

If the sampling has been done with replacement, then the number of reds in the sample of n would have obeyed the binomial distribution with probability of a red r/N on each draw. For the binomial distribution,

$$E(X) = np = nr/N$$

and

$$V(X) = npq = n \frac{r}{N} \left(1 - \frac{r}{N}\right).$$

Without replacement, we notice that $E(X)$ is still nr/N but the variance of X is $npq(N-n)/(N-1)$ which is smaller (except when $n = 1$, when they are equivalent). However, if N is large compared with n,

$$\frac{N-n}{N-1} = \frac{1-n/N}{1-1/N} \approx 1.$$

Hence, if n/N is small, sampling without replacement is roughly the same as sampling with replacement as would be expected. In sampling theory it is customary to neglect the effect of not replacing if $n/N < 1/10$.

Problem 3. Find $V(X)$ for the discrete random variables with the following p.d.f.s. Formulae should not be quoted, but the value of $E(X)$ should be taken from the corresponding problems on the expectations of a random variable (section 9.2).

(a) $\quad f(x) = \binom{5}{x} \binom{4}{3-x} / \binom{9}{3}, \qquad x = 0, 1, 2, 3.$

(b) $\quad f(x) = \binom{3}{x} \left(\frac{5}{9}\right)^x \left(\frac{4}{9}\right)^{3-x}, \qquad x = 0, 1, 2, 3.$

(c) $\quad f(x) = \frac{1}{3} \left(\frac{2}{3}\right)^{x-1}, \qquad\qquad x = 1, 2, \dots .$

Problem 4. Find $V(X)$ for the continuous random variables with p.d.f.s:

(a) $\quad f(x) = 1/2\sqrt{x}, \quad 0 < x < 1.$

(b) $\quad f(x) = 6x(1-x), \quad 0 < x < 1.$

(c) $\quad f(x) = \frac{1}{2}x^2 e^{-x}, \quad 0 \leqslant x \leqslant \infty.$

(d) $\quad f(x) = 1 - |1-x|, \quad 0 \leqslant x \leqslant 2.$

(e) $\quad f(x) = \lambda(\lambda x)^{\alpha-1} e^{-\lambda x}/\Gamma(\alpha), \quad 0 \leqslant x < \infty.$

(f) $\quad f(x) = (n+m-1)! \, x^{n-1}(1-x)^{m-1}/[(n-1)!(m-1)!], \quad 0 < x < 1.$

Problem 5. Let X be a random variable with mean μ and variance σ^2. Show that

$$\Pr\{|X-\mu| \geqslant t\} \leqslant \sigma^2/t^2, \quad (t > 0).$$

Show also that

$$\Pr\{X \geqslant t\} \leqslant E[(X+c)^2]/(t+c)^2, \quad (t > 0).$$

for every $c > 0$. If $\mu = 0$ show (by minimizing the right-hand side above with respect to c, or otherwise) that

$$\Pr\{X \geqslant t\} \leqslant \sigma^2/(\sigma^2 + t^2) \quad (t > 0).$$

Westfield College, 1982.

10.3 AN APPLICATION OF TCHEBYCHEV'S INEQUALITY

If X has the binomial distribution with parameters n, p then we have seen that $E(X) = np$, $V(X) = npq$. Applying Tchebychev's inequality

$$\Pr(|X - np| \geqslant k) \leqslant \frac{npq}{k^2} \text{ for } k > 0.$$

This has an interesting interpretation, for we also know that X is also the number of successes in a sequence of n independent trials in each of which the probability of a success is p. Consider the *proportion* of success, X/n, which has expectation p,

$$\Pr\left(\left|\frac{X}{n} - p\right| \geqslant k\right) = \Pr(|X - np| \geqslant nk) \leqslant \frac{V(X)}{n^2 k^2}$$

$$= \frac{pq}{nk^2}.$$

Now, for every fixed k, the limit of pq/nk^2, as n tends to infinity, is zero, that is

$$\lim_{n \to \infty} \left[\Pr\left(\left|\frac{X}{n} - p\right| \geqslant k\right) \right] = 0.$$

But k may be chosen as small as we please. Thus if n is large enough, the probability that the proportion of successes is far from p is negligible. This result is Bernoulli's form of the weak law of large numbers. It should be noted that this does not prove the phenomenon of statistical regularity, since the word probability appears in the limit. Since $0 < p < 1, q = 1 - p$, then

$$pq = p - p^2$$

$$= \frac{1}{4} - \left(\frac{1}{2} - p\right)^2 \leqslant \frac{1}{4}$$

for all p. Hence, we further deduce that

$$\Pr\left[\left|\frac{X}{n} - p\right| \geqslant k\right] \leqslant \frac{1}{4nk^2},$$

whatever the value of p.

    This may be used to estimate the number of trials required to bring this probability down to a required level. Thus if we need

$$\Pr\left(\left|\frac{X}{n} - p\right| \geqslant 0.1\right) \leqslant 0.05,$$

we have

$$\frac{1}{4n(0.1)^2} \leqslant 0.05 \quad \text{or} \quad n \geqslant \frac{1}{0.002} = 500.$$

The result is rather crude and may be considerably improved by an approximation to the distribution of $[(X/n) - p]$ for large n.

Problem 6. The continuous random variable X is said to have the **Pareto distribution** with positive parameters θ, n if its p.d.f. satisfies

$$f(x) = \frac{n\theta^n}{x^{n+1}}, \quad x \geqslant \theta$$

$$= 0 \text{ otherwise.}$$

Show that if n is sufficiently large,

$$E(X) = \frac{n\theta}{n-1}, \quad V(X) = \frac{\theta^2 n}{(n-1)^2 (n-2)}.$$

For the Pareto distribution with $\theta = 1, n = 3$, calculate $\Pr(X \geqslant 3.5)$.
Find also the upper bound set to this probability by Tchebychev's inequality.

Problem 7. X is a random variable for which $E(X) = \mu$ and $E[(X - \mu)^{2m}]$ exists where m is a positive integer. Prove that

$$\Pr(|X - \mu| \geqslant k) \leqslant E(X - \mu)^{2m}/k^{2m}.$$

If X has the normal distribution with parameters μ and σ, show that $E(X - \mu)^4 = 3\sigma^4$. By considering the cases $m = 1$, $m = 2$, find two upper bounds to $\Pr(|X - \mu| \geqslant 2\sigma)$ and find the result given by the table of the unit normal distribution.

Problem 8. X is a discrete random variable with p.d.f.

$$f(\mu - k\sigma) = \frac{1}{2k^2}$$

$$f(\mu) = 1 - \frac{1}{k^2}$$

$$f(\mu + k\sigma) = \frac{1}{2k^2}.$$

Show that $E(X) = \mu$, $V(X) = \sigma^2$ and that $\Pr(|X - \mu| \geqslant k\sigma) = 1/k^2$. We conclude that the Tchebychev's inequality cannot be improved if we only know that the variance exists.

Problem 9. Let X be a positive random variable and $g(x)$ a positive, strictly increasing function of x, such that $E[g(X)]$ exists. Prove that

$$\Pr(X \geqslant k) \leqslant \frac{E[g(X)]}{g(k)} \, .$$

If X has the exponential distribution with parameter $\lambda > 1$ and $g(x) = e^x$ show that,

$$\Pr(X \geqslant k) \leqslant \frac{\lambda \, e^{-k}}{\lambda - 1} \, .$$

10.4 MEAN AND VARIANCE OF A FUNCTION OF A CONTINUOUS RANDOM VARIABLE

We have seen that to find $E[g(X)]$ we require

$$\int g(x) f(x) \, dx \quad \text{or}$$

where $f(x)$ is the p.d.f. of the random variable X. Should the evaluation be dificult to carry out an approximation can be used. For example, we may be able to find a Taylor expansion of $g(x)$ about μ, the mean value of X. Provided the function is differentiable a sufficient number of times, we have

$$g(x) = g(\mu) + (x - \mu) g'(\mu) + \frac{(x - \mu)^2}{2!} g''(\mu) + R$$

where the remainder term is

$$\frac{(x - \mu)^3}{3!} g''[\mu + \theta_3 (x - \mu)], \quad 0 \leqslant \theta_3 \leqslant 1.$$

Now the expression is exact, and the only difficulty is the value of θ_3. For approximating purposes, the remainder is usually ignored and we have a representation which is good to some order of error for a stated interval of values of x. The difficulty, of course, is that if we use such an expression in the integration then we are letting x vary over its full range. If we do this, we obtain

$$E[g(X)] \approx \int \left[g(\mu) + (x - \mu) g'(\mu) + \frac{(x - \mu)^2}{2!} g''(\mu) \right] f(x) \, dx$$

$$= g(\mu) + \frac{g''(\mu) \, E(X - \mu)^2}{2!} \, ,$$

since the middle term vanishes

$$= g(\mu) + \frac{g''(\mu)}{2!} \, V(X).$$

Sometimes, indeed, $E[g(X)]$ is taken as $g(\mu)$.

$$V[g(X)] = E[g(X) - Eg(X)]^2 \, .$$

If we take $g(x) \approx g(\mu) + g'(\mu)(x - \mu)$

$$V[g(X)] \approx \int [g'(\mu)]^2 (x - \mu)^2 f(x)\, dx = [g'(\mu)]^2 V(X) \qquad (10.10)$$

Example 3
If X has the exponential distribution with parameter λ, $Y = \sin X$, then

$$E(Y) \approx \sin \frac{1}{\lambda} - \left(\frac{\sin 1/\lambda}{2\lambda^2} \right)$$

and, using (10.10),

$$V(Y) \approx \frac{1}{\lambda^2} \cos^2 \frac{1}{\lambda}. \qquad \blacksquare$$

The performance of the approximation is likely to prove pitiful in many cases. Suppose X has the uniform distribution over $(0, 1)$ then $E(X) = 1/2$, $V(X) = 1/12$. Consider the random variable $Y = -\log_e X$, this has the exponential distribution with parameter one and hence $V(Y) = 1$. The use of equation (10.10), however, gives

$$V(Y) \approx \left[\left(-\frac{1}{X} \right)^2 V(X) \right]_\mu = \frac{1}{3}.$$

10.5 TRUNCATED DISTRIBUTIONS

It may happen that a random variable represents a suitable model for a process except that by design or accident the values above or below certain fixed points are either discarded or not known. Such a distribution is said to be truncated. The truncation is said to be to the right of x_0 if all values $> x_0$ are discarded and is said to be to the left of x_0 if all values $< x_0$ are discarded.

Let X be a random variable with p.d.f. $f(x)$ and d.f. $F(x)$ and we require the distribution of X when it is truncated at the right of x_0.

$$\Pr(X \leqslant x \mid X \leqslant x_0) = \frac{\Pr(X \leqslant x \text{ and } X \leqslant x_0)}{\Pr(X \leqslant x_0)}.$$

When $x \leqslant x_0$, this reduces to

$$\frac{\Pr(X \leqslant x)}{\Pr(X \leqslant x_0)} = \frac{F(x)}{F(x_0)}$$

since $X \leqslant x$ implies $X \leqslant x_0$. While if $x > x_0$ we have

$$\frac{\Pr(X \leqslant x_0)}{\Pr(X \leqslant x_0)} = 1.$$

Thus the distribution function of the truncated random variable is $F(x)/F(x_0)$ when $x \leqslant x_0$ and unity otherwise, while its probability density function is $f(x)/F(x_0)$ when $x \leqslant x_0$ and zero otherwise. Naturally, when finding the moments we must take account of the modified p.d.f. Thus, in the continuous case,

$$E(X|X \leqslant x_0) = \frac{1}{F(x_0)} \int_{-\infty}^{x_0} x f(x) \, dx.$$

Example 4

Suppose a normal distribution with parameters μ, σ is truncated to the right at x_0, then

$$f(x|x \leqslant x_0) = \frac{\dfrac{1}{\sqrt{(2\pi)}\sigma} \exp\left[\dfrac{-\frac{1}{2}(x-\mu)^2}{\sigma^2}\right]}{\dfrac{1}{\sqrt{(2\pi)}\sigma} \int_{-\infty}^{x_0} \exp\left[\dfrac{-\frac{1}{2}(x-\mu)^2}{\sigma^2}\right] dx} = \frac{f(x)}{F(x_0)}.$$

$$E(X-\mu|X \leqslant x_0) = \frac{\dfrac{1}{\sqrt{(2\pi)}\sigma} \int_{-\infty}^{x_0} (x-\mu) \exp\left[\dfrac{-\frac{1}{2}(x-\mu)^2}{\sigma^2}\right] dx}{F(x_0)}$$

$$= -\frac{\sigma^2 \left\{ \exp\left[\dfrac{-\frac{1}{2}(x-\mu)^2}{\sigma^2}\right] \right\}_{-\infty}^{x_0}}{\sqrt{(2\pi)}\,\sigma F(x_0)}$$

$$= -\frac{\sigma \exp\left[\dfrac{-\frac{1}{2}(x_0-\mu)^2}{\sigma^2}\right]}{\sqrt{(2\pi)}\,F(x_0)}$$

$$= -\frac{\sigma^2 f(x_0)}{F(x_0)}.$$

Hence,

$$E(X|X \leqslant x_0) = \mu - \sigma^2 \frac{f(x_0)}{F(x_0)}.$$

Problem 10. The normal distribution with parameters μ, σ is truncated to the right at x_0. Show that

$$V(X|X \leqslant x_0) = \sigma^2 - (x_0 - \mu)\sigma^2 \frac{f(x_0)}{F(x_0)} - \left[\sigma^2 \frac{f(x_0)}{F(x_0)}\right]^2. \qquad \blacksquare$$

Similar methods may be employed with discrete distributions.

Example 5

For parents of a particular constitution, the probability that a child has a certain characteristic is p, independently for each child. The children of parents not of this type cannot have this characteristic. If we restrict our attention to a family of n children with at least one child having the characteristic, what is the probability that it has just r such children? The distribution of the number X of such children is binomial, truncated at $X = 0$.

$$\Pr(X = r \mid X > 0) = \frac{\Pr(X = r)}{\Pr(X > 0)}, \quad r = 1, 2, 3, \ldots, n$$

$$= \frac{\binom{n}{r} p^r q^{n-r}}{1 - q^n}.$$

The mean of the truncated random variable is

$$\frac{\sum_{r=1}^{n} r \binom{n}{r} p^r q^{n-r}}{1 - q^n} = \frac{np}{1 - q^n} > np.$$

Problem 11. A Poisson distribution with parameter λ is truncated at the origin. Find the p.d.f. of this distribution and its mean.

BRIEF SOLUTIONS AND COMMENTS ON THE PROBLEMS

Problem 1.

$$E[X(X - 1)] = \sum_{2}^{\infty} x(x - 1) \lambda^x e^{-\lambda}/x! = \lambda^2,$$

the rest is straightforward.

Problem 2. We build up $E[X(X - 1)]$ so that we can exploit the fact that

$$f(x) = 1 \quad \text{for all } r.$$

The essential part is

$$\sum_{2}^{\infty} x(x - 1) \binom{r + x - 1}{x} p^r q^x$$

$$= \frac{r(r + 1)q^2}{p^2} \sum_{2}^{\infty} \binom{r + 2 + x - 3}{x - 2} p^{r+2} q^{x-2}$$

$$= r(r + 1) q^2/p^2.$$

For this distribution, $E(X) = rq/p$.

Problem 3. The reader will find the expected values in the answer to Problem 1 in Chapter 9.

(a) $E(X^2) = \left[1^2 \binom{5}{1}\binom{4}{2} + 2^2 \binom{5}{2}\binom{4}{1} + 3^2 \binom{5}{3}\binom{4}{0} \right] / \binom{9}{3} = \dfrac{10}{3}.$

Hence $V(X) = \dfrac{10}{3} - \left(\dfrac{5}{3}\right)^2 = \dfrac{5}{9}.$

(b) $E(X^2) = \binom{3}{1}\left(\dfrac{5}{9}\right)\left(\dfrac{4}{9}\right)^2 + 4\binom{3}{2}\left(\dfrac{5}{9}\right)^2\left(\dfrac{4}{9}\right) + 9\binom{3}{3}\left(\dfrac{5}{9}\right)^3 = \dfrac{95}{27}.$

Hence, $V(X) = \dfrac{95}{27} - \left(\dfrac{5}{3}\right)^2 = \dfrac{20}{27}.$

(c) $E(X^2) = \displaystyle\sum_{1}^{\infty} \dfrac{x^2}{3}\left(\dfrac{2}{3}\right)^{x-1} = \sum_{1}^{\infty} x^2 \left(1 - \dfrac{2}{3}\right)\left(\dfrac{2}{3}\right)^{x-1}$

$= \displaystyle\sum x^2 \left(\dfrac{2}{3}\right)^{x-1} - \sum x^2 \left(\dfrac{2}{3}\right)^{x}$

$= \displaystyle\sum x^2 \left(\dfrac{2}{3}\right)^{x-1} - \sum (x-1)^2 \left(\dfrac{2}{3}\right)^{x-1}$

$= 2\displaystyle\sum x \left(\dfrac{2}{3}\right)^{x-1} - \sum \left(\dfrac{2}{3}\right)^{x-1} = 18 - 3 = 15.$

$V(X) = 15 - 3^2 = 6.$

Problem 4. The reader will find the expected values in the answer to Problem 2 in Chapter 9.

(a) $E(X^2) = \displaystyle\int_0^1 \dfrac{x^2}{2\sqrt{x}}\, dx = \dfrac{1}{2}\int_0^1 x^{3/2}\, dx = \dfrac{1}{5}.$

$V(X) = \dfrac{1}{5} - \left(\dfrac{1}{3}\right)^2 = \dfrac{4}{45}.$

(b) $E(X^2) = \displaystyle\int_0^1 (6x^3 - 6x^4)\, dx = 3/10.$

$V(X) = \dfrac{3}{10} - \left(\dfrac{1}{2}\right)^2 = \dfrac{1}{20}.$

(c) $E(X^2) = \dfrac{1}{2} \displaystyle\int_0^\infty x^4 \, e^{-x} \, dx = 12,$

$V(X) = 12 - 3^2 = 3.$

(d) $E(X^2) = \displaystyle\int_0^1 x^3 \, dx + \int_1^2 (2x^2 - x^3) \, dx = 7/6.$

$V(X) = (7/6) - 1^2 = 1/6.$

(e) $E(X^2) = \displaystyle\int_0^\infty \dfrac{\lambda^\alpha x^{\alpha+1}}{\Gamma(\alpha)} \, e^{-\lambda x} \, dx = \dfrac{\Gamma(\alpha + 2)}{\lambda^2 \Gamma(\alpha)} \int_0^\infty \dfrac{\lambda^{\alpha+2} x^{\alpha+1}}{\Gamma(\alpha + 2)} \, e^{-\lambda x} \, dx$

$\qquad = \dfrac{\alpha(\alpha + 1)}{\lambda^2}.$

Hence

$V(X) = [\alpha(\alpha + 1)/\lambda^2] - (\alpha/\lambda)^2 = \alpha/\lambda^2.$

(f) $E(X^2) = \displaystyle\int_0^1 \dfrac{(n + m - 1)!}{(n - 1)!(m - 1)!} \, x^{n+1} (1 - x)^{m-1} \, dx$

$\qquad = \dfrac{n(n + 1)}{(n + m)(n + m + 1)} \displaystyle\int_0^1 \dfrac{(n + m + 1)!}{(n + 1)!(m - 1)!} \, x^{n+1}(1 - x)^{m-1} \, dx$

$\qquad = \dfrac{n(n + 1)}{(n + m)(n + m + 1)}.$

Hence

$V(X) = \dfrac{n(n + 1)}{(n + m)(n + m + 1)} - \left(\dfrac{n}{n + m}\right)^2$

$\qquad = nm/[(n + m)^2 \, (n + m + 1)].$

Problem 5. The first result is Tchebychev's inequality

$$\Pr(X \geqslant t) = \Pr(X + c \geqslant t + c) \leqslant \Pr[(X + c)^2 \geqslant (t + c)^2]$$

$$\leqslant E(X + c)^2/(t + c)^2,$$

from Theorem 10.1. If $\mu = 0$, $E(X + c)^2 = \sigma^2 + c^2$ and $\sigma^2 + c^2/(t + c)^2$ is a minimum when $c = \sigma^2/t$. Substitute this value of c to obtain result.

Problem 6. The mean and variance are straightforward and exist if $n > 2$. When $\theta = 1, n = 3$,

$$\Pr(X \geqslant 3.5) = \int_{7/2}^\infty (3/x^4) \, dx = 8/343.$$

But

$$\Pr(X \geqslant 3.5) = \Pr(|X - 3/2| \geqslant 2),$$

which by Tchebychev's inequality is $\leqslant V(X)/2^2 = 3/16$.

Problem 7.

$$\Pr(|X - \mu| \geqslant k) = \Pr[(X - \mu)^2 \geqslant k^2]$$

$$= \Pr[(X - \mu)^{2m} \geqslant k^{2m}] \leqslant E[(X - \mu)^{2m}]/k^{2m}.$$

When $m = 1$, $E[(X - \mu)^2] = \sigma^2$ and for $k = 2\sigma$,

$$\Pr(|X - \mu| \geqslant 2\sigma) \leqslant \sigma^2/4\sigma^2 = 1/4.$$

When $m = 2$, $E[(X - \mu)^4] = 3\sigma^4$. Hence,

$$\Pr(|X - \mu| \geqslant 2\sigma) \leqslant 3\sigma^4/16\sigma^4 = 3/16.$$

From the table of the normal distribution,

$$\Pr(|X - \mu| \geqslant 2\sigma) \approx 0.045,$$

which is of course a good deal smaller.

Problem 8. All the probability is concentrated at three points.

$$E(X) = \frac{(\mu - k\sigma)}{2k^2} + \mu \left(1 - \frac{1}{k^2}\right) + \frac{(\mu + k\sigma)}{2k^2} = \mu.$$

$$V(X) = E(X - \mu)^2 = \frac{(-k\sigma)^2}{2k^2} + 0 \left(1 - \frac{1}{k^2}\right) + \frac{(+k\sigma)^2}{2k^2} = \sigma^2.$$

$$\Pr(|X - \mu| \geqslant k\sigma) = \Pr(X = \mu + k\sigma) + \Pr(X = \mu - k\sigma) = 1/k^2.$$

Problem 9.

$$g(x) \geqslant g(k) \leftrightarrow x \geqslant k.$$

Hence,

$$\Pr(X \geqslant k) = \Pr[g(X) \geqslant g(k)] \leqslant E[g(X)]/g(k),$$

from (10.4). If X is exponential, with parameter λ, $E(e^X) = \lambda/(\lambda - 1)$ and the result follows.

Problem 10. We must be careful since μ is *not* $E(X|X \leqslant x_0) = \mu^*$,

$$V(X|X \leqslant x_0) = E[(X - \mu^*)^2 \,|X \leqslant x_0]$$

$$= E[(X - \mu)^2 \,|X \leqslant x_0] - (\mu^* - \mu)^2.$$

$$E[(X - \mu)^2 \,|X \leqslant x_0] = [-(x_0 - \mu)\sigma^2 \, f(x_0)/F(x_0)] + \sigma^2 \,,$$

after one integration by parts, and the value of μ^* is displayed in Example 4.

Problem 11. $\Pr(X = 0) = e^{-\lambda}$, hence p.d.f. of truncated distribution is

$$f(x) = \lambda^x \, e^{-\lambda}/[x!(1 - e^{-\lambda})], \qquad x = 1, 2, \ldots,$$

with mean $\lambda/(1 - e^{-\lambda})$.

11

Moment Generating Functions

11.1 THE MOMENTS OF A DISTRIBUTION

The rth moment about the origin for a random variable X, denoted v_r', is defined as $v_r' = E(X^r)$ and is

$$\int x^r f(x)\, dx \quad \text{or} \quad \sum_x x^r f(x)$$

according as X is continuous or discrete. The rth central moment about the mean of a random variable X, denoted by v_r, is defined as $v_r = E[X - E(X)]^r$ and is

$$\int [x - E(X)]^r f(x)\, dx \quad \text{or} \quad \sum_x [x - E(X)]^r f(x)$$

according as X is continuous or discrete. In this notation, $E(X) = v_1'$, though it is also written μ. Also,

$$v_2 = V(X) = E[X - E(X)]^2 = E(X^2) - E^2(X) = v_2' - (v_1')^2,$$

though $V(X)$ is also written σ^2.

As to the number of moments which may exist, all possibilities may be found. Thus, for the Cauchy distribution, none exist, for the exponential distribution all moments exist, and for the Pareto distribution, a finite number, determined by the value of one of its parameters, exists. However, the 'moment' v_0' may, after all, be said to exist for all random variables, since $E(X^0) = 1$.

Problem 1. Show that $\nu_3 = \nu_3' - 3\nu_1'\nu_2' + 2(\nu_1')^3$.

Problem 2. Show for the exponential distribution with parameter λ, that $\nu_m' = (m/\lambda)\nu_{m-1}'$, and hence that $\nu_m' = m!/\lambda^m$.

Problem 3. Show for the normal distribution with parameters μ, σ that $\nu_m = (m-1)\sigma^2 \nu_{m-2}$, and deduce that $\nu_{2m+1} = 0, m = 0, 1, \ldots$ but

$$\nu_{2m} = \frac{(2m)!}{2^m\, m!}\,\sigma^{2m}, \quad m = 0, 1, 2, \ldots$$

Problem 4. Show for the Pareto distribution with p.d.f. $f(x) = n\theta^n/x^{n+1}, x \geqslant \theta$, that ν_k' exists if $k < n$, and then has value $n\theta^k/(n-k)$.

Problem 5. Show that $E(X^2) \geqslant E^2(|X|) \geqslant E^2(X)$. [Hint: Consider $V(|X|)$.] Verify that for the binomial distribution $V(X) < E(X)$, but if X has the geometric distribution and $Y = X - 1, V(Y) > E(Y)$.

11.2 SYMMETRY AND FLATNESS

Since for a distribution with a symmetric p.d.f. all odd central moments are zero, these are sometimes used as measures of asymmetry or skewness. In particular $\gamma_1 = \nu_3/\sigma^3$ is called the coefficient of skewness. Similarly, $\gamma_2 = (\nu_4/\sigma^4) - 3$ is employed to measure the degrees of flatness of a distribution. For a normal distribution $\gamma_2 = (3\sigma^4/\sigma^4) - 3 = 0$, and hence γ_2 is the coefficient of 'excess' (with respect to a normal distribution).

We shall not be much concerned with moments of higher order than the second. The evaluation of moments is tedious and for the case where moments of all positive integral orders exist, we shall proceed to discuss a function which generates them all at one blow.

11.3 MOMENT GENERATING FUNCTIONS

Definition. If X is a random variable, then its moment generating function, $M_X(t)$ is defined as

$$M_X(t) = E(e^{Xt})$$

provided that this expectation exists for some interval of t which includes the origin. That is, there is a positive number h, so that whenever $-h < t < h$, then $E(e^{Xt})$ is finite. For a continuous distribution, $M_X(t)$ is evaluated as

$$\int e^{xt} f(x)\, dx,$$

while for a discrete distribution, $M_X(t)$ is evaluated as

$$\sum_x e^{xt} f(x),$$

where the integration or summation is over those values of x for which $f(x) > 0$.

Example 1
X has the exponential distribution, for which $f(x) = \lambda e^{-\lambda x}, x > 0$.

$$M_X(t) = E(e^{Xt})$$

$$= \int e^{xt} f(x)\, dx = \int_0^\infty e^{xt} \lambda e^{-\lambda x}\, dx$$

$$= \lambda \int_0^\infty e^{x(t-\lambda)}\, dx$$

$$= \frac{\lambda[e^{x(t-\lambda)}]_0^\infty}{t - \lambda}.$$

For this to be finite we require

$$\lim_{x \to \infty} e^{x(t-\lambda)} = 0,$$

which will be the case if $t - \lambda < 0$. Thus if $t < \lambda$,

$$M_X(t) = \frac{\lambda}{\lambda - t}. \qquad (11.1)$$

Example 2
X has the binomial distribution, for which

$$f(x) = \binom{n}{x} p^x q^{n-x}, \quad x = 0, 1, \ldots, n.$$

$$M_X(t) = E(e^{Xt}) = \sum_x e^{xt} f(x)$$

$$= \sum_{x=0}^n e^{xt} \binom{n}{x} p^x q^{n-x}$$

$$= \sum_{x=0}^n \binom{n}{x} (e^t p)^x q^{n-x}$$

$$= (q + p\, e^t)^n. \qquad (11.2)$$

In this case, there is no restriction on t. ∎

In both examples, we note that we end up with a function of t and the parameters of the appropriate distribution. The variable t is in fact playing the part of a place-holder and restrictions of the type $|t| < h$ are not a source or embarrasment. Yet these examples contain no hint as to why we should consider any such function. If we trust intuition, then using the power series expansion for e^{Xt},

$$M_X(t) = E(e^{Xt}) = E\left(\sum_{r=0}^{\infty} \frac{X^r t^r}{r!}\right) = \sum_{r=0}^{\infty} \frac{t^r}{r!} E(X^r) = \sum_{r=0}^{\infty} \frac{t^r v_r'}{r!}.$$

This suggests that having found the function $M_X(t)$, we expand it as a power series in t, and then the coefficient of $t^r/r!$ will be the rth moment of X about the origin. This procedure is not always convenient and the effect of differentiating $M_X(t)$ r times with respect to t suggests and alternative.

$$\frac{d^r M(t)}{dt^r} = \frac{d^r}{dt^r}[E(e^{Xt})]$$

$$= E\left[\frac{d^r}{dt^r}(e^{Xt})\right]$$

$$= E(X^r e^{Xt}).$$

If this expectation is evaluated when $t = 0$, we have $M_X^{(r)}(0) = E(X^r) = v_r'$. We shall assume that both procedures are valid whenever the moment-generating function (m.g.f) exists.

Example 3

For the exponential distribution we have obtained,

$$M_X(t) = \frac{\lambda}{\lambda - t} = \frac{1}{1 - (t/\lambda)} = \left(1 - \frac{t}{\lambda}\right)^{-1}$$

$$= 1 + \frac{t}{\lambda} + \frac{t^2}{\lambda^2} + \ldots + \frac{t^r}{\lambda^r} + \ldots .$$

The binomial expansion is valid for $|t| < \lambda$, which also suffices for the existence of $M_X(t)$. The coefficient of $t^r/r!$ is $r!/\lambda^r$, and this is v_r'. In particular, $v_1' = 1/\lambda$, $v_2' = 2/\lambda^2$, in agreement with the values previously found. In this case finding the power series for $M_X(t)$ gives slightly less work than evaluating the rth derivative. We have, for instance,

$$\frac{d}{dt}M_X(t) = \frac{d}{dt}\left(1 - \frac{t}{\lambda}\right)^{-1} = \frac{1}{\lambda}\left(1 - \frac{t}{\lambda}\right)^{-2}$$

and when $t = 0$,

$$M_X^{(1)}(0) = \frac{1}{\lambda}.$$

Problem 6. Show for the exponential distribution

$$\frac{\mathrm{d}^r M(t)}{\mathrm{d}t^r} = \frac{r!}{\lambda^r} \left(1 - \frac{t}{\lambda}\right)^{-r-1}$$

and hence find ν_r'.

Problem 7. The random variable X has the rectangular distribution over the interval $(0, l)$. Show that $M_X(t) = (e^{lt} - 1)/lt$.

By expanding $M_X(t)$, show that $\nu_r' = l^r/(r + 1)$. Confirm the value ν_1' by differentiating $M_X(t)$. This last is somewhat harder — L'Hôpital's rule is of assistance.

Example 4

For the binomial distribution,

$$M_X(t) = (q + p\,e^t)^n = \sum_{x=0}^{n} \binom{n}{x} e^{tx} p^x q^{n-x}.$$

The coefficient of $t^r/r!$ in e^{tx} is x^r and hence the total coefficient of $t^r/r!$ is

$$\sum_{x=0}^{n} \binom{n}{x} x^r p^x q^{n-x},$$

which is not very enlightening as this is merely $E(X^r)$ as previously defined. Proceeding otherwise,

$$M_X(t) = \left[q + p\left(1 + t + \frac{t^2}{2!} + \ldots\right)\right]^n$$

$$= \left[1 + p\left(t + \frac{t^2}{2!} + \ldots\right)\right]^n$$

$$= 1 + np\left(t + \frac{t^2}{2!} + \ldots\right)$$

$$+ \frac{n(n-1)p^2}{2!}\left(t + \frac{t^2}{2!} + \ldots\right)^2 \ldots.$$

From which the coefficient of $t/1!$ is np and that of $t^2/2!$ is $n(n-1)p^2 + np$ are the values of ν_1' and ν_2'. For ν_1' the method of differentiating the m.g.f. yields

$$\frac{\mathrm{d}\,M(t)}{\mathrm{d}t} = \frac{\mathrm{d}}{\mathrm{d}t}(q + p\,e^t)^n = n(q + p\,e^t)^{n-1}\,p\,e^t$$

and

$$M_X^{(1)}(0) = n(q + p)^{n-1} p = np.$$

Problem 8. Show by differentiating the m.g.f. of the binomial distribution that $\nu_2' = np + n(n-1)p^2$.

Problem 9. The random variable X has the geometric distribution with p.d.f. $f(x) = pq^{x-1}$, $x = 1, 2, \ldots$. Show that the m.g.f. of X is $p\,e^t/(1 - e^t q)$. Verify that $\nu_1' = 1/p$, $\nu_2' = (q + 1)/p^2$.

Problem 10. The random variable X has the Poisson distribution with p.d.f.

$$f(x) = \frac{\lambda^x e^{-\lambda}}{x!}, \quad x = 0, 1, 2, \ldots .$$

Show that the m.g.f. of X is $e^{\lambda(e^t - 1)}$ and by differentiating show that $\nu_1' = \lambda$, $\nu_2' = \lambda(\lambda + 1)$.

Example 5
X has the normal distribution with p.d.f.

$$f(x) = \frac{1}{\sqrt{(2\pi)}\sigma} \exp\left[-\frac{1}{2}(x - \mu)^2/\sigma^2\right], \quad -\infty < x < \infty.$$

$$M_X(t) = E(e^{Xt}) = \int e^{xt} f(x)\, dx$$

$$= \int_{-\infty}^{+\infty} e^{xt} \frac{1}{\sqrt{(2\pi)}\sigma} \exp\left[-\frac{1}{2}(x - \mu)^2/\sigma^2\right] dx$$

$$= \int_{-\infty}^{+\infty} \frac{1}{\sqrt{(2\pi)}\sigma}$$

$$\exp\left\{-\frac{1}{2\sigma^2}\, [x^2 - 2x(\mu + t\sigma^2) + \mu^2]\right\} dx.$$

After 'completing the square' on x, we have

$$\int_{-\infty}^{+\infty} \frac{1}{\sqrt{(2\pi)}\sigma} \exp\left\{-\frac{1}{2\sigma^2}\, [(x - \mu - t\sigma^2)^2 - 2\mu t\sigma^2 - t^2\sigma^4]\right\} dx$$

$$= \exp\left(\mu t + \frac{1}{2}t^2\sigma^2\right)$$

$$\int_{-\infty}^{+\infty} \frac{1}{\sqrt{(2\pi)}\sigma} \exp\left\{-\frac{1}{2}\left[\frac{x - (\mu + t\sigma^2)}{\sigma}\right]^2\right\} dx.$$

But the definite integral is now of the form

$$\int_{-\infty}^{+\infty} \frac{1}{\sqrt{(2\pi)}\sigma} \exp\left[-\frac{1}{2}(x-\theta)^2/\sigma^2\right] dx,$$

where $\theta = \mu + t\sigma^2$, and thus is equal to unity. Finally,

$$M_X(t) = \exp\left(\mu t + \frac{1}{2}\sigma^2 t^2\right). \tag{11.3}$$

Problem 11. From the m.g.f., show, in two different ways, that for the normal distribution, $\nu_1' = \mu, \nu_2' = \sigma^2 + \mu^2$. ∎

At this stage it may be supposed that the m.g.f. ought to have been used to obtain the moments of distributions from the start. This idea may be resisted since the moment-generating function need not exist. This happens if there is a moment of some order which is not finite.

Example 6
X is a continuous random variable with p.d.f.

$$f(x) = \frac{2}{x^3}, \quad 1 \leqslant x < \infty$$

$$= 0 \text{ otherwise.}$$

ν_1' exists, since

$$E(X) = \int_1^\infty \frac{2}{x^2} dx = 2.$$

However,

$$\nu_2' = \int_1^\infty \frac{2}{x} dx = [2 \log_e x]_1^\infty$$

is unbounded, since $\log_e x$ exceeds any number if x is sufficiently large. Thus ν_2' does not exist; indeed, ν_r' does not exist if $r \geqslant 2$. In like fashion,

$$M_X(t) = \int_1^\infty e^{xt} \frac{2}{x^3} dx > \int_1^\infty \frac{x^3 t^3}{3!} \cdot \frac{2}{x^3} dx, \quad t > 0$$

and hence is not finite. This weakness of the m.g.f. can be remedied by considering instead the characteristic function of X, $\phi_X(t) = E(e^{iXt})$, which always exists. Further, it generates all those moments which exist.

11.4 FUNCTION OF A RANDOM VARIABLE

If X is a random variable and $Y = g(X)$ is also a random variable, then

$$M_Y(t) = E(e^{Yt}) = E[e^{g(X)t}] = M_{g(X)}(t)$$

and we can find the m.g.f. of Y by finding the appropriate expectation with respect to the distribution of X, and are thus relieved of the task of first finding the p.d.f. of Y.

Example 7
If X has the distribution $N(0, 1)$, find the m.g.f. of $Y = X^2$,

$$M_Y(t) = M_{X^2}(t) = E(e^{X^2 t})$$

$$= \int_{-\infty}^{+\infty} \frac{1}{\sqrt{(2\pi)}} e^{x^2 t} e^{-x^2/2} \, dx$$

$$= \int_{-\infty}^{+\infty} \frac{1}{\sqrt{(2\pi)}} e^{-(1/2)x^2(1-2t)} \, dx.$$

Now put $x(1 - 2t)^{1/2} = u$,

$$M_Y(t) = \frac{1}{(1 - 2t)^{1/2}} \int_{-\infty}^{+\infty} \frac{e^{-u^2/2}}{\sqrt{(2\pi)}} \, du = \frac{1}{(1 - 2t)^{1/2}}.$$

Problem 12. Find ν_1', ν_2' for the random variable Y in Example 7.

Problem 13. If X has the rectangular distribution over $(0, 1)$, show that the m.g.f. of $Y = -\log_e X$ is $M_Y(t) = 1/(1 - t)$. Identify the distribution of Y. ∎

A useful special case arises when $Y = a + bX$, when

$$M_Y(t) = M_{a+bX}(t) = E[e^{(a+bX)t}] = E[e^{at} e^{X(bt)}] = e^{at} M_X(bt).$$

In particular, if $a = -\mu$, $b = 1$, $M_{X-\mu}(t) = e^{-\mu t} M_X(t)$. Also,

$$M_{X-\mu}(t) = E[e^{(X-\mu)t}]$$

$$= E\left\{ \sum_{r=0}^{\infty} [(X - \mu)^r t^r/r!] \right\}$$

$$= \sum_{r=0}^{\infty} \left\{ E[(X - \mu)^r t^r/r!] \right\}$$

$$= \sum_{r=0}^{\infty} (\nu^r t^r/r!)$$

and the coefficient of $t^r/r!$ in $e^{-\mu t} M_X(t)$ is ν_r, the central moment of X of order r. Should this prove inconvenient we may alternatively evaluate the rth derivative of $e^{-\mu t} M_X(t)$ at $t = 0$. Either method presupposes that μ is known.

Example 8
If X has the Poisson distribution, then $M_X(t) = e^{\lambda e^t - \lambda}$ and $\nu_1' = \lambda$. Thus

$$M_{X-\lambda}(t) = e^{-\lambda t} M_X(t) = e^{\lambda e^t - \lambda - \lambda t},$$

$$\frac{d}{dt}[M_{X-\lambda}(t)] = (-\lambda + \lambda e^t) e^{\lambda e^t - \lambda - \lambda t} = (-\lambda + \lambda e^t) M_{X-\lambda}(t)$$

and when $t = 0$, the value of this derivative is zero, which is always the case for the first central moment.

$$\frac{d^2}{dt^2}[M_{X-\lambda}(t)] = \frac{d}{dt}[(-\lambda + \lambda e^t) M_{X-\lambda}(t)]$$

$$= (-\lambda + \lambda e^t) \frac{d}{dt}[M_{X-\lambda}(t)] + (\lambda e^t) M_{X-\lambda}(t).$$

When $t = 0$, the first of these terms is again zero, the second has value $\lambda M_{X-\lambda}(0) = \lambda$.

Problem 14. If X is distributed $N(\mu, \sigma^2)$ show that $M_{X-\mu}(t) = e^{+\sigma^2 t^2/2}$. Hence deduce that $\nu_{2r+1} = 0, \nu_{2r} = (2r!) \, \sigma^{2r}/(r! \, 2^r)$.

11.5 PROPERTIES OF MOMENT GENERATING FUNCTIONS

It would be convenient if given a m.g.f. it were possible to recover the probability density function of the random variable. In some cases, when X is discrete, this is possible. Thus if

$$M_X(t) = \sum_{x=0}^{\infty} f(x) e^{xt} = \frac{1}{3} e^t + \frac{2}{3} e^{2t}$$

for all t, then by picking out the coefficients of e^{xt}, we have $f(1) = 1/3$, $f(2) = 2/3$ and $f(x) = 0$ otherwise. Interest evaporates on noting that such a method is useless for continuous random variables. Indeed, there is no general elementary method for recovering the p.d.f. from the m.g.f. However, we may be able to spot the type of distribution which yields a particular variety of m.g.f. Thus if $M_X(t) = e^{2t^2}$, then this is of the form $e^{\mu t + (1/2)\sigma^2 t^2}$, with $\mu = 0$ and $\sigma = 2$. We are prompted to claim that X has the distribution $N(0, 4)$. Now, while it is certainly the case that this normal distribution has e^{2t^2} as its m.g.f., we need assuring that this is the only possibility. This security is provided by the uniqueness theorem, which is stated without proof.

Theorem 11.1. If X and Y are random variables, each with the same m.g.f. $M(t)$, for all values of t for which $M(t)$ is defined, then X and Y have the same p.d.f.

This result may allow us to discover the p.d.f. of a random variable with a given m.g.f. by inspecting the list of m.g.f.s derived from known distributions. The theorem is only relevant when the m.g.f. exists, for it can happen that two different distributions have a finite number of moments in common. Nor does it provide an answer as to whether a given infinite sequence of numbers 1, $v_1', v_2', \ldots$ can be the moments of some possible distribution.

Problem 15. Let X be a random variable such that $v_r' = (n+r-1)!/[\lambda^r(n-1)!]$ $r = 1, 2, \ldots$. Show that $M_X(t) = \lambda^n(\lambda - t)^{-n}$, and verify that the p.d.f. of X is $f(x) = \lambda(\lambda x)^{n-1} e^{-\lambda x}/(n-1)!, 0 < x < \infty$.   ■

We have already seen that, under certain conditions, the limiting form of a binomial distribution with parameters n, p is a Poisson distribution with parameter $\lambda = np$. We next look at the behaviour of the corresponding sequence of m.g.f.s as n increases subject to the same conditions. For the binomial distribution

$$M_X(t) = (q + p e^t)^n$$

Now put $p = \lambda/n, q = 1 - (\lambda/n)$:

$$M_X(t) = \left(1 - \frac{\lambda}{n} + \frac{\lambda e^t}{n}\right)^n$$

$$M_X(t) = \left(1 + \frac{\lambda e^t - \lambda}{n}\right)^n$$

$$\lim_{n \to \infty} [M_X(t)] = e^{\lambda e^t - \lambda}.$$

But *this* is the m.g.f. of a Poisson distribution with parameter λ. We state the following theorem on sequences of random variables, again without proof.

Theorem 11.2. Let $f_n(x)$ and $M_{X_n}(t)$ be respectively the probability density function and the m.g.f. of a random variable X_n. Then, if

$$\lim_{n \to \infty} M_{X_n}(t)$$

exists and is $M(t)$, then

$$\lim_{n \to \infty} f_n(x)$$

exists and is, say, $f(x)$, where $f(x)$ is the p.d.f. of a random variable X which has the m.g.f. $M(t)$.

Problem 16. The random variable X_n has the exponential distribution with parameter $n\lambda$. Show that

$$\lim_{n\to\infty} M_{X_n}(t) = 1.$$

Verify that correspondingly, for every $k > 0$,

$$\lim_{n\to\infty} \Pr(X_n < k) = 1.$$

Prove, however, that if $Y_n = nX_n$ then the m.g.f. of Y_n does not depend on n and deduce the distribution of Y_n.

BRIEF SOLUTIONS AND COMMENTS ON THE PROBLEMS

Problem 1

$$\begin{aligned} E(X-\mu)^3 &= E(X^3 - 3X^2\mu + 3X\mu^2 - \mu^3) \\ &= E(X^3) - 3E(X^2)\mu + 3E(X)\mu^2 - \mu^3 \\ &= v_3' - 3v_1' v_2' + 2(v_1')^3. \end{aligned}$$

Problem 2

$$v_m' = \int_0^\infty x^m \, \lambda e^{-\lambda x} \, dx$$

$$= (m/\lambda) \int_0^\infty x^{m-1} \, \lambda e^{-\lambda x} \, dx = (m/\lambda)v_{m-1}',$$

after one integration by parts. The result is obtained by continued application.

Problem 3

$$v_m = \int_{-\infty}^\infty (x-\mu)^m f(x) \, dx = \int_{-\infty}^{+\infty} (x-\mu)^{m-1} (x-\mu) f(x) \, dx,$$

and, after integration by parts,

$$v_m = (m-1)\sigma^2 v_{m-2}.$$

The different results obtain because m is stepped down by two.

$$v_{2m+1} = 2m\sigma^2 v_{2m-1} = 2m(2m-2)\sigma^4 v_{2m-3} \cdots,$$

and result terminates at v_1. But

$$v_1 = E(X-\mu) = E(X) - \mu = 0. \qquad v_{2m} = (2m-1)\sigma^2 v_{2m-2}$$

and, after repeated application, stops at $v_2 = \sigma^2$. Write $(2m-1)(2m-3)\ldots 3$ as

$$\frac{2m(2m-1)(2m-2)\ldots 3.2}{2m(2m-2)\ldots 2} = \frac{(2m)!}{2^m m!} \;.$$

Problem 4

$$E(X^k) = \int_\theta^\infty (n\theta^n x^{k-n-1})\, dx$$

$$= \left(\frac{n\theta^n x^{k-n}}{k-n}\right)_\theta^\infty,$$

which if $k < n$, is $n\theta^k/(n-k)$. If $k \geqslant n$, the moment does not exist.

Problem 5

$$V(|X|) = E(|X|^2) - (E|X|)^2$$
$$= E(X^2) - (E|X|)^2 \geqslant 0.$$

Hence $E(X^2) \geqslant (E|X|)^2$. Also $E(|X|) \geqslant E(X)$ because $|X| \geqslant X$. For the binomial distribution, $V(X) = npq < np = E(X)$. For the geometric distribution,

$$V(Y) = V(X) = q/p^2 > q/p = E(Y) = E(X) - 1 = 1/p - 1 = q/p,$$

so that $V(Y) > E(Y)$.

Problem 6. Presents no difficulty, either directly or by induction.

Problem 7

$$M_X(t) = \int_0^l \frac{e^{xt}}{l}\, dx = \frac{e^{lt} - 1}{lt}$$

$$= \sum_1^\infty (lt)^{r-1}/r!.$$

Hence the coefficient of t^r is $l^r/(r+1)!$ and of $t^r/r!$ is thus $l^r/(r+1)$.

$$M_X^{(1)}(t) = [lt(l\,e^{lt}) - (e^{lt} - 1)\,l]/(lt)^2.$$

Application of L'Hôpital's rule leads to $M_X^{(1)}(0) = l/2$.

Problem 8

$$M_X^{(2)}(t) = n(n-1)(q + p\,e^t)^{n-2} p^2 e^{2t} + n(q + p\,e^t)^{n-1} p\,e^t.$$
$$v_2' = M_X^{(2)}(0) = n(n-1)p^2 + np.$$

Problem 9

$$M_X(t) = \sum_1^\infty e^{xt} p q^{x-1} = p\, e^t/(1 - q\, e^t)$$

$$= (-p/q) + (p/q)/(1 - q\, e^t).$$

Hence $M_X^{(1)}(t) = p\, e^t/(1 - q\, e^t)^2$ and $M_X^{(1)}(0) = 1/p$. $M_X^{(2)}(t)$ involves no new difficulty.

Problem 10

$$M_X(t) = \sum_0^\infty e^{tx} \lambda^x\, e^{-\lambda}/x! = e^{-\lambda} \sum_0^\infty (\lambda\, e^t)^x/x!$$

$$= e^{-\lambda}\, e^{\lambda e^t}.$$

$$M_X^{(1)}(t) = e^{-\lambda} e^{\lambda e^t} \lambda\, e^t = \lambda\, e^t M_X(t),$$

which is rather convenient.

$$M_X^{(1)}(0) = \lambda M_X(0) = \lambda \quad \text{and} \quad M_X^{(2)}(t) = \lambda\, e^t M_X(t) + \lambda\, e^t M_X^{(1)}(t),$$

whence

$$M_X^{(2)}(0) = \lambda + \lambda^2.$$

Problem 11

$$M_X^{(1)}(t) = (\mu + t\sigma^2)\, M_X(t),$$

hence

$$M_X^{(2)}(t) = \sigma^2 M_X(t) + (\mu + t\sigma^2)\, M_X^{(1)}(t).$$

Thus $M_X^{(1)}(0) = \mu$, $M_X^{(2)}(0) = \sigma^2 + \mu^2$. Alternatively by expanding as a power series, $M_X(t) = 1 + \mu t + (1/2)\,(\sigma^2 + \mu^2)t^2 + \dots$.

Problem 12

$$M_X^1(t) = 1/(1 - 2t)^{3/2}, \qquad \nu_1' = M^1(0) = 1.$$
$$M_X^2(t) = 3/(1 - 2t)^{5/2}, \qquad \nu_2' = M^2(0) = 3.$$

Problem 13

$$E[e^{(-\log_e X)t}] = E(X^{-t})$$

$$= \int_0^1 x^{-t}\, dx = 1/(1 - t).$$

This is the m.g.f. of an exponential distribution with parameter 1.

Problem 14

$$M_{X-\mu}(t) = e^{-\mu t} e^{\mu t + (\sigma^2 t^2/2)} = e^{\sigma^2 t^2/2}$$

$$= \sum (\sigma^2 t^2/2)^r/r!$$

The coefficient of $t^{2r}/2r!$ is ν_{2r} and is as shown.

Problem 15. The coefficient of t^r must be

$$\binom{n+r-1}{r} \left(\frac{1}{\lambda}\right)^r = a_r .$$

Hence

$$\sum_{r=0}^{\infty} a_r t^r = (1 - t/\lambda)^{-n}$$

as required. Evaluate

$$E(e^{Xt}) = \int_0^\infty e^{xt} \, f(x) \, dx$$

to show that this is also $(1 - t/\lambda)^{-n}$.

Problem 16

$$M_X(t) = n\lambda/(n\lambda - t) \to 1 \text{ as } n \to \infty,$$

$$\Pr(X < k) = 1 - e^{-n\lambda k} \to 1 \text{ as } n \to \infty.$$

These indicate that all the probability is being forced towards the origin.

$$M_{Y_n}(t) = E(e^{Y_n t}) = E(e^{nX_n t})$$

$$= E(e^{X_n nt}) = M_{X_n}(nt) = n\lambda/(n\lambda - nt) = \lambda/(\lambda - t),$$

which is the m.g.f. of an exponential distribution with parameter λ.

12

Moments of Bivariate Distributions

12.1 CONDITIONAL AND UNCONDITIONAL EXPECTATIONS

If X_1, X_2 have a joint distribution with p.d.f. $f(x_1, x_2)$, then we define $E(X_1)$, $E(X_2)$ as

$$E(X_1) = \iint x_1 f(x_1, x_2)\, dx_1\, dx_2,$$

$$E(X_2) = \iint x_2 f(x_1, x_2)\, dx_2\, dx_1$$

in the continuous case, and as

$$E(X_1) = \sum_{x_1} \sum_{x_2} x_1 f(x_1, x_2), \quad E(X_2) = \sum_{x_2} \sum_{x_1} x_2 f(x_1, x_2)$$

in the discrete case, provided the integrals and sums are absolutely convergent. With this latter requirement the order of integration or summation is immaterial, the same result being obtained in either case. It may, of course, be easier to evaluate in one particular order. Thus,

$$E(X_1) = \sum_{x_1} \sum_{x_2} x_1 f(x_1, x_2) = \sum_{x_1} x_1 \left[\sum_{x_2} f(x_1, x_2) \right]$$

$$= \sum_{x_1} x_1 f_1(x_1),$$

where $f_1(x_1)$ is the marginal p.d.f. of X_1, and agrees with $E(X_1)$ where X_1 is regarded as having a univariate distribution. Similarly

$$E(X_2) = \sum_{x_2} x_2 f_2(x_2).$$

For the continuous case

$$E(X_1) = \int x_1 f_1(x_1) dx_1, \quad E(X_2) = \int x_2 f_2(x_2) dx_2.$$

On the other hand,

$$E(X_1) = \sum_{x_1} \sum_{x_2} x_1 f(x_1, x_2) = \sum_{x_2} \sum_{x_1} x_1 f(x_1|x_2) f_2(x_2)$$

$$= \sum_{x_2} f_2(x_2) \left[\sum_{x_1} x_1 f(x_1|x_2) \right] = E[E_1(X_1|X_2)]$$

where we term $E_1(X_1|x_2)$, the **conditional** expectation of X_1 given $X_2 = x_2$. $E_1(X_1|x_2)$ is a particular value of the random variable $E_1(X_1|X_2)$. This conditional expectation is, of course, a function of x_2, and is also known as the regression function of X_1 on X_2. The graph of $x_1 = E_1(X_1|x_2)$ is the regression curve of X_1 on X_2. Similarly,

$$E(X_2) = \sum_{x_1} f_1(x_1) \left[\sum_{x_2} x_2 f(x_2|x_1) \right] = E[E_2(X_2|X_1)]$$

$$\text{and} \quad E_2(X_2|x_1)$$

is the conditional mean of X_2 given $X_1 = x_1$ and is the regression function of X_2 on X_1. There are corresponding conditional means for the continuous case. In general, if X_1, X_2 have a bivariate distribution and $g(X_1, X_2)$ is a random variable then $E[g(X_1, X_2)]$ is

$$\iint g(x_1, x_2) f(x_1, x_2) dx_1 dx_2 \quad \text{or} \quad \sum_{x_1} \sum_{x_2} g(x_1, x_2) f(x_1, x_2)$$

according as the distribution of X_1, X_2 is continuous or discrete.

Handling bivariate distributions for which $f(x_1, x_2) > 0$ in a finite region which is not rectangular may require a little care.

Example 1

$$f(x_1, x_2) = 2, \quad 0 < x_1 < x_2 < 1$$

$$= 0 \text{ otherwise.}$$

The p.d.f. is positive within the shaded triangle in Fig. 12.1. The total probability is, of course, 1, since it is equal to the area of the triangle, times the constant value of $f(x_1, x_2) = 2$. For the marginal distributions we have

$$f_1(x_1) = \int f(x_1, x_2) dx_2 = \int_{x_1}^{1} 2. dx_2 = 2(1 - x_1); \quad 0 < x_1 < 1,$$

$$f_2(x_2) = \int f(x_1, x_2) dx_1 = \int_{0}^{x_2} 2 dx_1 = 2x_2, \quad 0 < x_2 < 1.$$

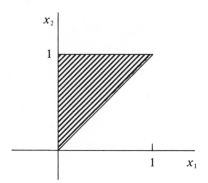

Fig. 12.1

Thus,

$$E(X_1) = \int x_1 f_1(x_1)\, dx_1 = \int_0^1 x_1(2 - 2x_1)\, dx_1$$

$$= \int_0^1 (2x_1 - 2x_1^2)\, dx_1 = \frac{1}{3},$$

$$E(X_2) = \int x_2 f_2(x_2)\, dx_2 = \int_0^1 x_2 . 2x_2\, dx_2 = \int_0^1 2x_2^2\, dx = \frac{2}{3}.$$

We remark in passing that since $f_1(x_1) f_2(x_2) \neq f(x_1, x_2)$, X_1, X_2 are not independent. Also,

$$f(x_1 | x_2) = \frac{f(x_1, x_2)}{f_2(x_2)} = \frac{2}{2x_2} = \frac{1}{x_2}, \quad 0 < x_1 < x_2 < 1.$$

Hence the **conditional** mean of X_1 given $X_2 = x_2$ is

$$\int_0^{x_2} x_1 f(x_1 | x_2)\, dx_1 = \int_0^{x_2} \frac{x_1}{x_2}\, dx_1 = \frac{1}{x_2} \int_0^{x_2} x_1\, dx_1 = \frac{x_2}{2},$$

$$0 < x_2 < 1.$$

From this we can regain the **unconditional** mean of X_1 by averaging over the distribution of X_2

$$E(X_1) = E[E_1(X_1 | X_2)] = E\left(\frac{X_2}{2}\right) = \int_0^1 \frac{x_2}{2} f_2(x_2)\, dx_2$$

$$= \int_0^1 x_2^2\, dx_2 = \frac{1}{3}.$$

Similarly, the conditional mean of X_2 given $X_1 = x_1$ is

$$\int_{x_1}^{1} x_2\, f(x_2|x_1)\, dx_2 \;=\; \int_{x_1}^{1} x_2\, \frac{f(x_1,x_2)}{f_1(x_1)}\, dx_2$$

$$=\; \int_{x_1}^{1} x_2\, \frac{2}{2(1-x_1)}\, dx_2 \;=\; \frac{1}{2}(1+x_1)$$

and then to recover the **unconditional** mean,

$$E(X_2) = E[E_2(X_2|X_1)] \;=\; E[(1/2)(1+X_1)]$$
$$= 2/3.$$

Thus the regression curve of X_1 on X_2 is $x_1 = x_2/2$, and of X_2 on X_1 is $x_2 = (1/2)(1+x_1)$. Thse are both straight lines, though this will not be the case in general.

We now consider the variances of X_1, X_2 in a bivariate distribution.

$$V(X_1) = E[X_1 - E(X_1)]^2$$
$$= \int\int [x_1 - E(X_1)]^2\, f(x_1,x_2)\, dx_1\, dx_2$$
$$= \int [x_1 - E(X_1)]^2\, f_1(x_1)\, dx_1 = E(X_1^2) - E^2(X_1),$$

which agrees with the one-dimentional case. Similarly,

$$V(X_2) = \int [x_2 - E(X_2)]^2\, f_2(x_2)\, dx_2 = E(X_2^2) - E^2(X_2).$$

Thus, in example (1) we have

$$E(X_1^2) = \int x_1^2\, f_1(x_1)\, dx_1 = \int_0^1 x_1^2(2-2x_1)\, dx_1$$

$$= \int_0^1 (2x_1^2 - 2x_1^3)\, dx_1 = \frac{1}{6}.$$

Hence

$$V(X_1) = E(X_1^2) - E^2(X_1) = \frac{1}{6} - \left(\frac{1}{3}\right)^2 = \frac{1}{18}.$$

We have earlier discussed two important bivariate distributions to which we apply our present results. We are entitled to use our available information about the marginal and conditional distributions.

Example 2
The multinomial distribution

$$f(x_1,x_2) = \frac{n!}{x_1!x_2!(n-x_1-x_2)!}\; p_1^{x_1} p_2^{x_2} (1-p_1-p_2)^{n-x_1-x_2}$$

for $0 \leqslant x_1 + x_2 \leqslant n$. The marginal distribution of X_1 is binomial with parameters n and p_1. Hence $E(X_1) = np_1$ and $V(X_1) = np_1 q_1$. Similarly, $E(X_2) =$
np_2, $V(X_2) = np_2 q_2$. Also, the conditional distribution of X_1 given $X_2 = x_2$ is
again binomial with parameters $n - x_2$ and $p_1/(1 - p_2)$. Hence the conditional
mean of X_1 given $X_2 = x_2$ is $(n - x_2)p_1/(1 - p_2)$ and the conditional variance
is $(n - x_2)(p_1/q_2)(1 - p_1/q_2)$ with similar results for the conditional distribution of X_2 given $X_1 = x_1$.

Example 3
The bivariate normal distribution,

$$f(x_1, x_2) = \frac{1}{2\pi\sigma_1\sigma_2\sqrt{(1 - \rho^2)}} \exp\left\{ -\frac{1}{2(1 - \rho^2)}\left[\left(\frac{x_1 - \mu_1}{\sigma_1}\right)^2 \right.\right.$$

$$\left.\left. - 2\rho\left(\frac{x_1 - \mu_1}{\sigma_1}\right)\left(\frac{x_2 - \mu_2}{\sigma_2}\right) + \left(\frac{x_2 - \mu_2}{\sigma_2}\right)^2\right]\right\}.$$

The marginal distribution of X_1 is $N(\mu_1, \sigma_1^2)$ and hence $E(X_1) = \mu_1$, $V(X_1) = \sigma_1^2$.
Similarly, $E(X_2) = \mu_2$, $V(X_2) = \sigma_2^2$. The conditional distribution of X_1 given
$X_2 = x_2$ is $N\{[\mu_1 + \rho\sigma_1(x_2 - \mu_2)/\sigma_2], \sigma_1^2(1 - \rho^2)\}$ and hence the conditional
mean of X_1 is $\mu_1 + \rho\sigma_1(x_2 - \mu_2)/\sigma_2$ and the conditional variance $\sigma_1^2(1 - \rho^2)$.
There are corresponding values for the conditional distribution of X_2 given
$X_1 = x_1$.

For the problems any marginal or conditional probability density functions
for which the means and variances are known may be utilized.

Problem 1. If X_1, X_2 are discrete random variables with joint p.d.f.

$$f(x_1, x_2) = \frac{\lambda^{x_2} e^{-2\lambda}}{x_1!(x_2 - x_1)!}, \quad x_1 = 0, 1, \ldots, x_2,$$

$$x_2 = 0, 1, 2, \ldots, \lambda > 0$$

evaluate (a) $E_1(X_1|x_2)$, (b) $V(X_1|x_2)$, (c) $E_2(X_2|x_1)$, (d) $V(X_2|x_1)$. {See
Example 3, Chapter 8.}

Problem 2. If X_1, X_2 are discrete random variables with joint p.d.f.

$$f(x_1, x_2) = q^2 p^{x_2 - 2}, \quad 0 \leqslant x_1 \leqslant x_2 - 1, \quad 0 < p < 1,$$

$$x_2 = 2, 3, \ldots, \quad q = 1 - p$$

find (a) $E_1(X_1|x_2)$, (b) $V(X_1|x_2)$, (c) $E_2(X_2|x_1)$, (d) $V(X_2|x_1)$. {See solution
Problem 5, Chapter 8.}

Problem 3. If X, Y are continuous random variables with point p.d.f.

$$f(x, y) = \frac{3}{2x^3\,y^2}\,, \quad \frac{1}{x} \leqslant y \leqslant x, \quad 1 \leqslant x < \infty$$

show that

$$E(X|y) = \frac{2}{y} \text{ if } 0 < y < 1$$

$$= 2y \text{ if } 1 \leqslant y < \infty$$

but that $V(X|y)$ does not exist. {See solution to Problem 10, Chapter 8.}

12.2 EXPECTATIONS OF FUNCTIONS

Sums of functions

The following result covers a wide class of instances.

$$E[ag(X_1, X_2) + bh(X_1, X_2)] = aE[g(X_1, X_2)] + bE[h(X_1, X_2)]$$

$$(12.1)$$

providing the expectations exist. For in the continuous case,

$$aE[g(X_1, X_2)] + bE[h(X_1, X_2)]$$

$$= a\iint g(x_1, x_2)\,f(x_1, x_2)\,dx_1\,dx_2$$

$$+ b\iint h(x_1, x_2)\,f(x_1, x_2)\,dx_1\,dx_2$$

$$= \iint [ag(x_1, x_2) + bh(x_1, x_2)]\,f(x_1, x_2)\,dx_1\,dx_2$$

$$= E[ag(X_1, X_2) + bh(X_1, X_2)]\,.$$

The heart of the matter being that the 'sum of the integrals is the integral of the sum'. The same result holds for discrete distributions.

If $g(X_1, X_2) = X_1$, $h(X_1, X_2) = X_2$, then in particular

$$E(aX_1 + bX_2) = aE(X_1) + bE(X_2) \tag{12.2}$$

$$E(X_1 + X_2) = E(X_1) + E(X_2). \tag{12.3}$$

Nothing in this section assumes anything as to the independence of the random variables.

Products of functions

Theorem 12.1. If $Y_1 = g(X_1, X_2)$, $Y_2 = h(X_1, X_2)$ are independent random variables, then $E(Y_1, Y_2) = E(Y_1)\,E(Y_2)$ or equivalently

$$E[g(X_1, X_2)\,h(X_1, X_2)] = E[g(X_1, X_2)]\,E[h(X_1, X_2)]\,.$$

For suppose the joint p.d.f. of Y_1, Y_2 is $\phi(y_1, y_2)$ then, since they are independent, $\phi(y_1, y_2) = \phi_1(y_1)\phi_2(y_2)$. Hence,

$$E(Y_1\ Y_2) = \iint y_1 y_2\ \phi(y_1, y_2)\,dy_2\,dy_1$$

$$= \iint y_1 y_2\ \phi(y_1)\,\phi_2(y_2)\,dy_2\,dy_1$$

$$= \int y_1 \left[\int y_2 \phi_2(y_2)\,dy_2\right]\phi_1(y_1)\,dy_1$$

$$= \int y_1 [E(Y_2)]\phi_1(y_1)\,dy_1$$

$$= E(Y_2)\int y_1\ \phi_1(y_1)\,dy_1 = E(Y_2)\,E(Y_1).$$

The converse of this result is neither asserted nor need be true. If X_1, X_2 are *independent*, then also

$$E[g(X_1)\,h(X_2)] = E[g(X_1)]\,E[h(X_2)]\,.$$

In particular,

$$E(X_1 X_2) = E(X_1)\,E(X_2).$$

12.3 COVARIANCE

We next consider a measure of association between two random variables. The covariance of X_1, X_2, written $C(X_1, X_2)$, is defined as

$$C(X_1, X_2) = E\big\{[X_1 - E(X_1)][X_2 - E(X_2)]\big\}. \tag{12.4}$$

This expression may be expanded as

$$E[X_1\ X_2 - X_2 E(X_1) - X_1 E(X_2) + E(X_1)E(X_2)]$$

$$= E(X_1\ X_2) - E(X_2)E(X_1) - E(X_1)E(X_2) + E(X_1)E(X_2)$$

$$= E(X_1\ X_2) - E(X_1)E(X_2). \tag{12.5}$$

We note at once that if X_1, X_2 are independent, then $E(X_1 X_2) = E(X_1)E(X_2)$ and hence $C(X_1, X_2) = 0$. The converse need not be true, that is if $C(X_1, X_2) = 0$, it does not follow that X_1, X_2 are independent.

Problem 4. X is uniformly distributed over $(-1, +1)$, $Y = X^2$. Show that $C(X, Y) = 0$ although X, Y are not independent. ∎

For continuous distributions we evaluate $E(X_1 X_2)$ as

$$\iint x_1 x_2\ f(x_1, x_2)\,dx_1\,dx_2$$

and for discrete distributions as

$$\sum_{x_1} \sum_{x_2} x_1 x_2 f(x_1, x_2).$$

It is seen that $C(X_1, X_2) = C(X_2, X_1)$ and that $C(X_1, X_1) = V(X_1)$.

Example 4

The continuous random variables X_1, X_2 have a joint p.d.f.

$$f(x_1, x_2) = 2, \quad 0 < x_1 < x_2 < 1$$
$$= 0 \text{ otherwise.}$$

$$\begin{aligned}
E(X_1 X_2) &= \int_0^1 \int_{x_1}^1 x_1 x_2 \, 2 dx_1 \, dx_2 \\
&= \int_0^1 x_1 \left(\int_{x_1}^1 2x_2 \, dx_2 \right) dx_1 \\
&= \int_0^1 x_1 (1 - x_1^2) \, dx_1 \\
&= \left[\frac{x_1^2}{2} - \frac{x_1^4}{4} \right]_0^1 = \frac{1}{4}.
\end{aligned}$$

We have already evaluated $E(X_1) = 1/3$, $E(X_2) = 2/3$, thus $C(X_1, X_2) = 1/36$.

Problem 5. Verify Example 1 by integrating with respect to x_1 first.

Example 5

X_1, X_2 have the trinomial distribution with parameters n, p_1, p_2.

$$\begin{aligned}
E(X_1 X_2) &= \sum_{0 \leqslant x_1 + x_2 \leqslant n} \sum x_1 x_2 \frac{n!}{x_1! x_2! (n - x_1 - x_2)!} \\
&\quad \times p_1^{x_1} p_2^{x_2} (1 - p_1 - p_2)^{n - x_1 - x_2} \\
&= \sum_{2 \leqslant x_1 + x_2 \leqslant n} \sum \frac{n!}{(x_1 - 1)! (x_2 - 1)! (n - x_1 - x_2)!} \\
&\quad \times p_1^{x_1} p_2^{x_2} (1 - p_1 - p_2)^{n - x_1 - x_2} \\
&= n(n-1) p_1 p_2 \sum_{2 \leqslant x_1 + x_2 \leqslant n} \sum \frac{(n-2)!}{(x_1 - 1)! (x_2 - 1)! (n - x_1 - x_2)!} \\
&\quad \times p_1^{x_1 - 1} p_2^{x_2 - 1} (1 - p_1 - p_2)^{n - 2 - (x_1 - 1) - (x_2 - 1)} \\
&= n(n-1) p_1 p_2 [p_1 + p_2 + (1 - p_1 - p_2)]^{n-2} \\
&= n(n-1) p_1 p_2.
\end{aligned}$$

Hence, $C(X_1, X_2) = n(n-1) p_1 p_2 - n p_1 n p_2 = -n p_1 p_2$.

Problem 6. X_1, X_2 have the trinomial distribution with parameters n, p_1, p_2, by evaluating $E(X_1 X_2)$ as $E\{X_2[E_1(X_1|X_2)]\}$, show that $C(X_1, X_2) = -np_1 p_2$. {Use results on the trinomial distribution from Chapter 8.}

Problem 7. X_1, X_2 have the bivariate normal distribution. Use the independence of $X_2 - \rho(\sigma_2/\sigma_1)X_1$ and X_1 to show that $C(X_1, X_2) = \rho\sigma_1 \sigma_2$. {Use results on the bivariate normal distribution from Chapter 8.}

12.4 CORRELATION COEFFICIENT

If the units are changed, so does the covariance. This defect is removed by considering instead, the **product moment correlation coefficient**, $\rho(X_1, X_2)$, which is defined as

$$\rho(X_1, X_2) = \frac{C(X_1, X_2)}{\sqrt{[V(X_1) V(X_2)]}} \tag{12.6}$$

provided both $V(X_1) > 0$, $V(X_2) > 0$. It will be seen that $\rho(X_1, X_2)$ is, in fact, the covariance of the random variables $Z_1 = (x_1 - \mu_1)/\sqrt{[V(X_1)]}$, $Z_2 = (X_2 - \mu_2)/\sqrt{[V(X_2)]}$. Such random variables with zero means and unit standard deviations are said to be standardized.

We next derive an unexpected property of the correlation coefficient, namely $-1 \leqslant \rho(X_1, X_2) \leqslant +1$. For consider the random variable

$$[\lambda(X_1 - \mu_1) + (X_2 - \mu_2)]^2$$

which, for real λ, is non-negative. Hence this random variable must have a non-negative expectation. Thus

$$E[\lambda(X_1 - \mu_1) + (X_2 - \mu_2)]^2 \geqslant 0$$

$$E[\lambda^2(X_1 - \mu_1)^2 + 2\lambda(X_1 - \mu_1)(X_2 - \mu_2) + (X_2 - \mu_2)^2] \geqslant 0$$

$$\lambda^2 E(X_1 - \mu_1)^2 + 2\lambda E[(X_1 - \mu_1)(X_2 - \mu_2)] + E(X_2 - \mu_2)^2 \geqslant 0$$

$$V(X_1)\lambda^2 + 2C(X_1, X_2)\lambda + V(X_2) \geqslant 0.$$

Now this expression is a quadratic form in λ and since the coefficient of λ^2 is positive we must have

$$C^2(X_1, X_2) \leqslant V(X_1)V(X_2) \quad \text{or} \quad \rho^2(X_1, X_2) \leqslant 1 \tag{12.7}$$

Example 6
For the trinomial distribution we have obtained

$$V(X_1) = np_1 q_1, \quad V(X_2) = np_2 q_2, \quad C(X_1, X_2) = -np_1 p_2,$$

hence

$$\rho(X_1, X_2) = -\sqrt{(p_1 p_2 / q_1 q_2)}.$$

Example 7

For the bivariate normal distribution we have obtained $V(X_1) = \sigma_1^2$, $V(X_2) = \sigma_2^2$, $C(X_1, X_2) = \rho\sigma_1\sigma_2$, hence $\rho(X_1, X_2) = \rho$. That is, the parameter ρ is the correlation coefficient (appropriately enough!).

In the following problems, if the marginal distribution of a random variable is recognized, its mean and variance may be quoted.

Problem 8. X, Y are continous random variables whose joint p.d.f. is

$$f(x, y) = 6x, \quad 0 < x < y < 1$$

$$= 0 \text{ otherwise.}$$

Find $E(X), E(Y), V(X), V(Y), \rho(X, Y)$.

Problem 9. X, Y are continuous random variables whose point p.d.f. is

$$f(x, y) = \frac{(m + n + 2)!}{m!\,n!} (1 - x)^n y^m, \quad 0 \leqslant y \leqslant x \leqslant 1.$$

By considering $E[(1 - X)Y]$, or otherwise, show that

$$C(X, Y) = \frac{(m + 1)(n + 1)}{(m + n + 3)^2 (m + n + 4)}$$

12.5 RELATION BETWEEN CORRELATION COEFFICIENT AND REGRESSION CURVES

We begin with a certain minimal property of the regression curves. Suppose that X_1, X_2 have a joint continuous distribution and knowing that $X_2 = x_2$ we wish to predict the (undisclosed!) value of X_1. To do this we can choose any function $R(x_2)$ and the discrepancy in the prediction will be $x_1 - R(x_2)$. A good choice of $R(.)$ will be such that the expected value of some function of the discrepancy is as small as possible. The classical choice of function has been the square of the discrepancy. Thus we require

$$E[X_1 - R(X_2)]^2$$

to be a minimum. But,

$$E[X_1 - R(X_2)]^2 = E\ E_1\{[X_1 - R(X_2)]^2 \,|\, X_2\}$$

$$= E\ E_1\{[X_1 - R(X_2)]^2 \,|\, X_2\}$$

and $E_1\{[X_1 - R(X_2)]^2 \,|\, X_2 = x_2\}$ is minimized when $R(x_2) = E_1(X_1|x_2)$, (see section 10.1).

An interesting further result is obtained if we consider only those functions $R(x_2) = a + bx_2$. We have

$$E(X_1 - a - bX_2)^2 = E[X_1 - \mu_1 - b(X_2 - \mu_2) + (\mu_1 - a - b\mu_2)]^2,$$

where $E(X_1) = \mu_1$ and $E(X_2) = \mu_2$. After expanding and taking expectations we arrive at

$$\sigma_1^2 + b^2 \sigma_2^2 - 2b\rho\sigma_1\sigma_2 + (\mu_1 - a - b\mu_2)^2,$$

$$= \sigma_1^2(1 - \rho^2) + (\rho\sigma_1 - b\sigma_2)^2 + (\mu_1 - a - b\mu_2)^2,$$

where $\sigma_1^2 = V(X_1)$, $\sigma_2^2 = V(X_2)$, $\rho = \mathrm{corr}(X_1, X_2)$. Thus $E(X_1 - a - bX_2)^2$ is minimized with respect to choice of a, b if

$$b = \rho\sigma_1/\sigma_2,$$

$$a = \mu_1 - b\mu_2.$$

Thus the 'best straight line' predictor of x_1 for any x_2 is

$$a + bx_2 = \mu_1 + \rho\sigma_1(x_2 - \mu_2)/\sigma_2,$$

and with this choice, the expected value of the square of the discrepancy collapses to $\sigma_1^2(1 - \rho^2)$. We observe that $\sigma_1^2(1 - \rho^2)$ becomes neglible as $|\rho|$ approaches one, in which case $E(X_1 - a - bX_2)^2$ is small, that is, X_1 is close to $a + bX_2$. It is in this sense that ρ is sometimes viewed as a measure of the strength of the linear association between X_1 and X_2.

We conclude this section with a useful result which connects the variance of a random variable to its conditional mean and variance given some other random variable. Suppose then that X_1, X_2 have a joint distribution such that their variances exist.

$$V(X_1) = E[X_1 - E(X_1)]^2$$

$$= E[X_1 - E_1(X_1|X_2) + E_1(X_1|X_2) - E(X_1)]^2$$

$$= E\{[X_1 - E(X_1|X_2)]^2 + [E_1(X_1|X_2) - E(X_1)]^2\},$$

since the cross-product term vanishes. Since $E[E_1(X_1|X_2)] = E(X_1)$, we may write

$$E\{[X_1 - E(X_1|X_2)]^2\} = E\,E_1\{[X_1 - E(X_1|X_2)]^2 |X_2\}$$

$$= E\,V(X_1|X_2)$$

so that

$$V(X_1) = E[V(X_1|X_2)] + V[E_1(X_1|X_2)]. \quad (12.8)$$

Notice that the 'unconditional' $V(X_1)$ is *not* just the expected value of the conditional variance. Our first illustrative example will scarcely strike the reader as a recommendation for the formula, though the stages of the calculation are easily verified.

Example 8

The variance of X_1 in Example 1 is reconsidered in the light of the formula (12.8). We have $E_1(X_1|X_2) = X_2/2, E_1(X_1^2|X_2) = X_2^2/3$. Hence

$$V(X_1) = E\left[\frac{X_2^2}{3} - \left(\frac{X_2}{2}\right)^2\right] + V\left(\frac{X_2}{2}\right).$$

Unfortunately the evaluation of the alternative expression requires more computation than tackling $V(X_1)$ directly! The formula is seen to rather better advantage in the next example.

Example 9
X_1 has the binomial distribution with parameters N, p_1 where N has the geometric distribution with parameter p_2. From known results for these standard distributions,

$$V(X_1|n) = np_1q_1, \quad E(X_1|n) = np_1.$$

Hence

$$V(X_1) = E(Np_1q_1) + V(Np_1)$$

$$= p_1q_1E(N) + p_1^2 V(N)$$

$$= \frac{p_1q_1}{p_2} + \frac{p_1^2 q_2}{p_2^2}.$$

Problem 10. Verify formula (12.8) for Problem 1, using the partial results there made available.

Problem 11. X, Y are continuous random variables with joint probability density function (p.d.f.) given by

$$f_{X,Y}(x, y) = 12xy \quad \text{within} \quad A, 0 \text{ outside } A, \tag{1}$$

where A is the area bounded by $y = 0, x = 1$ and $y = x^2$.

(i)   Show that the regression of Y on X is given by

$$E(Y|X = x) = \frac{2}{3}x^2. \tag{2}$$

and that the conditional variance of Y given $X = x$ increases from 0 at $x = 0$ to $1/18$ at $x = 1$. Why must this conditional variance necessarily be zero at $x = 0$?

(ii)   Prove that the 'best' linear prediction relationship for Y given $X = x$ (i.e. best in the sense of minimizing the expected squared error of prediction) is

$$Y_{\text{pred}} = \frac{28}{27}x - \frac{7}{18} \tag{3}$$

$$\left(\text{You may use the formula } Y_{\text{pred}} = E(Y) + \frac{\text{cov}(X, Y)}{\text{var } X} [x - E(X)].\right)$$

Comment on the relative merits of equations (2) and (3) for predicting Y from an X-value between 0 and 3/8.

(iii)   In equation (1), $12xy$ factorizes into (function of x) $\times$ (function of y); explain why this does *not* imply that X, Y are independent.

University of Hull, 1978.

12.6 MOMENT GENERATING FUNCTIONS: BIVARIATE DISTRIBUTIONS

The moments of a bivariate distribution are defined by

$$v'_{j,k} = E(X_1^j X_2^k) \tag{12.9}$$

and for the corresponding moments about the means,

$$v_{j,k} = E[(X_1 - \mu_1)^j(X_2 - \mu_2)^k], \tag{12.10}$$

where $\mu_1 = E(X_1) = v'_{1,0}, \mu_2 = E(X_2) = v'_{0,1}$.

In this notation, $V(X_1) = v_{2,0}$, $V(X_2) = v_{0,2}$, $C(X_1, X_2) = v_{1,1}$. We now proceed to a brief discussion of the corresponding m.g.f.

Definition. If X_1, X_2 have a joint distribution with p.d.f. $f(x_1, x_2)$, then their m.g.f. $M_{X_1,X_2}(t_1, t_2)$ is defined as

$$M_{X_1,X_2}(t_1, t_2) = E(e^{X_1 t_1 + X_2 t_2})$$

provided this expectation exists for values of t_1, t_2 in an interval which includes $t_1 = 0, t_2 = 0$. For convenience, the suffixes X_1, X_2 may be omitted.

$$M(t_1, t_2) = \iint e^{x_1 t_1 + x_2 t_2} f(x_1, x_2) \, dx_1 \, dx_2$$

in the continuous case and

$$M(t_1, t_2) = \sum_{x_1} \sum_{x_2} e^{x_1 t_1 + x_2 t_2} f(x_1, x_2)$$

in the discrete case. We first show that this function is appropriate in the sense that there are methods for recovering the moments.

Method of expanding the m.g.f.

$$E(e^{X_1 t_1 + X_2 t_2}) = E\left[\sum_{r=0}^{\infty} \frac{(X_1 t_1 + X_2 t_2)^r}{r!}\right]$$

$$= \sum_{r=0}^{\infty} \left[E \frac{(X_1 t_1 + X_2 t_2)^r}{r!}\right]$$

$$
= \sum_{r=0}^{\infty} \left\{ E \frac{\displaystyle\sum_{j=0}^{r} \binom{r}{j} (X_1^j X_2^{r-j} t_1^j t_2^{r-j})}{r!} \right\}
$$

$$
= \sum_{r=0}^{\infty} \left\{ \sum_{j=0}^{r} \frac{\left[\binom{r}{j} t_1^j t_2^{r-j} E(X_1^j X_2^{r-j}) \right]}{r!} \right\}
$$

$$
= \sum_{r=0}^{\infty} \left[\sum_{j=0}^{r} \left(\frac{v'_{j,r-j}}{j!(r-j)!} t_1^j t_2^{r-j} \right) \right]. \qquad (12.11)
$$

Hence the coefficient of $(t_1^j t_2^{r-j})/[j!(r-j)!]$ is the **mixed moment** $v'_{j,r-j}$.

Differentiating the m.g.f.
Consider differentiating $M(t_1, t_2)$, k times with respect to t_2, then j times with respect to t_1 and finally setting $t_1 = t_2 = 0$. It is clear that the only non-zero term will be that involving $t_1^j t_2^k$ in the expansion of $M(t_1, t_2)$. But

$$
\frac{\partial^{j+k}}{\partial t_1^j \, \partial t_2^k} \left(\frac{v'_{j,k}}{j!k!} t_1^j t_2^k \right) = v'_{j,k}.
$$

Hence,

$$
\left\{ \frac{\partial^{j+k}}{\partial t_1^j \, \partial t_2^k} [M_{X_1,X_2}(t_1, t_2)] \right\}_{t_1=t_2=0} = v'_{j,k}.
$$

Example 10
The bivariate normal distribution. We employ a technique to avoid performing the double integration. We have seen for this distribution that although with $\rho \neq 0$, X_1, X_2 are dependent, the random variables X_2, $X_1 - \rho\sigma_1 X_2/\sigma_2$ are independent and normally distributed.

$$
\begin{aligned}
M(t_1, t_2) &= E(e^{X_1 t_1 + X_2 t_2}) \\
&= E[e^{(X_1 - \rho\sigma_1 X_2/\sigma_2)t_1 + X_2(t_2 + \rho\sigma_1 t_1/\sigma_2)}] \\
&= E[e^{(X_1 - \rho\sigma_1 X_2/\sigma_2)t_1} e^{X_2(t_2 + \rho\sigma_1 t_1/\sigma_2)}] \\
&= E[e^{(X_1 - \rho\sigma_1 X_2/\sigma_2)t_1}] E[e^{X_2(t_2 + \rho\sigma_1 t_1/\sigma_2)}],
\end{aligned}
$$

using the independence just remarked.
We now that for the normal distribution $N(\mu, \sigma^2)$, $M(t) = e^{\mu t + (1/2)\sigma^2 t^2}$. But the marginal distribution of X_2 is $N(\mu_2, \sigma_2^2)$. Hence,

$$
\begin{aligned}
E[e^{X_2(t_2 + \rho\sigma_1 t_1/\sigma_2)}] &= M_{X_2}(t_2 + \rho\sigma_1 t_1/\sigma_2) \\
&= e^{\mu_2(t_2 + \rho\sigma_1 t_1/\sigma_2) + (1/2)\sigma_2^2(t_2 + \rho\sigma_1 t_1/\sigma_2)^2}.
\end{aligned}
$$

$$
(12.12)
$$

$X_1 - \rho\sigma_1 X_2/\sigma_2$ is distributed $N[\mu_1 - \rho\sigma_1\mu_2/\sigma_2, \sigma_1^2(1-\rho^2)]$, hence

$$E[e^{(X_1 - \rho\sigma_1 X_2/\sigma_2)t_1}] = e^{(\mu_1 - \rho\sigma_1\mu_2/\sigma_2)t_1 + (1/2)\sigma_1^2(1-\rho^2)t_1^2}.$$

(12.13)

Finally, combining the results of (12.12) and (12.13), and after a little cancelling,

$$M_{X_1,X_2}(t_1, t_2) = e^{\mu_1 t_1 + \mu_2 t_2 + (1/2)\sigma_1^2 t_1^2 + (1/2)\sigma_2^2 t_2^2 + \rho\sigma_1\sigma_2 t_1 t_2}$$

$$= e^{Q(t_1, t_2)}.$$

(12.14)

We notice that if $\rho = 0$, then $M_{X_1,X_2}(t_1, t_2) = M_{X_1}(t_1)M_{X_2}(t_2)$. The second and third terms of the expansion of $M(t_1, t_2)$ are

$$\frac{Q(t_1, t_2)}{1!} + \frac{[Q(t_1, t_2)]^2}{2!}$$

from which we readily pick out the coefficients of $t_1, t_2, t_1^2/2!, t_2^2/2!, t_1 t_2/1!1!$ as $\mu_1, \mu_2, \sigma_1^2 + \mu_1^2, \sigma_2^2 + \mu_2^2, \rho\sigma_1\sigma_2 + \mu_1\mu_2$ which are the moments $\nu'_{1,0}, \nu'_{0,1}, \nu'_{2,0}, \nu'_{0,2}, \nu'_{1,1}$.

Alternatively, in order to evaluate ν'_{11} by differentiation

$$\frac{\partial^2 M(t_1, t_2)}{\partial t_1 \partial t_2} = \frac{\partial}{\partial t_1}[(\mu_2 + \sigma_2^2 t_2 + \rho\sigma_1\sigma_2 t_1)M(t_1, t_2)]$$

$$= M(t_1, t_2)(\rho\sigma_1\sigma_2) + (\mu_2 + \sigma_2^2 t_2 + \rho\sigma_1\sigma_2 t_1)$$

$$\times (\mu_1 + \sigma_1^2 t_1 + \rho\sigma_1\sigma_2 t_2)M(t_1, t_2).$$

Now set $t_1 = t_2 = 0$, and since $M(0, 0) = 1$, we are left with $\rho\sigma_1\sigma_2 + \mu_1\mu_2$.

Problem 12. By differentiating the m.g.f., verify that $\nu'_{1,0} = \mu_1, \nu'_{2,0} = \sigma_1^2 + \mu_1^2$ for the bivariate normal distribution. ∎

The last example provides us with a generalization. We have

$$M_{a_1 X_1, a_2 X_2}(t_1, t_2) = M_{X_1,X_2}(a_1 t_1, a_2 t_2) = \exp(\mu_1 a_1 t_1 + \mu_2 a_2 t_2$$

$$+ (1/2)\sigma_1^2 a_1^2 t_1^2 + (1/2)\sigma_2^2 a_2^2 t_2^2$$

$$+ \rho\sigma_1\sigma_2 a_1 a_2 t_1 t_2).$$

(12.15)

We deduce that $a_1 X_1, a_2 X_2$ also have a bivariate normal distribution. Furthermore, if we put $t_1 = t_2 = t$ then

$$M_{X_1,X_2}(a_1 t, a_2 t) = e^{(a_1\mu_1 + a_2\mu_2)t + (1/2)(a_1^2\sigma_1^2 + 2\rho a_1 a_2 \sigma_1 \sigma_2 + a_2^2\sigma_2^2)t^2}.$$

(12.16)

But this is the m.g.f. of a normal distribution with parameters,

$$\mu = a_1\mu_1 + a_2\mu_2, \sigma = (a_1^2\sigma_1^2 + 2\rho a_1 a_2 \sigma_1 \sigma_2 + a_2^2\sigma_2^2)^{1/2}.$$

Moreover,

$$M_{X_1,X_2}(a_1 t, a_2 t) = E(e^{a_1 X_1 t + a_2 X_2 t})$$

$$= M_{a_1 X_1 + a_2 X_2}(t).$$

Hence the distribution of $a_1 X_1 + a_2 X_2$ is

$$N[a_1\mu_1 + a_2\mu_2, (a_1^2\sigma_1^2 + 2\rho a_1 a_2 \sigma_1 \sigma_2 + a_2^2\sigma_2^2].$$

Thus if X_1, X_2 have the bivariate normal distribution, any linear combination $a_1 X_1 + a_2 X_2$ has a normal distribution.

Problem 13. X_1, X_2 have the trinomial distribution with parameters, n, p_1, p_2. Show that

$$M_{X_1,X_2}(t_1, t_2) = [p_1 e^{t_1} + p_2 e^{t_2} + (1 - p_1 - p_2)]^n$$

and verify that $\nu'_{1,0} = np_1, \nu'_{2,0} = n(n-1)p_1^2 + np_1, \nu'_{1,1} = n(n-1)p_1 p_2$. Find the distribution of $X_1 + X_2$.

12.7 BIVARIATE MOMENT GENERATING FUNCTIONS AND INDEPENDENCE

We conclude this chapter with a useful result relating independence of random variables and factorization of their joint m.g.f, when this exists.

Theorem 12.2. Let X_1, X_2 have joint m.g.f. $M(t_1, t_2)$, then X_1, X_2 are independent if and only if

$$M(t_1, t_2) = M(t_1, 0) M(0, t_2). \tag{12.17}$$

For if X_1, X_2 are independent,

$$M(t_1, t_2) = E(e^{X_1 t_1 + X_2 t_2}) = E(e^{X_1 t_1} e^{X_2 t_2})$$

$$= E(e^{X_1 t_1}) E(e^{X_2 t_2}).$$

In the continuous case,

$$E(e^{X_1 t_1}) = \iint e^{(x_1 t_1)} f(x_1, x_2)\, dx_1\, dx_2$$

$$= \iint \lim_{t_2 \to 0} e^{(x_1 t_1 + x_2 t_2)} f(x_1, x_2)\, dx_1\, dx_2$$

and, assuming that the limiting operation and integration can be interchanged,

$$E(e^{X_1 t_1}) = \lim_{t_2 \to 0} \iint e^{(x_1 t_1 + x_2 t_2)} f(x_1, x_2)\, dx_1\, dx_2$$

$$= \lim_{t_2 \to 0} [M(t_1, t_2)] = M(t_1, 0).$$

Similarly, $E(e^{X_2 t_2}) = M(0, t_2)$. Conversely, if $M(t_1, t_2) = M(t_1, 0) M(0, t_2)$,

$$\iint e^{(x_1 t_1 + x_2 t_2)} f(x_1, x_2) \, dx_1 \, dx_2$$

$$= \left[\iint e^{(x_1 t_1)} f(x_1, x_2) \, dx_1 \, dx_2 \right] \left[\iint e^{(x_2 t_2)} f(x_1, x_2) \, dx_1 \, dx_2 \right]$$

$$= \left[\int e^{(x_1 t_1)} f_1(x_1) \, dx_1 \right] \left[\int e^{(x_2 t_2)} f_2(x_2) \, dx_2 \right]$$

$$= \iint e^{(x_1 t_1 + x_2 t_2)} f_1(x_1) f_2(x_2) \, dx_1 \, dx_2 .$$

But since m.g.f.s are unique, this implies that

$$f(x_1, x_2) = f_1(x_1) f_2(x_2),$$

which is a sufficient condition for X_1, X_2 to be independent.

Problem 14. X_1, X_2 are independent random variables, each distributed $N(0, 1)$. Show that the m.g.f. of the random variable $X_1 - X_2, X_1 + X_2$ is $e^{t_1^2 + t_2^2}$.

Problem 15. The discrete random variables X_1, X_2 have joint p.d.f.

$$f(x_1, x_2) = \frac{\lambda^{x_2} e^{-2\lambda}}{x_1! (x_2 - x_1)!}, \quad \begin{aligned} & x_1 = 0, 1, \ldots, x_2, \\ & x_2 = 0, 1, 2, \ldots \end{aligned}$$

Show that the m.g.f. of X_1, X_2 is

$$M(t_1, t_2) = e^{-2\lambda + \lambda(\exp(t_2) + \exp(t_1 + t_2))}$$

Hence recover the distributions of X_1, X_2.

BRIEF SOLUTIONS AND COMMENTS ON THE PROBLEMS

Problem 1. From Example 3, Chapter 8, the distribution of X_1 given $X_2 = x_2$ is binomial with parameters x_2 and $1/2$. Hence $E_1(X_1 | x_2) = x_2/2$ and $V(X_1 | x_2) = x_2/4$. From the same example, the conditional p.d.f. of X_2 given $X_1 = x_1$ is

$$[\lambda^{x_2 - x_1} e^{-\lambda}] / (x_2 - x_1)!], \quad x_2 \geqslant x_1.$$

By inspection we see that the distribution of $X_2 - x_1$ given $X_1 = x_1$ is Poisson with parameter λ. Hence $E_2[X_2 - x_1 | x_1] = \lambda \Rightarrow E_2[X_2 | x_1] = \lambda + x_1$; and $V[X_2 | x_1] = V[X_2 - x_1 | x_1] = \lambda$.

Problem 2. From the solution to Problem 5, Chapter 8, the p.d.f. of the conditional distribution of X_1 given $X_2 = x_2$ is $1/(x_2 - 1)$ for $1 \leqslant x_1 \leqslant x_2 - 1$. Hence

$$E_1(X_1|x_2) = \sum_1^{x_2-1} x_1/(x_2-1) = (x_2/2)(x_2-1)/(x_2-1) = x_2/2.$$

For $V(X_1|x_2)$, we need the well known identity

$$\sum_1^k r^2 = k(k+1)(2k+1)/6.$$

We have,

$$E_1[X_1^2|x_2] = \sum_1^{x_2-1} x_1^2/(x_2-1) = x_2(2x_2-1)/6$$

and $V(X_1|x_2) = x_2(x_2-2)/12$. The p.d.f. of X_2 given $X_1 = x_1$ is $qp^{x_2-x_1-1}$ for $x_2 \geqslant x_1 + 1$. Thus, the distribution of $X_2 - x_1$ given x_1 is geometric with parameter q. $E(X_2|x_1) = x_1 + 1/q$, $V(X_2|x_1) = p/q^2$.

Problem 3. When $0 < y < 1$, the conditional p.d.f. of X given y is $2/x^3y^2$ for $x \geqslant 1/y$.

$$E(X|y) = \int_{1/y}^{\infty} \frac{2}{x^2y^2}\,dx = 2/y,$$

and $E(X^2|y)$ does not exist. On the other hand if $1 < y < \infty$, the conditional p.d.f. of X given y is $2y^2/x^3$, for $x \geqslant y$.

$$E(X|y) = \int_y^{\infty} \frac{2y^2}{x^2}\,dx = 2y$$

and again $E(X^2|y)$ does not exist. See Problem 10, Chapter 8.

Problem 4

$$E(XY) = E(X^3) = \int_{-1}^{+1} x^3\,dx = 0$$

and $E(X) = 0$, hence $C(X, Y) = 0$, though Y is a function of X.

Problem 5

$$\int_0^1 \left(\int_0^{x_2} 2x_1x_2\,dx_1 \right) dx_2 = \int_0^1 x_2^3\,dx_2 = \frac{1}{4}.$$

Problem 6. X_1 given x_2 has the binomial distribution with parameters $n - x_2$ and $p_1/(1-p_2)$. X_2 has the binomial distribution with parameters n and p_2. Hence

$$E(X_1 | x_2) = (n - x_2)p_1/(1 - p_2)$$

$$E(X_2(n - X_2)p_1/(1 - p_2)) = [E(nX_2) - E(X_2^2)]p_1/(1 - p_2)$$

$$= \{n^2 p_2 - [np_2(1 - p_2) + n^2 p_2^2]\} p_1/(1 - p_2)$$

$$= n(n - 1)p_1 p_2,$$

so that

$$C(X_1, X_2) = n(n - 1)p_1 p_2 - np_1 np_2 = -np_1 p_2.$$

Problem 7. X_1 is $N(\mu_1, \sigma_1^2) \Rightarrow E(X_1) = \mu_1$, $V(X_1) = \sigma_1^2$.

$$E[(X_2 - \rho\sigma_2 X_1/\sigma_1)X_1] = E(X_2 X_1) - \rho\sigma_2 E(X_1^2)/\sigma_1.$$

$$E(X_2 - \rho\sigma_2 X_1/\sigma_1) = \mu_2 - \rho\sigma_2\mu_1/\sigma_1.$$

Now use

$$E[(X_2 - \rho\sigma_2 X_1/\sigma_1)X_1] = E(X_2 - \rho\sigma_2 X_1/\sigma_1)E(X_1)$$

to obtain result (see Chapter 8 for the independence of $X_2 - \rho\sigma_2 X_1/\sigma_1$ and X_1).

Problem 8. All the integrations are straightforward and we merely list the main results.

$$f_1(x) = 6x(1 - x), \quad 0 < x < 1. \quad E(X) = 1/2, \quad V(X) = 1/20.$$

$$f_2(y) = 3y^2, \quad 0 < y < 1. \quad E(Y) = 3/4, \quad V(Y) = 3/80.$$

$$E(XY) = 6 \int_0^1 \left[\int_0^y x^2 y \, dx \right] dy = 2/5, \quad C(X, Y) = 1/40,$$

hence

$$\rho(X, Y) = 1/\sqrt{3}.$$

Problem 9. Note that $f(x, y)$ is a p.d.f. for integral m, n.

$$E[(1 - X)Y] = \int_0^1 \left[\int_0^x \frac{(m + n + 2)!}{m! n!} (1 - x)^{n+1} y^{m+1} \, dy \right] dx$$

$$= \frac{(n + 1)(m + 1)}{(n + m + 4)(n + m + 3)},$$

after 'building up' the integrand. $E[(1 - X)Y] = E(Y) - E(XY)$. Show

$$E(Y) = (m + 1)/(n + m + 3), \quad E(X) = (m + 2)/(n + m + 3),$$

and then evaluate $E(XY) - E(X)E(Y)$.

Problem 10. X_1 given $X_2 = x_2$ has binomial distribution with parameters x_2 and $1/2$.

$E(X_1|x_2) = x_2/2$, hence $V[E(X_1|X_2)] = V(X_2/2) = V(X_2)/4 = \lambda/2$, since X_2 has a Poisson disttribution with parameter 2λ. Also

$$V(X_1|x_2) = x_2(1/2)(1/2) = x_2/4, \quad \text{and} \quad E(X_2/4) = \lambda/2.$$

Thus sum is λ and *is* $V(X_1)$ since X_1 has a Poisson distribution with parameter λ.

Problem 11.

$$f_1(x) = \int_0^{x^2} 12xy \, dy = 6x^5, \quad 0 \leqslant x \leqslant 1.$$

$$f(y|x) = (12xy)/6x^5 = 2y/x^4, \quad 0 < y \leqslant x^2.$$

$$E(Y|x) = \int_0^{x^2} (2y^2/x^4) \, dy = 2x^2/3.$$

$$E(Y^2|x) = \int_0^{x^2} (2y^3/x^4) \, dy = x^4/2,$$

hence $V(Y|x) = x^2/18$.

When $x = 0, f_1(0) = 0$, hence conditional distribution has no spread of probability. As an alternative to the given formula for Y_{pred}, the reader may care to consider

$$E[(Y - ax - b)^2 |x]^2$$
$$= V(Y|x) + [E(Y|x) - ax - b]^2$$
$$= \frac{1}{18}x^4 + \left[\frac{2}{3}x^2 - ax - b\right]^2.$$

Now average this over the distribution of X and minimize with respect to a, b.

For $0 < x < 3/8$, we have $Y_{\text{pred}} < 0$, which does not appear to be desirable!

It is easily shown that $f(y|x) \neq f_2(y)$, so that X, Y cannot be independent. In any case, the region in which $f(x, y) > 0$ is not a rectangle.

Problem 12.

$$\frac{\partial M(t_1, t_2)}{\partial t_1} = (\mu_1 + \sigma_1^2 t_1 + \rho \sigma_1 \sigma_2 t_2)M(t_1, t_2),$$

$$\frac{\partial^2 M(t_1, t_2)}{\partial t_1^2} = \sigma_1^2 M(t_1, t_2) + (\mu_1 + \sigma_1^2 t_1 + \rho \sigma_1 \sigma_2 t_2)\frac{\partial M(t_1, t_2)}{\partial t_1}.$$

Put $t_1 = t_2 = 0$ to obtain required result.

Problem 13

$$M(t_1, t_2) = E[\exp(X_1 t_1 + X_2 t_2)]$$

$$= \sum_{x_2=0}^{n} \sum_{x_1=0}^{n-x_2} \frac{n!}{x_1! x_2! (n - x_1 - x_2)!}$$

$$(p_1 e^{t_1})^{x_1} (p_2 e^{t_2})^{x_2} (1 - p_1 - p_2)^{n-x_1-x_2}$$

$$= [p_1 e^{t_1} + p_2 e^{t_2} + (1 - p_1 - p_2)]^n = [G(t_1, t_2)]^n.$$

$$\frac{\partial M(t_1, t_2)}{\partial t_1} = np_1 e^{t_1} G^{n-1}, \quad v'_{1,0} = np_1$$

$$\frac{\partial^2 M(t_1, t_2)}{\partial t_1^2} = np_1 e^{t_1} G^{n-1} + n(n-1)p_1^2 e^{2t_1} G^{n-2},$$

$$v'_{2,0} = np_1 + n(n-1)p_1^2.$$

$$\frac{\partial^2 M(t_1, t_2)}{\partial t_2 \partial t_1} = np_1 e^{t_1} (n-1)p_2 e^{t_2} G^{n-2}, \quad v'_{1,1} = n(n-1)p_1 p_2.$$

Put $t_1 = t_2 = t$, then $M(t, t) = [(p_1 + p_2)e^t + (1 - p_1 - p_2)]^n$ which is the m.g.f. of the binomial distribution with parameters n and $p_1 + p_2$.

Problem 14. Let $Y_1 = X_1 - X_2$, $Y_2 = X_1 + X_2$.

$$M_{Y_1, Y_2}(t_1, t_2) = E[e^{(X_1 - X_2)t_1 + (X_1 + X_2)t_2}]$$

$$= E[e^{X_1(t_2 + t_1) + X_2(t_2 - t_1)}]$$

$$= E[e^{X_1(t_2 + t_1)}] E[e^{X_2(t_2 - t_1)}],$$

since X_1, X_2 independent,

$$= M_{X_1}(t_1 + t_2) M_{X_2}(t_2 - t_1)$$

$$= e^{(t_2 + t_1)^2/2} e^{(t_2 - t_1)^2/2} = e^{t_1^2 + t_2^2} = e^{t_1^2} e^{t_2^2}.$$

As a bonus we recover Y_1, Y_2 are independent and each is distributed $N(0, 2)$.

Problem 15

$$M(t_1, t_2) = e^{-2\lambda} \sum_{x_2=0}^{\infty} \frac{\lambda^{x_2} e^{x_2 t_2}}{x_2!} \left[\sum_{x_1=0}^{x_2} \binom{x_2}{x_1} e^{t_1 x_1} \right]$$

$$= e^{-2\lambda} \sum_{x_2=0}^{\infty} \frac{\lambda^{x_2} e^{x_2 t_2}}{x_2!} (1 + e^{t_1})^{x_2}$$

$$= e^{-2\lambda} e^{\lambda \exp(t_2)[1 + \exp(t_1)]}.$$

$M(0, t_2) = e^{-2\lambda} e^{2\lambda \exp(t_2)}$, hence the distribution of X_2 is Poisson with parameter 2λ.

$M(t_1, 0) = e^{-\lambda} e^{\lambda \exp(t_1)}$, hence the distribution of X_1 is Poisson with parameter λ. Be it noted that X_1, X_2 are not independent since

$$M(t_1, 0) M(0, t_2) \neq M(t_1, t_2).$$

13

Probability Generating Functions

13.1 INTRODUCTION

In the case of discrete random variables, which have positive probability only on the non-negative integers, we can construct a probability generating function.

Definition. If $\Pr(X = x) = f(x), x = 0, 1, 2, \ldots$, then the **probability generating function**, $G_X(t)$, is

$$G_X(t) = E(t^X) = \sum_{x=0}^{\infty} t^x f(x). \tag{13.1}$$

Example 1
$f(x) = 1/6, x = 1, 2, \ldots, 6.$

$$G_X(t) = \sum_{1}^{6} t^x/6 = (t - t^7)/[(1 - t)6].$$

Example 2

$$f(x) = \binom{n}{x} p^x q^{n-x}, \quad x = 0, 1, \ldots, n, \quad q = 1 - p.$$

$$G_X(t) = \sum_{0}^{n} \binom{n}{x} (pt)^x q^{n-x} = (q + pt)^n. \qquad \blacksquare$$

In the two examples above, the existence of the probability generating function (p.g.f.) was not in doubt since only a finite number of terms entered the sum. However, $G(t)$ always exists for $|t| < 1$. This is because for a discrete distribution $f(x) \leqslant 1$ and hence

$$|G(t)| \leqslant \sum_1^\infty |t|^x$$

which is convergent for $|t| < 1$. In fact for a proper distribution,

$$G(1) = \sum_0^\infty f(x) = 1$$

and $G(t)$ exists for $|t| \leqslant 1$. Indeed, for some distributions, the radius of convergence will exceed 1. Subject to some restriction on the domain of the dummy variable t, we can say that the probability generating function $G(t)$ always exists. This is more than could be claimed for moment generating functions.

Problem 1. If X has the Poisson distribution with parameter λ, show that $G_X(t) = e^{\lambda t - \lambda}$.

Problem 2. If X has the geometric distribution with parameter p, show that $G_X(t) = pt/(1 - qt)$, $|t| < 1/q$ where $q = 1 - p$.

13.2 EVALUATION OF MOMENTS

A probability generating function can be used to obtain any moments which exist. Thus since a power series may be differentiated term by term within its radius of convergence.

$$G'(t) = \frac{\mathrm{d}}{\mathrm{d}t}\left[\sum t^x f(x)\right]$$

$$= \sum \frac{\mathrm{d}}{\mathrm{d}t}\,[t^x f(x)]$$

$$= \sum xt^{x-1}f(x),$$

at least for $|t| < 1$. Now to say $E(X)$ exists implies that $\sum xf(x)$ is convergent and so

$$E(X) = \lim_{t \to 1}\left[\sum xt^{x-1}f(x)\right] = G'(1). \tag{13.2}$$

Example 3

For the binomial distribution with $G(t) = (q + pt)^n$, we have

$$G'(t) = np(q + pt)^{n-1} \quad \text{and} \quad G'(1) = np.$$

Problem 3. Use the probability generating function to calculate $E(X)$ for:

(a)   the Poisson distribution with parameter λ,
(b)   the geometric distribution with parameter p.

Example 4

For the binomial distribution, $G(t) = (q + pt)^n$, hence

$$G''(t) = n(n-1)p^2(q + pt)^{n-2} \quad \text{and} \quad G''(1) = n(n-1)p^2.$$

We can readily find the variance since

$$V(X) = E(X^2) - E^2(X) = E[X(X-1)] + E(X) - E^2(X)$$
$$= n(n-1)p^2 + np - (np)^2 = np(1-p).$$

13.3 SUMS OF INDEPENDENT RANDOM VARIABLES

If X_1, X_2 are independent random variables then

$$E(t^{X_1 + X_2}) = E(t^{X_1} t^{X_2}) = E(t^{X_1}) E(t^{X_2}).$$

If further, X_1, X_2 are discrete and assume only non-negative integral values with positive probability, an equivalent formulation is

$$G_{X_1 + X_2}(t) = G_{X_1}(t) G_{X_2}(t). \tag{13.3}$$

This result can be used to identify the distribution of sums of such independent random variables.

Example 5

X_1, X_2 have independent Poisson distributions with parameters λ_1, λ_2.

$$G_{X_1 + X_2}(t) = G_{X_1}(t) G_{X_2}(t) = e^{\lambda_1 t - \lambda_1} e^{\lambda_2 t - \lambda_2}$$
$$= e^{(\lambda_1 + \lambda_2)t - (\lambda_1 + \lambda_2)}.$$

But this is the probability generating function of another Poisson distribution with parameter $\lambda_1 + \lambda_2$.

Problem 4. X_1, X_2, ... X_n are independent random variables each having the geometric distribution with parameter p. Find the probability distribution of

$$\sum_1^n X_i.$$

Problem 5. If X_1 is binomial n_1, p, and X_2 is binomial n_2, p show that if X_1, X_2 are independent then $X_1 + X_2$ has the binomial distribution with parameters $n_1 + n_2$ and p.

Example 6
A fair dice has its sides numbered one to six. If it is rolled twice then the probability of a total score of, say five, is easily found by enumeration. The favourable outcomes are $(1, 4), (4, 1), (3, 2), (2, 3)$, and the required probability is $4/6^2 = 1/9$. If, however the dice is rolled thrice and we seek the probability of a total score of twelve, then the method of enumeration becomes more arduous and we might be grateful for the assistance of a probability generating function approach. For each roll,

$$G(t) = \sum_1^6 t^x/6.$$

Hence for the total of three rolls, is

$$\left[\sum_1^6 t^x/6 \right]^3$$

$$= \frac{1}{6^3} \left(\frac{t - t^7}{1 - t} \right)^3$$

$$= t^3 (1 - t^6)^3 \, (1 - t)^{-3}/6^3.$$

The coefficient of t^{12} is easily found to be $(55 - 30)/6^3 = 25/216$.

Problem 6. If X has the binomial distribution with parameters n and p, find the limiting form of $G_X(t)$ if $n \to \infty$ and $p \to 0$ in such a way that np remains constant and equal to λ.

Problem 7. The random variable R can take all non-negative integer values and its probability generating function is

$$G(t) = p_0 + p_1 t + p_2 t^2 + \dots$$

(a)   Show that $E(R)$ is $G'(1)$ and obtain a formula for $V(R)$ in terms of values of the derivatives of $G(t)$.

(b)   Show that the probability that R is an odd integer is

$$[1 - G(-1)]/2$$

(c)   An unbiased six-sided die is thrown repeatedly. Show that the probability generating function for the number of throws needed to obtain a six is

$$t/(6 - 5t).$$

Write down the probability generating function for the number of throws needed to obtain two sixes.

(d) Two players throw this die in turn. The player who throws the *second* six to appear is the winner. By using (b) and (c) or otherwise, find the probability that the game is won by the player who made the first throw.

Oxford & Cambridge, A.L., 1981.

Problem 8. The probability generating function for the random variable R is $G(t) = p_0 + p_1 t + \ldots$. Find $G(1)$ and show that the mean and variance of R are $G'(1)$ and $G''(1) + G'(1) - [G'(1)]^2$ respectively. If $q_r = \Pr(R > r)$ for all r, show that

$$(1 - t) \sum_{r=0}^{\infty} q_r t^r = 1 - G(t).$$

If the random variable M is defined to be $\min(r_1, r_2 \ldots, r_k)$ where $r_1, r_2, \ldots, r_k$ are k independent values of R, prove that

$$\Pr(M = r) = (q_{r-1})^k - (q_r)^k.$$

If $\Pr(R = r) = \alpha \theta^r$ where α and θ are constants such that $0 < \theta < 1$, find α and q_r in terms of θ and show that

$$G(t) = \frac{(1 - \theta)}{1 - \theta t}.$$

Give as simple an expression as you can for the probability generating function for M, where R is as given, find the mean value and the variance of M.

Oxford & Cambridge, A.L., 1979.

Problem 9. In Chapter 5, we obtained the following differential equations for the random stream of events:

$$p_n'(t) = \lambda_{n-1}(t) - \lambda p_n(t), \quad n \geqslant 0, \quad p_{-1}(t) \equiv 0,$$

where $p_n(t)$ was the probability of just n events in $(0, t)$. If

$$G(t) = \sum_{n=0}^{\infty} p_n(t) \theta^n,$$

show that

$$\frac{\partial G}{\partial t} = \lambda(\theta - 1)G.$$

Hence show that $G = e^{\lambda(\theta - 1)t}$.

13.4 USE OF RECURRENCE RELATIONS

Calculating the probability of an event by enumerating all the possible sequences of outcomes in a series of trials is not always feasible. Sometimes it is possible to relate the probability of an event on a particular trial to the corresponding probabilities on previous trials. In that case a probability generating function may be of great service, though our first examples would scarcely warrant such an elaborate process.

Example 7
In a series of n independent trials, the probability of a success on each trial is p. It is required to find the probability of just r successes by relating this event to the outcome of the first trial. Let $\phi_n(r)$ be the required probability, then since the first trial is a success with probability p or a failure with probability $q(= 1 - p)$ we have, for $n \geqslant 2$

$$\phi_n(r) = p\phi_{n-1}(r-1) + q\phi_{n-1}(r), \quad 1 \leqslant r \leqslant n.$$

If

$$G_n(t) = \sum_{r=0}^{n} \phi_n(r)t^r,$$

multiplying by t^r and summing r from 1 to n, we have

$$G_n(t) - \phi_n(0) = tp \sum_{1}^{n} \phi_{n-1}(r-1)t^{r-1} + q \sum_{1}^{n} \phi_{n-1}(r)t^r$$

$$= tp\, G_{n-1}(t) + q[G_{n-1}(t) - \phi_{n-1}(0)].$$

But $\phi_n(0) = q\phi_{n-1}(0)$, hence

$$G_n(t) = (pt + q)G_{n-1}(t), \quad n \geqslant 2.$$

By repeated application, we arrive at

$$G_n(t) = (pt + q)^{n-1}G_1(t).$$

Evidently, $G_1(t) = \phi_1(0) + \phi_1(1)t = q + pt$ and we conclude that

$$G_n(t) = (pt + q)^n.$$

Problem 10. In a sequence of independent trials the probability of a success on each trial is p. By considering the outcome of the first trial show that $G_r(t)$, the probability generating function of the number of trials required to achieve the rth success, satisfies

$$G_r(t) = ptG_{r-1}(t) + qtG_r(t),$$

and hence obtain $G_r(t)$. ∎

Our next example illustrates more forcibly the power of the method.

Example 8

In a sequence of independent trials, the probability of a success on each trial is p. It is required to find the probability generating function for the first trial on which the second of two consecutive successes is obtained.

Let ϕ_i be the probability of this event on trial i, then clearly $\phi_1 = 0, \phi_2 = p^2$ and $\phi_3 = qp^2$ where $q = 1 - p$. When $k \geqslant 4$, the *last three* trials must result in a failure followed by two successes, with probability qp^2. The previous $k - 3$ trials must not display the event and this is the case with probability $1 - \phi_1 - \phi_2 \ldots - \phi_{k-3}$. Hence

$$\phi_k = \left[1 - \sum_{i=1}^{k-3} \phi_i \right] qp^2, \quad k \geqslant 4.$$

We obtain the probability generating function, $G(t)$, for the ϕ_k by multiplying by t^k and summing for $k \geqslant 4$.

$$\sum_{4}^{\infty} \phi_k t^k = qp^2 \left[\sum_{4}^{\infty} t^k - \sum_{k=4}^{\infty} t^k \sum_{i=1}^{k-3} \phi_i \right],$$

$$= qp^2 \left[\sum_{4}^{\infty} t^k - \sum_{i=1}^{\infty} \phi_i \left(\sum_{k=i+3}^{\infty} t^k \right) \right]$$

$$= qp^2 \left[\frac{t^4}{1-t} - \sum_{i=1}^{\infty} \phi_i \frac{t^{i+3}}{1-t} \right], \quad |t| < 1,$$

That is

$$G(t) - \phi_3 t^3 - \phi_2 t^2 = qp^2 \left[\frac{t^4}{1-t} - \frac{t^3 G(t)}{1-t} \right],$$

and after collecting terms,

$$G(t) = p^2 t^2 (1 - pt)/(1 - t + t^3 qp^2)$$

$$= p^2 t^2 /(1 - qt - pqt^2).$$

The reader should verify that $G(1) = 1$ and that the expected number of trials needed to obtain two consecutive successes is $(1 + p)/p^2$. The next problem is concerned with the nuts and bolts of Example 8 for a particular value of p.

Problem 11. For Example 8 calculate ϕ_i when $p = 2/3$.

13.5 COMPOUND DISTRIBUTIONS

Suppose that the number of carriers of a disease has a Poisson distribution and that the numbers of persons infected by each carrier have independent Poisson distributions. Then the total number of persons infected is the sum of a random number of random variables. More precisely let $X_1, X_2, \ldots$ be independent discrete random variables and

$$Y = \sum_1^N X_i$$

where N is an integer-valued random variable, independent of the X_i, then Y is said to have a compound distribution.

If the p.g.f. of each X_i is $G(t)$, then when $N = n$, the p.g.f. of

$$\sum_1^n X_i \quad \text{is} \quad \prod_1^n G(t) = [G(t)]^n.$$

Now suppose $\Pr(N = n) = \phi_n$, then

$$E(t^Y) = \sum_1^\infty E(t^Y | n)\phi_n$$

$$= \sum_1^\infty [G(t)]^n \phi_n. \tag{13.4}$$

But if the p.g.f. of N is $H(t)$, the form of (13.4) is $H[G(t)]$.

Example 9
X_i has a Poisson distribution with parameter λ and N has a Poisson distribution with parameter θ

$$G_X(t) = e^{-\lambda + \lambda t}$$

$$H_N(t) = e^{-\theta + \theta t}.$$

Hence

$$Y = \sum_1^N X_i$$

has p.g.f. $H[G(t)] = e^{-\theta + \theta \, [\exp(-\lambda + \lambda t)]}$.

Problem 12. The number of insect colonies in an area has a Poisson distribution with parameter λ. The probability that any particular colony contains just $k(\geqslant 1)$ insects is $p^{k-1}(1 - p)$ where $0 < p < 1$. If the numbers in different colonies are independent of each other, show that the probability generating function of the total number of insects present in the area is

$$e^{[\lambda(1-p)t/(1-pt)]} e^{-\lambda}.$$

Problem 13. Suppose that each X_i, $i = 1, 2, \ldots$ has p.g.f. $G(t)$ and N has p.g.f. $H(t)$, then from (13.4),

$$Y = \sum_1^N X_i$$

has p.g.f. $H[G(t)]$. Show that, if they exist,

(a) $E(Y) = E(N) E(X)$;
(b) $V(Y) = V(N) E^2(X) + E(N) V(X)$.

13.6 BIVARIATE PROBABILITY GENERATING FUNCTIONS

Suppose that X_1, X_2 have a joint discrete distribution but only assume non-negative integral values with positive probability. Then their joint probability generating function, $G(t_1, t_2)$, is defined as

$$G(t_1, t_2) = E(t_1^{X_1} t_2^{X_2}). \tag{13.5}$$

It is clear that $G(t_1, 1)$, $G(1, t_2)$ are the p.g.f.s for the marginal distributions of X_1, X_2.

Example 10
X_1, X_2 have the trinomial distribution with parameters n, p_1, p_2. Thus the p.d.f. of X_1, X_2 is

$$\frac{n!}{x_1!x_2!(n-x_1-x_2)!} p_1^{x_1} p_2^{x_2} (1-p_1-p_2)^{n-x_1-x_2},$$

$$0 \leqslant x_1 + x_2 \leqslant n,$$

and

$$E(t_1^{X_1} t_2^{X_2})$$

$$= \sum_{x_1=0}^{n} \sum_{x_2=0}^{n-x_1} \frac{n!}{x_1!x_2!(n-x_1-x_2)!}$$

$$(p_1 t_1)^{x_1} (p_2 t_2)^{x_2} (1-p_1-p_2)^{n-x_1-x_2}$$

$$= [p_1 t_1 + p_2 t_2 + (1-p_1-p_2)]^n.$$

We note that $G(t_1, 1) = [p_1 t_1 + (1-p_1)]^n$, $G(1, t_2) = [p_2 t_2 + (1-p_2)]^n$ which confirms that the marginal distributions of X_1, X_2 are binomial with parameters n, p_1 and n, p_2 respectively. Moreover

$$G(t, t) = [(p_1 + p_2)t + (1 - p_1 - p_2)]^n,$$

and hence the distribution of $X_1 + X_2$ is binomial with parameters $n, p_1 + p_2$.

Problem 14. The joint p.d.f. of X_1, X_2 is

$$f(x_1, x_2) = \frac{\lambda^{x_2} e^{-2\lambda}}{x_1!(x_2 - x_1)!}, \quad 0 \leqslant x_1 \leqslant x_2, \quad x_2 = 0, 1, 2, \ldots.$$

Show that the joint p.g.f. of X_1, X_2 is

$$G(t_1, t_2) = e^{-2\lambda + \lambda t_2 + \lambda t_1 t_2}.$$

Hence derive the distributions of (1) X_1, (2) X_2, (3) $X_2 - X_1$.

Problem 15. Show that the coefficient of $\alpha^x \beta^y$ in the expansion of

$$\prod(\alpha, \beta) = e^{\lambda(\alpha-1) + \mu(\beta-1) + \nu(\alpha-1)(\beta-1)} \tag{1}$$

in powers of α and β is

$$e^{-\lambda - \mu - \nu} \sum_{j=0}^{k} \frac{\nu^j}{j!} \frac{(\lambda - \nu)^{x-j}}{(x-j)!} \frac{(\mu - \nu)^{y-j}}{(y-j)!},$$

where $k = \min(x, y)$, and hence that, when

$$\lambda \geqslant \nu, \quad \mu \geqslant \nu, \quad \nu \geqslant 0. \tag{2}$$

$\prod(\alpha, \beta)$ satisfies the conditions to be a bivariate probability generating function (p.g.f.).

The discrete random variables X, Y have $\prod(\alpha, \beta)$ as their bivariate p.g.f., with λ, μ, ν satisfying the conditions (2). Show that the marginal distributions of X and Y are Poisson with respective means λ and μ.

Show that the conditional p.g.f. of Y given $X = x$ is

$$\frac{\text{coefficient of } \alpha^x \text{ in } \prod(\alpha, \beta)}{\text{coefficient of } \alpha^x \text{ in } \prod(\alpha, 1)}$$

$$= \left[1 + \frac{\nu}{\lambda}(\beta - 1)\right]^x e^{(\mu - \nu)(\beta - 1)}.$$

Comment on this result.

Part question, University of Hull, 1978.

13.7 GENERATING SEQUENCES

The methods presented in this chapter may be employed more widely. We shall say that $H(t)$ is a generating function for the sequence $a_0, a_1, \ldots$ if

$$H(t) = a_0 + a_1 t + \ldots ,$$

and the series converges for some interval $|t| < t_0$. It is a generating function in the sense that the coefficient of t^n is a_n.

Example 11
(i) $H(t) = 1/(1 - t) = 1 + t + t^2 + \ldots , |t| < 1$, so that $1/(1 - t)$ generates the sequence $1, 1, 1, \ldots$.
(ii) $H(t) = 1/(1 - t^2) = 1 + t^2 + t^4 + \ldots , |t| < 1$, so that $1/(1 - t^2)$ generates the sequence $1, 0, 1, 0, \ldots$.
(iii) $H(t) = \exp(t) = 1 + t + t^2/2! \ldots$ so that $\exp(t)$ generates the sequence $1, 1, 1/2!, 1/3!, \ldots$.

Naturally for such general sequences, characteristics such as mean and variance have no relevance. However, there are some sequences related to discrete random variables which can usefully be derived from generating functions.

Example 12
Suppose X has probability generating function $G(t)$ and we wish to find a function $H(t)$ which generates the sequence $h(x) = \Pr(X > x)$. Now

$$\Pr(X = x) = h(x - 1) - h(x), \quad x \geqslant 1,$$

hence

$$G(t) = \sum_{x=0}^{\infty} t^x \Pr(X = x)$$

$$= \sum_{1}^{\infty} t^x [h(x - 1) - h(x)] + \Pr(X = 0)$$

$$= t \sum_{1}^{\infty} t^{x-1} h(x - 1) - \sum_{1}^{\infty} t^x h(x) + \Pr(X = 0)$$

$$= t H(t) - [H(t) - h(0)] + \Pr(X = 0)$$

$$= (t - 1) H(t) + \Pr(X > 0) + \Pr(X = 0)$$

$$= (t - 1) H(t) + 1.$$

That is,

$$H(t) = [1 - G(t)]/(1 - t). \tag{13.6}$$

Problem 16. In Example 12 show that if $\lim\limits_{t \to 1} H(t)$ exists then it is equal to $E(X)$.

Hence calculate $E(X)$ for the geometric dsitribution.

Example 13

In a sequence of independent trials, the probability of a success on each trial is p. Every time a trial is the second of two consecutive successes since the last such event (the sequence being assumed to restart after each such event) we shall say that the recurrent event ϵ has taken place. Thus the sequence $S\,F\,S\,S\,S\,S\,\ldots$ displays a first appearance of ϵ on trial 4 and a second on trial 6. We seek the probability, u_n, of ϵ on trial n. Suppose successes occur on trials $n-1$ and n, with probability p^2. Then either ϵ occurs on trial n, with probability u_n, or ϵ occurs on trial $n-1$ and is followed by a success, with probability $u_{n-1}\ p$. Hence

$$u_{n-1}p + u_n = p^2, \quad n \geqslant 2 \tag{13.7}$$

and $u_2 = p^2$, $u_1 = 0$. We form the generating function

$$H(t) \;=\; \sum_{2}^{\infty} u_n\, t^n$$

by multiplying equation (13.7) by t^n and summing over $n \geqslant 2$. This yields

$$ptH(t) + H(t) = p^2 \sum_{2}^{\infty} t^n = p^2 t^2/(1-t),$$

or

$$H(t) \;=\; \frac{p^2 t^2}{(1+pt)\,(1-t)}\,, \quad |t| < 1.$$

$H(t)$ may now, after splitting into partial fractions, be expanded as a power series. The reader should verify that the coefficient of t^n is

$$p^2 \left[1 + (-1)^n p^{n-1} \right]/(1+p), \quad n \geqslant 2.$$

Problem 17. For Example 13, if f_r is the probability that ϵ *first* appears on trial r, show that

$$u_n = \sum_{r=2}^{n-1} f_r u_{n-r} + f_n, \quad n \geqslant 3.$$

Hence deduce that if $G(t)$ is the probability generating function of $f_2, f_3, \ldots$, then

$$G(t) \;=\; H(t)/[1 + H(t)] = p^2 t^2/(1 - qt - qpt^2).$$

13.8 A NOTE ON DEFECTIVE DISTRIBUTIONS

We remarked that for a proper discrete distribution, the probability generating function, $G_X(t)$, satisfies $G_X(1) = 1$. By proper we here mean that

$$\sum \Pr(X = x) = 1.$$

If this property does not hold, then the distribution of X is said to be defective. Such a feature may be present even in seemingly simple situations. Suppose in a sequence of independent games, player A has probability p of winning any particular game, and the corresponding probability for player B is $q = 1 - p$. Player A resolves to quit immediately he has a positive gain on the basis of unit stake per game. It can be shown [1] that the generating function of the probabilities of quitting in an infinite sequence of games is

$$H(t) = [1 - (1 - 4pqt^2)^{1/2}]/2qt,$$

$$H(1) = (1 - |p - q|)/2q.$$

Hence if $p \geq q$, $|p - q| = p - q$ and $H(1) = 1$. If, however, $p < q$, $|p - q| = q - p$ and $H(1) = p/q$. Thus, if $p < 1$, there is a non-zero probability that player A never obtains a profit.

REFERENCE

[1]   W. Feller, *An Introduction to Probability Theory and its Applications*, Vol. I, John Wiley.

BRIEF SOLUTIONS AND COMMENTS ON THE PROBLEMS

Problem 1

$$\sum_{0}^{\infty} t^x \lambda^x e^{-\lambda}/x! = e^{-\lambda} \sum (\lambda t)^x/x! = e^{-\lambda} e^{\lambda t}.$$

Converges for all t.

Problem 2

$$G_X(t) = \sum_{1}^{\infty} t^x q^{x-1} p = pt \sum (qt)^{x-1} = pt/(1 - qt), \quad |qt| < 1,$$

here the radius of convergence exceeds 1 for $0 < p < 1$.

Problem 3
(a) $G(t) = \exp(\lambda t - \lambda), G'(t) = \lambda \exp(\lambda t - \lambda). G'(1) = \lambda.$
(b) $G(t) = pt/(1 - qt) = \{[1/(1 - qt)] - 1\}p/q.$ Hence
      $G'(t) = p/(1 - qt)^2$, hence $G'(1) = p/(1 - q)^2 = 1/p.$ Similarly, differentiating $G(t)$ twice,

$$G''(t) = \sum_{x=2}^{\infty} x(x - 1) t^{x-2} f(x)$$

and if $E(X^2)$ exists, then

$$G''(t) = \sum x(x-1) f(x) = E[X(X-1)].$$

Problem 4. The p.g.f. of each X_i is $pt/(1-qt)$. Hence the p.g.f. of

$$\sum_1^n X_i$$

is $[pt/(1-qt)]^n$, $|t| < 1/q$. We are to recover

$$\Pr\left[\sum_1^n X_i = r\right].$$

That is to say, we need the coefficient of t^r in the expansion of $p^n t^n [1-qt]^{-n}$ as a power series. Since $|qt| < 1$, we have

$$(1-qt)^{-n} = 1 + nqt + \frac{n(n+1)}{2!} q^2 t^2 \ldots$$

$$\frac{n(n+1)\ldots(n+k-1)}{k!} q^k t^k$$

and the coefficient of t^{r-n} in *this* expansion is

$$\frac{n(n+1)\ldots(n+r-n-1)}{(r-n)!} q^{r-n} = \binom{r-1}{n-1} q^{r-n}.$$

Hence

$$\Pr\left(\sum X_i = r\right) = \binom{r-1}{n-1} p^n q^{r-n}, \quad r = n, \quad n+1, \ldots.$$

Problem 5. $G_{X_1}(t) G_{X_2}(t) = (q+pt)^{n_1} (q+pt)^{n_2} = (q+pt)^{n_1+n_2}$ which is of the required form.

Problem 6

$$G_X(t) = (q+pt)^n = \left[1 - \frac{\lambda}{n} + \frac{\lambda}{n} t\right]^n$$

$$= \left[1 + \frac{\lambda(t-1)}{n}\right]^n \to e^{\lambda t - \lambda}.$$

The limit is the p.g.f. of a Poisson distribution with parameter λ.

Problem 7

Ans. (c) $H(t) = [t/(6-5t)]^2$.

Hint. Probability of first six on the rth throw is $(5/6)^{r-1}(1/6)$.

(d) $(1/2)[1-H(-1)]$.

Problem 8. Hint. $\Pr(M=r) = \Pr(M \geqslant r) - \Pr(M \geqslant r+1)$

$$= q_{r-1}^k - q_r^k.$$

$\alpha = (1-\theta)/(1-\alpha t), \quad q_r = \theta^{r+1}$.

Required p.g.f. $(1-\theta^k)/(1-\theta^k t)$.

Problem 9. Compare coefficients of θ^n both in $\partial G/\partial t$ and $\lambda(\theta-1)G$ and check that they are equal in virtue of the differential equation. Write the partial differential equation as

$$\frac{\partial}{\partial t}[\log_e G(t)] = \lambda(\theta-1),$$

and integrate with respect to time.

Problem 10. If $\phi_r(n)$ is the probability that the nth trials is the rth success, then either:

(a)   the first trial is a success and there remain $(n-1)$ trials to reach a further $(r-1)$ successes; or

(b)   the first trial is a failure and all r successes must be obtained in $(n-1)$ trials. Thus $\phi_r(n) = p\phi_{r-1}(n-1) + q\phi_r(n-1)$ for $2 \leqslant r \leqslant n$. Multiply equation by t^n and sum n from r to ∞

$$G_r(t) = ptG_{r-1}(t) + qtG_r(t),$$

$$\Rightarrow G_r(t) = \frac{qt}{1-qt} G_{r-1}(t)$$

$$= \left(\frac{pt}{1-qt}\right)^{r-1} G_1(t),$$

by repeated application.

But

$$G_1(t) = pt + qpt^2 + \ldots q^{k-1}pt^k \ldots$$

$$= pt/(1-qt),$$

so that

$$G_r(t) = [pt/(1-qt)]^r.$$

Problem 11. From Example 8,

$$G(t) = \frac{4t^2}{(3-2t)(3+t)}$$

$$= \frac{4t^2}{9} \left[\frac{2}{3-2t} + \frac{1}{3+t} \right]$$

$$= \frac{4t^2}{9} \left[\frac{2}{3} \left(1 - \frac{2t}{3}\right)^{-1} + \frac{1}{3} \left(1 + \frac{t}{3}\right)^{-1} \right].$$

We require the coefficient of t^i, which is

$$\frac{4}{9} \left[\frac{2}{3} \left(\frac{2}{3}\right)^{i-2} + \frac{1}{3} \left(-\frac{1}{3}\right)^{i-2} \right], \quad i \geqslant 2.$$

Problem 12. If N is the number of colonies, then the total number of insects is zero if $N = 0$. If $N > 0$, the total number of insects is

$$\sum_1^N X_i$$

where X_i is the number in the ith colony. $G(t) = t(1-p)/(1-pt)$. Now $\Pr(N = n) = \lambda^n e^{-\lambda}/n!$, and hence the required p.g.f. is

$$e^{-\lambda} + \sum_1^\infty [t(1-p)/(1-pt]^n \lambda^n e^{-\lambda}/n!$$

$$= e^{\lambda(1-p)t/(1-pt)} e^{-\lambda}.$$

Problem 13. If $K(t) = H[G(t)]$, then $K'(t) = H'[G(t)]G'(t)$, hence

$$K'(1) = H'[G(1)] G'(1) = H'(1) G'(1) = E(N) E(X).$$

Differentiating a second time with respect to t

$$K''(t) = H''[G(t)][G'(t)]^2 + H'[G(t)]G''(t),$$

hence

$$K''(1) = H''(1) [G'(1)]^2 + H'(1) G''(1)$$

$$= E[N(N-1))] [E(X)]^2 + E(N) E[X(X-1)].$$

But $K''(1) = E[Y(Y-1)]$. After collecting terms and using

$$V(N) = E[N(N-1)] + E(N) - [E(N)]^2$$

and similarly for $V(X)$, we obtain the result.

Problem 14

$$G(t_1, t_2) = \sum_{x_1=0}^{\infty} \sum_{x_2=x_1}^{\infty} \frac{\lambda^{x_2} e^{-2\lambda}}{x_1!(x_2-x_1)!} t_1^{x_1} t_2^{x_2}$$

$$= \sum_{x_1=0}^{\infty} \frac{e^{-2\lambda}}{x_1!} (\lambda t_2 t_1)^{x_1} \left\{ \sum_{x_2=x_1}^{\infty} \frac{(\lambda t_2)^{x_2-x_1}}{(x_2-x_1)!} \right\}$$

$$= e^{-2\lambda} \sum_{x_1=0}^{\infty} \frac{(\lambda t_2 t_1)^{x_1}}{x_1!} \; [e^{\lambda t_2}]$$

$$= e^{-2\lambda} e^{\lambda t_2 t_1} e^{\lambda t_2}$$

$$= e^{-2\lambda + \lambda t_2 + \lambda t_2 t_1}.$$

Now $G(t_1, 1) = e^{-\lambda + \lambda t_1}$ and this is the p.g.f. of a Poisson distribution with parameter λ. Also $G(1, t_2) = e^{-2\lambda + 2\lambda t_2}$, another Poisson distribution but with parameter 2λ. Finally we put $t_1 = 1/t_2$ when $E(t_1^{X_1} t_2^{X_2}) = E(t_2^{X_2-X_1})$. But $G(1/t_2, t_2) = e^{-\lambda + \lambda t_2}$. Hence $X_2 - X_1$ has a Poisson distribution with parameter λ.

Problem 15. $\Pi(\alpha, \beta) = e^{\alpha(\lambda - \nu + \nu\beta)} e^{\beta(\mu - \nu)} e^{-\lambda - \mu + \nu}$, coefficient of α^x in $e^{[\alpha(\lambda - \nu + \nu\beta)]} e^{[\beta(\mu-\nu)]}$ is

$$\frac{(\lambda - \nu + \nu\beta)^x}{x!} e^{\beta(\mu-\nu)},$$

in which the coefficient of β^y is

$$\frac{1}{x!} \sum_{j=0}^{k} \binom{x}{j} (\nu\beta)^j (\lambda - \nu)^{x-j} (\mu - \nu)^{y-j}/(y-j)!$$

and hence result. $\Pi(\alpha, 1) = e^{\lambda(\alpha-1)}$ and X is Poisson parameter λ, while $\Pi(1, \beta) = e^{\mu(\beta-1)}$ and Y is Poisson parameter μ. Conditional p.g.f. of Y given x is of the form

$$\sum_{y} \Pr(X=x \quad \text{and} \quad Y=y)\, \beta^y / \Pr(X=x) = \sum(\text{coefficient } \alpha^x \beta^y)\beta^y /$$
$$\text{coefficient of } \alpha^x$$

in $\Pi(\alpha, 1)$ and the numerator is the coefficient of α^x in $\Pi(\alpha, \beta)$.

Problem 16. If $\lim_{t\to 1} H(t)$ exists then so does $\lim_{t\to 1} [1 - G(t)]/(1 - t)$ which is $G'(1) = E(X)$. Hence

$$H(1) = \sum_{0}^{\infty} h(x).$$

If $\Pr(X = x) = q^{x-1}p$, then $\Pr(X > x) = q^x$ and

$$\sum_{x=0}^{\infty} q^x = 1/p.$$

Problem 17. If ϵ occurs on trial n, then if it appeared on trial r for the first time it is observed again $(n - r)$ trials later $(2 \leqslant r \leqslant n - 1)$. However, trial n may also be the first occasion. Hence the result follows. Multiply the equation by t^n, sum over $n \geqslant 3$, and interchange the order of summation. Thus

$$\sum_{n=3}^{\infty} u_n t^n = \sum_{n=3}^{\infty} \left(\sum_{r=2}^{n-1} f_r u_{n-r} \right) t^n + \sum_{n=3}^{\infty} f_n t^n,$$

$$H(t) - u_2 t^2 = \sum_{r=2}^{\infty} \left\{ t^r f_r \sum_{n=r+1}^{\infty} u_{n-r} t^{n-r} \right\} + G(t) - f_2 t^2,$$

$$H(t) - p^2 t^2 = G(t) H(t) + G(t) - p^2 t^2,$$

to obtain the result. Finally substitute for $H(t)$ from Example 13. Compare with the solution in Example 8 for the same problem.

14

Sums of Random Variables

14.1 VARIANCES AND COVARIANCES

Exploring the characteristics of a population or a distribution via a sample of values is an activity of critical importance in statistics. In particular, the sample average is used as a guide to the population mean and an aid to predicting the average of a future sample. The sample average is a random variable and its properties are described by its p.d.f. We shall first tackle the problem of finding the mean and variance of the distribution of the sample average, *assuming these to exist*.

If X_1, X_2 are random variables, using the result in equation (12.2),

$$E(X_1 + X_2) = E(X_1) + E(X_2),\tag{14.1}$$

$$E(aX_1 + bX_2) = aE(X_1) + bE(X_2)\tag{14.2}$$

whether or not X_1, X_2 are independent.

$$\begin{aligned}
V(aX_1 + bX_2) &= E[(aX_1 + bX_2) - E(aX_1 + bX_2)]^2\\
&= E\{a[X_1 - E(X_1)] + b[X_2 - E(X_2)]\}^2\\
&= E\{a^2[X_1 - E(X_1)]^2\\
&\quad + 2ab[X_1 - E(X_1)][X_2 - E(X_2)]\\
&\quad + b^2[X_2 - E(X_2)]\}
\end{aligned}$$

$$= a^2 E[X_1 - E(X_1)]^2 + 2abE\{[X_1 - E(X_1)]$$

$$[X_2 - E(X_2)]\} + b^2 E[X_2 - E(X_2)]^2$$

$$= a^2 V(X_1) + 2abC(X_1, X_2) + b^2 V(X_2). \tag{14.3}$$

If, further, X_1, X_2 are independent, $C(X_1, X_2) = 0$ and

$$V(aX_1 + bX_2) = a^2 V(X_1) + b^2 V(X_2). \tag{14.4}$$

In particular,

$$V(X_1 + X_2) = V(X_1 - X_2) = V(X_1) + V(X_2). \tag{14.5}$$

By repeated application of (14.1) and (14.5), if X_1, $X_2, \ldots, X_n$ are independent random variables,

$$E\left(\sum_{i=1}^{n} X_i\right) = \sum_{i=1}^{n} E(X_i), \tag{14.6}$$

$$V\left(\sum_{i=1}^{n} X_i\right) = \sum_{i=1}^{n} V(X_i). \tag{14.7}$$

Example 1

If X_1, X_2 are independent,

(a) $V(-X_1) = (-1)^2 V(X_1) = V(X_1)$.

(b) $V(X_1 + X_1) = V(2X_1) = 2^2 V(X_1) = 4V(X_1)$.

(c) $V(3X_1 - 4X_2) = 3^2 V(X_1) + 4^2 V(X_2) = 9V(X_1) + 16V(X_2)$.

(d) $V(3X_1 + 2) = 3^2 V(X_1) = 9V(X_1)$.

In the case where the X_i are independent and have the same distribution, then $E(X_i) = \mu$, $V(X_i) = \sigma^2$ for all i and

$$E\left(\sum_{i=1}^{n} X_i\right) = n\mu \tag{14.8}$$

$$V\left(\sum_{i=1}^{n} X_i\right) = n\sigma^2. \tag{14.9}$$

Since the sample mean

$$\bar{X} = \sum_{i=1}^{n} X_i / n,$$

$$E(\bar{X}) = \frac{1}{n} E\left(\sum_{i=1}^{n} X_i\right) = \mu, \tag{14.10}$$

$$V(\bar{X}) = V\left[\sum_{i=1}^{n} (X_i)/n\right] = \frac{1}{n^2} V\left(\sum_{i=1}^{n} X_i\right)$$

$$= \frac{n\sigma^2}{n^2} = \frac{\sigma^2}{n}.$$ (14.11)

The standard deviation of $\bar{X}$, $\sigma/\sqrt{n}$, is also known as the **standard error** of the mean.

As n increases, $V(\bar{X})$ decreases, hence the larger the sample drawn the more likely it is that the sample mean is near the population mean. More precisely, by Tchebychev's inequality, equation (10.5),

$$\Pr(|\bar{X} - \mu| \geqslant k) \leqslant \frac{V(\bar{X})}{k^2} = \frac{\sigma^2}{nk^2}.$$ (14.12)

Since σ^2/nk^2 tends to zero as n tends to infinity, the limit of the $\Pr(|\bar{X}-\mu| \geqslant k)$ also tends to zero. We also have a crude estimate of this probability given n.

Results for covariances similar to equation (14.3) for variances may also be obtained. Thus

$$C(aX_1 + bX_2, cY_1 + dY_2)$$
$$= E\{[aX_1 + bX_2 - E(aX_1 + bX_2)][cY_1 + dY_2 - E(cY_1 + dY_2)]\}$$
$$= E\{[aX_1 - aE(X_1) + bX_2 - bE(X_2)][cY_1 - cE(Y_1)$$
$$\qquad + dY_2 - dE(Y_2)]\}$$
$$= E\{ac[X_1 - E(X_1)][Y_1 - E(Y_1)] + ad[X_1 - E(X_1)][Y_2 - E(Y_2)]$$
$$\qquad + bc[X_2 - E(X_2)][Y_1 - E(Y_1)] +$$
$$\qquad + bd[X_2 - E(X_2)][Y_2 - E(Y_2)]\}$$
$$= acC(X_1, Y_1) + adC(X_1, Y_2) + bcC(X_2, Y_1) + bdC(X_2, Y_2).$$

(14.13)

The mode of expansion for the covariance of two linear combinations of random variables is evident from (14.13).

Example 2
If X_1, X_2 are independent,

(a) $C(X_1, X_1 + X_2) = C(X_1, X_1) + C(X_1, X_2) = V(X_1) + 0.$

(b) $C(3X_1 - 4X_2, 2X_1 + 5X_2) = C(3X_1, 2X_1) + C(3X_1, 5X_2)$
$$\qquad\qquad + C(-4X_2, 2X_1) + C(-4X_2, 5X_2)$$
$$\qquad\qquad = 6C(X_1, X_1) + 0 + 0 - 20C(X_2, X_2)$$
$$\qquad\qquad = 6V(X_1) - 20V(X_2).$$

(c) $C(X_1, \bar{X}) = C[X_1, (X_1 + X_2)/2] = C(X_1, X_1/2) = V(X_1)/2.$

Example 3
If $V(X_1) = V(X_2) = \sigma^2, C(X_1, X_2) = \sigma^2/2.$

(a) $C(3X_1 + 2, 2X_2 - 1) = C(3X_1, 2X_2) + C(3X_1, -1)$
$$+ C(2, 2X_2) + C(2, -1)$$
$$= 6C(X_1, X_2) + 0 + 0 + 0 = 3\sigma^2.$$

Thus constants may be 'dropped'.

(b) $C(X_1 + X_2, X_1 - X_2) = C(X_1, X_1) + C(X_1, -X_2) + C(X_2, X_1)$
$$+ C(X_2, -X_2)$$
$$= \sigma^2 - \sigma^2/2 + \sigma^2/2 - \sigma^2 = 0.$$

Problem 1. Show that

$$C\left(\sum_{i=1}^n a_i X_i, \sum_{j=1}^m b_j X_j\right) = \sum_{i=1}^n \sum_{j=1}^m a_i b_j C(X_i, X_j).$$

Problem 2. Show that

$$V\left(\sum_{i=1}^n a_i X_i\right) = \sum_{i=1}^n a_i^2 V(X_i) + \sum\sum_{i \neq j} a_i a_j C(X_i, X_j).$$

Problem 3. If $X_1, X_2, \ldots, X_n$ are such that $V(X_i) = \sigma^2$, $C(X_i, X_j) = \rho\sigma^2$, show that

$$V(\bar{X}) = \sigma^2\left(\frac{1}{n} + \frac{n-1}{n}\rho\right).$$

This result will apply in sampling a finite population *without* replacement.

14.2 SUMS OF INDEPENDENT RANDOM VARIABLES, DISTRIBUTIONS

The task of this section is to consider ways of evaluating the distribution of $X_1 + X_2$ where X_1, X_2 are independent random variables with known distributions. The first method to be considered does not require the existence of any moments and hence is very general, though sometimes tedious to apply.

Theorem 14.1. If X_1, X_2 are independent discrete random variables with p.d.f.s $f_1(x_1), f_2(x_2)$ then the p.d.f. of the random variable $Y = X_1 + X_2$ is $\phi(y)$, where

$$\phi(y) = \sum_{x_1} f_1(x_1) f_2(y - x_1) \tag{14.14}$$

where the summation is over those values of x_1 such that $f_1(x_1) > 0$ and $f_2(y - x_1) > 0$.

Proof

$$\phi(y) = \Pr(Y = y) = \Pr(X_1 + X_2 = y)$$

$$= \sum_{x_1} [\Pr(X_1 = x_1 \text{ and } X_2 = y - x_1)]$$

$$= \sum_{x_1} \Pr(X_1 = x_1) \Pr(X_2 = y - x_1),$$

since X_1, X_2 are independent,

$$= \sum_{x_1} f_1(x_1) f_2(y - x_1).$$

There is an important special case when X_1, X_2 have the same p.d.f., $f(x)$, when

$$\phi(y) = \sum_x f(x) f(y - x) \qquad (14.15)$$

Example 4

X_1, X_2 have independent Poisson distributions with parameters λ_1, λ_2. Thus

$$f_1(x_1) = \frac{\lambda_1^{x_1} e^{-\lambda_1}}{x_1!}, \quad x_1 = 0, 1, 2, \dots, f_2(x_2) = \frac{\lambda_2^{x_2} e^{-\lambda_2}}{x_2!},$$

$$x_2 = 0, 1, 2, \dots .$$

If $Y = X_1 + X_2$,

$$\phi(y) = \sum_{x_1} f_1(x_1) f_2(y - x_1)$$

where the limits of summation must be carefully considered, $f_1(x_1) > 0$ if $x_1 \geqslant 0$, while $f_2(y - x_1) \geqslant 0$ if $y - x_1 \geqslant 0$, that is $x_1 \leqslant y$. Both conditions are satisfied if $0 \leqslant x_1 \leqslant y$, and accordingly the summation should be from 0 to y inclusive.

$$\phi(y) = \sum_{x_1 = 0}^{y} \frac{\lambda_1^{x_1} e^{-\lambda_1}}{x_1!} \cdot \frac{\lambda_2^{y - x_1} e^{-\lambda_2}}{(y - x_1)!}$$

$$= \frac{e^{-(\lambda_1 + \lambda_2)}}{y!} \sum_{x_1 = 0}^{y} \frac{y!}{x_1! (y - x_1)!} \lambda_1^{x_1} \lambda_2^{y - x_1}$$

$$= \frac{e^{-(\lambda_1 + \lambda_2)}}{y!} (\lambda_1 + \lambda_2)^y.$$

Rather unexpectedly, we see that $\phi(y)$ is the p.d.f. of another Poisson distribution with parameter $\lambda_1 + \lambda_2$. By continued application, the sum of any number of independent Poisson distributions is again a Poisson distribution, with parameter equal to the sum of the separate parameters.

Example 5
X_1, X_2, X_3 have independent Poisson distributions with parameters $\lambda_1 = 1$, $\lambda_2 = 2$, $\lambda_3 = 3$ respectively. Calculate $\Pr(X_1 + X_2 + X_3 \geqslant 3)$. $X_1 + X_2 + X_3$ has a Poisson distribution with parameter $1 + 2 + 3 = 6$.

$$\Pr(X_1 + X_2 + X_3 \geqslant 3) = 1 - \Pr(X_1 + X_2 + X_3 \leqslant 2)$$
$$= 1 - (e^{-6} + 6e^{-6} + 6^2 e^{-6}/2)$$
$$= 1 - 25\, e^{-6}.$$

Suppose further that $X_1 + X_2 + X_3 = 2$, then,

$$\Pr(X_1 = 1 | X_1 + X_2 + X_3 = 2) = \frac{\Pr(X_1 = 1 \text{ and } X_1 + X_2 + X_3 = 2)}{\Pr(X_1 + X_2 + X_3 = 2)}$$
$$= \frac{\Pr(X_1 = 1 \text{ and } X_2 + X_3 = 1)}{\Pr(X_1 + X_2 + X_3 = 2)}$$
$$= \frac{\Pr(X_1 = 1) \Pr(X_2 + X_3 = 1)}{\Pr(X_1 + X_2 + X_3 = 2)}$$
$$= \frac{e^{-1}(5\,e^{-5})}{6^2\, e^{-6}/2!} = \frac{5}{18},$$

since X_1 is independent of $X_2 + X_3$, which has a Poisson distribution with parameter $2 + 3 = 5$.

Problem 4. X_1, X_2, X_3, X_4 have independent Poisson distributions with corresponding parameters $\lambda_1 = 1, \lambda_2 = 1, \lambda_3 = 1, \lambda_4 = 2$. Evaluate

(a) $\Pr(X_1 = 1, X_2 = 0, X_3 = 1, X_4 = 0 | X_1 + X_2 + X_3 + X_4 = 2)$.
(b) $\Pr(X_1 + X_2 = 1 | X_1 + X_2 + X_3 + X_4 = 2)$.

Problem 5. If X_1, X_2 have independent binomial distributions with parameters n_1, p and n_2, p respectively, show that $X_1 + X_2$ has also a binomial distribution with parameters $n_1 + n_2, p$. Deduce the sum of n independent random variables each having a Bernoulli distribution with parameter p.

Problem 6. If X_i has a Poisson distribution with parameter $\lambda_i, i = 1, 2, \ldots, n$ and these random variables are independent, show that the joint distribution of the X_i given that their sum is m, is multinomial with parameters $m, p_i = \lambda_i/\Sigma\lambda_i$.

Theorem 14.2. If X_1, X_2 are independent continuous random variables with p.d.f. $f_1(x_1)$ and $f_2(x_2)$, then the p.d.f. $\phi(y)$ of the random variable $Y = X_1 + X_2$ is

$$\phi(y) = \int f_1(x_1) f_2(y - x_1) \, dx_1. \qquad (14.16)$$

The integration over x_1 being for those intervals for which $f_1(x_1) > 0$ and $f_2(y - x_1) > 0$. Although Y, X_1 are *not* independent, their point p.d.f. $\phi(y, x_1)$ may be expressed as $\phi(y, x_1) = \phi(y|x_1) f_1(x_1)$. To obtain a suitable alternative expression for $\phi(y|x_1)$, we observe

$$
\begin{aligned}
\Pr(Y \leqslant y | X_1 = x_1) &= \Pr(X_1 + X_2 \leqslant y | X_1 = x_1) \\
&= \Pr(x_1 + X_2 \leqslant y | X_1 = x_1) \\
&= \Pr(X_2 \leqslant y - x_1 | X_1 = x_1) \\
&= \Pr(X_2 \leqslant y - x_1)
\end{aligned}
$$

since X_2, X_1 *are* independent.

Hence $\phi(y|x_1) = f_2(y - x_1)$. Finally,

$$\phi(y) = \int \phi(y, x_1) \, dx_1 = \int f_2(y - x_1) f_1(x_1) \, dx_1.$$

Example 6

X_1, X_2 have independent exponential distributions with parameters λ_1, λ_2. We have $f_1(x_1) = \lambda_1 e^{-\lambda_1 x_1}$, $x_1 \geqslant 0$ and $f_2(x_2) = \lambda_2 e^{-\lambda_2 x_2}$, $x_2 \geqslant 0$. As in example 1, we must have $x_1 \geqslant 0$ and $y - x_1 \geqslant 0$ or $0 \leqslant x_1 \leqslant y$. Hence

$$\phi(y) = \int_0^y \lambda_1 e^{-\lambda_1 x_1} \lambda_2 e^{-\lambda_2 (y - x_1)} \, dx_1$$

$$= \lambda_1 \lambda_2 e^{-\lambda_2 y} \int_0^y e^{x_1 (\lambda_2 - \lambda_1)} \, dx_1.$$

These are two cases to consider. If $\lambda_1 \neq \lambda_2$

$$\phi(y) = \lambda_1 \lambda_2 e^{-\lambda_2 y} \left[\frac{e^{x_1 (\lambda_2 - \lambda_1)}}{\lambda_2 - \lambda_1} \right]_0^y$$

$$= \frac{\lambda_1 \lambda_2}{\lambda_2 - \lambda_1} (e^{-\lambda_1 y} - e^{-\lambda_2 y}), \quad y \geqslant 0.$$

While if $\lambda_1 = \lambda_2$

$$\phi(y) = \lambda_1^2 e^{-\lambda_1 y} \int_0^y 1 \, dx_1$$

$$= \lambda_1^2 y \, e^{-\lambda_1 y}, \quad y \geqslant 0.$$

This last is in fact the p.d.f. of a gamma distribution with parameters $2, \lambda_1$.

Problem 7. X_1, X_2 have independent rectangular distributions over the interval $(0, l)$. Show that $Y = X_1 + X_2$ has p.d.f. $\phi(y) = y/l^2$ if $0 \leqslant y \leqslant l$, $\phi(y) = (2l - y)/l^2$ if $l \leqslant y \leqslant 2l$, and is zero otherwise. (Y is said to have the triangular distribution. Hint: If $0 \leqslant y \leqslant l$, then $0 \leqslant x_1 \leqslant y$, but if $l \leqslant y \leqslant 2l$, then $y - l \leqslant x_1 \leqslant l$).

Problem 8. X_1 has the distribution $N(\mu_1, \sigma_1^2)$, X_2 has the distribution $N(\mu_2, \sigma_2^2)$ and X_1, X_2 are independent. Show that the distribution of $X_1 + X_2$ is $N[\mu_1 + \mu_2, \sigma_1^2 + \sigma_2^2]$.

14.3 SUMS OF INDEPENDENT RANDOM VARIABLES, MOMENT GENERATING FUNCTIONS

We now consider another method of obtaining the distribution of the sum of two or more independent random variables, which will apply when their m.g.f.s exist. This method will thus not be so widely applicable as the method of convoluting their p.d.f.s. On the other hand, the method frequently needs less work.

Theorem 14.3. If X_1, X_2 are independent random variables, discrete or continuous, with m.g.f.s $M_{X_1}(t), M_{X_2}(t)$, then the m.g.f. of $X_1 + X_2$ is

$$M_{X_1 + X_2}(t) = M_{X_1}(t) M_{X_2}(t).$$

For

$$M_{X_1 + X_2}(t) = E[e^{(X_1 + X_2)t}]$$
$$= E(e^{X_1 t} e^{X_2 t})$$
$$= E(e^{X_1 t}) E(e^{X_2 t}),$$

since X_1, X_2 are independent,

$$= M_{X_1}(t) M_{X_2}(t). \tag{14.17}$$

Hence the m.g.f. of the sum of any finite number of independent random variables is the product of their separate m.g.f.s. The problem has now been reduced to identifying the distribution which has a m.g.f. equal to this product. At this stage we rely on the uniqueness theorem.

Example 7

If X_1 is distributed $N(\mu_1, \sigma_1^2)$, X_2 is independently distributed $N(\mu_2, \sigma_2^2)$ then

$$M_{X_1}(t) = e^{\mu_1 t + (1/2)\sigma_1^2 t^2},$$
$$M_{X_2}(t) = e^{\mu_2 t + (1/2)\sigma_2^2 t^2},$$

and we have

$$M_{X_1+X_2}(t) = M_{X_1}(t)\,M_{X_2}(t)$$

$$= e^{\mu_1 t + (1/2)\sigma_1^2 t^2}\, e^{\mu_2 t + (1/2)\sigma_2^2 t^2}$$

$$= e^{(\mu_1+\mu_2)t + (1/2)(\sigma_1^2+\sigma_2^2)t^2}. \qquad (14.18)$$

But this is the m.g.f. of a normal distribution with mean $\mu_1 + \mu_2$ and variance $\sigma_1^2 + \sigma_2^2$. This is, the distribution of $X_1 + X_2$ is $N[\mu_1 + \mu_2, \sigma_1^2 + \sigma_2^2]$. We knew, of course, in advance that $E(X_1 + X_2) = E(X_1) + E(X_2)$ and that since X_1, X_2 are independent $V(X_1 + X_2) = V(X_1) + V(X_2) = \sigma_1^2 + \sigma_2^2$. The *new* information is that $X_1 + X_2$ is also normally distributed.

Example 8

X_1 is distributed $N(1, 9)$, X_2 is distributed $N(2, 16)$ and X_1, X_2 are independent. Calculate $\Pr(X_1 + X_2 > 5)$, $\Pr(X_1 > X_2)$.

$X_1 + X_2$ is distributed $N(\mu, \sigma^2)$ where $\mu = \mu_1 + \mu_2 = 1 + 2 = 3$ and $\sigma^2 = \sigma_1^2 + \sigma_2^2 = 25$. Thus $Z = (X_1 + X_2 - 3)/5$ is distributed $N(0, 1)$. Thus

$$\Pr(X_1 + X_2 > 5) = \Pr[(X_1 + X_2 - 3)/5 > (5 - 3)/5],$$

$$\Pr(Z > 0.4) \qquad = 1 - \Pr(Z < 0.4) = 1 - 0.6554 = 0.3446.$$

We recast $\Pr(X_1 > X_2)$ as $\Pr(X_1 - X_2 > 0)$. We have $X_1 - X_2$ is distributed $N[1 - 2, 3^2 + 4^2]$, that is, $N(-1, 25)$. Hence

$$\Pr(X_1 - X_2 > 0) = \Pr[(X_1 - X_2 + 1)/5 > 1/5].$$

Since $(X_1 - X_2 + 1)/5$ is distributed $N(0, 1)$, this probability is 0.421. It should be noted that we cannot treat $\Pr[(X_1/X_2) > 1]$ *immediately* in this way, since there are two cases to consider, according as X_2 is positive or negative.

Problem 9. X_1, X_2, X_3 are independent and distributed $N(1, 9)$, $N(2, 16)$, $N(3, 144)$ respectively, calculate $\Pr(X_1 + X_2 + X_3 < 19)$.

Problem 10. X_1, X_2 are independent and each is distributed $N(0, 1)$. From the table of the unit normal distribution we have $\Pr(X_1 > 1.64) = 0.05$, and thus also $\Pr(X_2 > 1.64) = 0.05$. Explain why $\Pr[(X_1 + X_2)/2 > 1.64] \neq 0.05$, and calculate this probability. Calculate also $\Pr(6X_1 + 8X_2 < 7)$.

Problem 11. X_1, X_2, $\ldots$, X_9 are independent and each is distributed $N(\mu, \sigma^2)$. Find, in terms of μ, σ, numbers c_1 and c_2 such that

(a) $\Pr\left(\sum_{i=1}^{9} X_i \geqslant c_1\right) = 0.05$

(b) $\Pr\left(\left|\sum_1^9 X_i\right| \geqslant c_2\right) = 0.05$.

Problem 12. $X_1, X_2, \ldots, X_n$ are independent and each is distributed $N(\mu, 9)$. Find the least positive integer n, such that for some number c,

$$\Pr\left(\sum_{i=1}^n X_i \geqslant c \mid \mu = 1\right) = 0.05,$$

and

$$\Pr\left(\sum_{i=1}^n X_i \leqslant c \mid \mu = 2\right) = 0.05.$$

Problem 13. X_1, X_2 have independent normal distributions with parameters μ_1, σ_1 and μ_2, σ_2. Show that the distribution of $a_1 X_1 + a_2 X_2$ is

$$N[a_1\mu_1 + a_2\mu_2, (a_1^2 \sigma_1^2 + a_2^2 \sigma_2^2)].$$

Consider the special cases (i) $a_1 = 1, a_2 = -1$, (ii) $a_1 = 0$.

Problem 14. An engine part with two critical dimensions x mm. and y mm, is is produced by a machine. The values of x and y are independent of each other and of the corresponding measurements on other parts and are normally distributed with variances 0.02 mm^2 and 0.01 mm^2 respectively. From a sample of n parts, the quantity $\bar{z}$ is found where

$$\bar{z} = \bar{x} - 2\bar{y}$$

and where $\bar{x}$ and $\bar{y}$ are the sample means. If $\bar{z}$ is greater than a limit l, the machine is stopped for adjustment. Find l and the smallest value of n such that, when the mean m of the distribution of $\bar{z}$ is zero, the probability P of stopping the machine is less than 0.02, but when $m = 0.2, P$ is greater than 0.9.
   Show that, for this value of n,

$$P = \frac{1}{2} + \frac{1}{\sqrt{(2\pi)}} \int_0^u e^{-t^2/2} \, dt,$$

where $u \approx 16.8 \, (m - l)$, and sketch the graph of P as a function of m.

Oxford and Cambridge, A.L. 1981. ∎

   By repeated application of Example 7, if X_i is distributed $N(\mu_i, \sigma_i^2), i = 1, 2,$ $\ldots, n$ and the X_i are independent, then the distribution of

$$\sum_1^n X_i$$

is

$$N\left(\sum_{1}^{n} \mu_i, \sum_{1}^{n} \sigma_i^2\right).$$

This is an important result, which will be frequently used. In particular, if $\mu_i = \mu$, $\sigma_i = \sigma$, $i = 1, 2, \ldots, n$, then the distribution of

$$\sum_{1}^{n} X_i$$

is $N(n\mu, n\sigma^2)$ and that of

$$\bar{X} = \sum_{i=1}^{n} X_i/n$$

is $N(\mu, \sigma^2/n)$. That is, the mean of a random sample of n from a distribution $N(\mu, \sigma^2)$ has the distribution

$$N(\mu, \sigma^2/n).$$

14.4 CENTRAL LIMIT THEOREM

Suppose a random sample is drawn from a distribution. If the distribution has mean μ and a variance σ^2, then $E(\bar{X}) = \mu$ and $V(\bar{X}) = \sigma^2/n$. Thus we might estimate μ by the value assumed by the sample mean $\bar{X}$ and, if n is large enough, then it is very probable that $\bar{X}$ is near to μ. The exact probability can be found if we know the distribution of $\bar{X}$. There is a remarkable theorem which states that, if n is large enough, then the random variable

$$Z_n = \sqrt{(n)} \, (\bar{X} - \mu)/\sigma$$

has approximately the distribution $N(0, 1)$. This will be the case if the X_i are independent and the moments μ, σ^2 exist. We prove the result making much stronger assumptions, namely that the moment generating function $M_X(t)$ exists.

Theorem 14.4. If $X_1, \ldots, X_n$ is a sequence of independent random variables, each with the same distributions, for which $M_X(t)$ exists and $Z_n = \sqrt{(n)} \, (\bar{X} - \mu)/\sigma$, where $E(X_i) = \mu$, $V(X_i) = \sigma^2$ and

$$\bar{X} = \sum_{i=1}^{n} X_i/n$$

then the limiting distribution of Z_n is $N(0, 1)$.

Proof

$$M_{Z_n}(t) = M_{\sqrt{(n)}(\bar{X}-\mu)/\sigma}(t)$$

$$= E[e^{\sqrt{(n)}(\bar{X}-\mu)t/\sigma}]$$

$$= E\left[e^{\sum_{i=1}^{n}(X_i-\mu)t/(\sigma\sqrt{n})}\right]$$

$$= E\left[\prod_{i=1}^{n} e^{(X_i-\mu)t/(\sigma\sqrt{n})}\right]$$

since the X_i are independent,

$$M_{Z_n}(t) = \prod_{i=1}^{n} E\left[e^{(X_i-\mu)t/(\sigma\sqrt{n})}\right]$$

$$= \prod_{i=1}^{n} M_{X_i-\mu}[t/(\sigma\sqrt{n})]$$

$$= \left\{M_{X-\mu}[t/(\sigma\sqrt{n})]\right\}^n,$$

since the X_i all have the same m.g.f.

Now $M_{X-\mu}(t)$ has a full expansion as a power series in t; indeed, it generates the central moments

$$M_{X-\mu}(t) = 1 + \frac{\nu_1 t}{1!} + \frac{\nu_2 t^2}{2!} + \frac{\nu_3 t^3}{3!} + \ldots$$

$$= 1 + \frac{\nu_2 t^2}{2!} + \frac{\nu_3 t^3}{3!} + \ldots$$

(since $\nu_1 = 0$). This full expression is not needed, we only require the Maclaurin expansion for the first three terms, namely

$$M_{X-\mu}(t) = 1 + \frac{\nu_2 t^2}{2!} + \frac{R_3 t^3}{3!}$$

where R_3 is bounded. Hence

$$M_{x-\mu}\left(\frac{t}{\sigma\sqrt{n}}\right) = 1 + \frac{\nu_2 t^2}{2n\sigma^2} + \frac{R_3 t^3}{3!\,\sigma^3\sqrt{n^3}}.$$

Substituting this in the expression for $M_{Z_n}(t)$, we have, noting that $\nu_2 = \sigma^2$,

$$M_{Z_n}(t) = \left[1 + \left(\frac{t^2}{2} + \frac{R_3 t^3}{3!\sigma^3\sqrt{n}}\right)\bigg/ n\right]^n.$$

Now there is a theorem in analysis which states that if $b_n \to 0$ as $n \to \infty$, then

$$\lim_{n \to \infty} \left(1 + \frac{a + b_n}{n}\right)^n = e^a.$$

In our case, as R_3 is bounded,

$$\lim_{n \to \infty} \frac{R_3 t^3}{3! \sigma^3 \sqrt{n}} = 0$$

and hence

$$\lim_{n \to \infty} M_{Z_n}(t) = e^{t^2/2}.$$

But this is the m.g.f. of a normal distribution with paramters $\mu = 0$, $\sigma = 1$. Hence the limiting distribution Z_n is $N(0, 1)$. That is, for sufficiently large n, $\sqrt{(n)}(\bar{X} - \mu)/\sigma$ is distributed approximately $N(0, 1)$. This is indeed a remarkable result, since the approximate distribution does not depend on the original distribution, subject to the m.g.f. existing. The following comments should be noted

(a)   We need only the mean and variance of the original distribution in order to use the approximation.

(b)   The result is the same, whether or not the original is continuous or discrete. In the latter case, a small 'continuity correction' is often used to improve the approximation.

(c)   The proof gives no estimate as to how large n should be for the approximation to be 'good'.

(d)   The theorem discusses the behaviour of $\sqrt{n}(\bar{X} - \mu)/\sigma$. The random variable $\bar{X}$ has a limiting distribution which is degenerate — it converges in probability to the value μ.

(e)   We shall assume the Central Limit Theorem (C.L.T.) also applies when it is only known that the mean and variance of the distribution exists.

Example 9
$X_1, X_2, \ldots, X_n$ is a random sample of n values from the exponential distribution with parameter λ. In this case, $E(X_i) = 1/\lambda$ and $V(X_i) = 1/\lambda^2$. Hence by the C.L.T., $Z_n = \sqrt{(n)}\lambda(\bar{X} - 1/\lambda)$ is distributed approximately $N(0, 1)$ for large n.

If $\lambda = 1$, $n = 100$, suppose we require to calculate $\Pr(\bar{X} \geqslant 1.1)$. In this case, $Z = 10(\bar{X} - 1)$ is distributed approximately $N(0, 1)$. Now $\bar{X} \geqslant 1.1$ when $10(\bar{X} - 1) \geqslant 1$. From the tables,

$$\Pr(Z \geqslant 1) = 1 - \Pr(Z \leqslant 1) = 1 - 0.8413 = 0.1587 = 0.16.$$

[For the exponential distribution, the exact distribution of

$$\sum_{i=1}^{100} X_i$$

can be found. It is a gamma distribution with parameters 100 and 1.]

Problem 15. If $\bar{X}$ is the mean of a random sample of 48 values from the rectangular distributions over 0, 1, calculate approximately

$$\Pr(0.48 < \bar{X} < 0.52).$$ ∎

If $X_1, \ldots, X_n$ is a random sample of n values from the Bernoulli distribution with parameter p, then $E(X_i) = p$, $V(X_i) = pq$. Hence $E(\bar{X}) = p$, $V(\bar{X}) = pq/n$. By the C.L.T., $\sqrt{(n)}\,(\bar{X} - p)/\sqrt{(pq)}$ has approximately the $N(0, 1)$ distribution for sufficiently large n. Alternatively,

$$\frac{\left[\left(\sum_{i=1}^{n} X_i\right) - np\right]}{\sqrt{(npq)}}$$

has approximately the $N(0, 1)$ distribution. This is a result of some practical importance. In the first form, it allows us to approximate the proportion of successes in a series of independent trials, while the second version performs the same service for the total number of successes. A reinterpretation is also available, namely that since

$$\sum_{i=1}^{n} X_i$$

has in this instance the binomial distribution, then if X has a binomial distribution with parameters n and p, we have $(X - np)/\sqrt{(npq)}$ has approximately the $N(0, 1)$ distribution. This result may also be proved by considering the limiting behaviour of the p.d.f. of $(X - np)/\sqrt{npq}$ where X has the binomial distribution.

Example 10
If X has the binomial distribution with parameters $n = 300$, $p = 1/4$, calculate $\Pr(X \geqslant 85)$. Since

$$np = 300 \times 1/4 = 75, npq = 300 \times 1/4 \times 3/4, \sqrt{(npq)} = 15/2.$$

If X has the binomial distribution with parameters n and p, $Z = 2(X - 75)/15$ is distributed approximately $N(0, 1)$.

$$\Pr(X \geqslant 85) = \Pr\left[\frac{2(X-75)}{15} \geqslant \frac{2(85-75)}{15}\right]$$

$$= \Pr(Z \geqslant 1.33)$$

$$= 1 - \Pr(Z \leqslant 1.33)$$

$$= 1 - 0.9082 \approx 0.092.$$

The question of when n is sufficiently large has so far been avoided. In this respect, the binomial distribution has received close attention. Unfortunately, it is difficult to give other than conservative rules, since the minimum value of n depends not only on the degree of accuracy required but also on the value of p. One such rule is, that if $0.1 \leqslant p \leqslant 0.9$ then $n \geqslant 50$. In fact if p is very close to 0.5, two-decimal accuracy is obtained for $n \geqslant 10$. Outside those values of p, the Poisson distribution should also be considered.

When a continuous distribution is used to approximate a discrete distribution then a continuity correction can be applied. We suppose, as in the binomial distribution, that the discrete random variable assumes integral values with positive probability. Then if x is an integer, in the approximating continuous distribution every value in an interval will correspond to x in the discrete distribution. In the present case over the interval $[x - (1/2), x + (1/2)]$. Thus to approximate $\Pr(X \geqslant x)$ in the discrete distribution, we evaluate $\Pr[X \geqslant x - (1/2)]$, while for $\Pr(X \leqslant x)$ we evaluate $\Pr[X \leqslant x + (1/2)]$. This continuity correction is not worth while for $n \geqslant 50$.

Example 11
Calculate the probability of obtaining between three and six heads (inclusive) in 10 tosses of a fair coin. The number of heads has the binomial distribution with parameters $n = 10, p = 1/2$.

$$\Pr(3 \leqslant X \leqslant 6) = \sum_{x=3}^{6} \binom{10}{x} \left(\frac{1}{2}\right)^x \left(\frac{1}{2}\right)^{10-x}$$

$$= \frac{15}{128} + \frac{105}{512} + \frac{63}{256} + \frac{105}{512} = \frac{99}{128} = 0.7734.$$

Suppose now we use the normal approximation, including the continuity correction, $np = 5, \sqrt{(npq)} \approx 1.58$

$$\Pr(2.5 \leqslant X \leqslant 6.5) = \Pr\left(\frac{2.5-5}{1.58} \leqslant \frac{X-5}{1.58} \leqslant \frac{6.5-5}{1.58}\right)$$

$$= \Pr(-1.58 \leqslant Z \leqslant 0.95)$$

$$= \Pr(Z \leqslant 0.95) - \Pr(Z \leqslant -1.58)$$

$$= \Pr(Z \leqslant 0.95) - [1 - \Pr(Z \leqslant 1.58)]$$

$$= 0.8289 + 0.9429 - 1$$

$$= 0.7718.$$

At this stage we have grounds for supposing that for the Poisson distribution, which has mean λ and variance λ, the random variable $(X - \lambda)/\sqrt{\lambda}$ is approximately distributed $N(0, 1)$ if λ is sufficiently large. Our reasons for so thinking are on the one hand that the limiting form of a binomial distribution under certain conditions is a Poisson distribution and on the other that the standardized binomial random variable tends to the unit normal distribution. The supposition is indeed correct but there is a lurking suspicion that the sets of conditions under which these two limits take place may interfere with each other. It is perhaps worth while to investigate the Poisson distribution directly.

Example 12

If the random variable X has the Poisson distribution with parameter λ, then the limiting distribution of $Z = (X - \lambda)/\sqrt{\lambda}$ as λ tends to infinity is $N(0, 1)$. The m.g.f. of X is $M_X(t) = \exp[\lambda(e^t - 1)]$. Hence

$$
\begin{aligned}
M_Z(t) &= E[e^{(X-\lambda)t/\sqrt{\lambda}}] \\
&= e^{-\sqrt{(\lambda)}t} E(e^{Xt/\sqrt{\lambda}}) \\
&= e^{-\sqrt{(\lambda)}t} M_X(t/\sqrt{\lambda}) \\
&= e^{-\sqrt{(\lambda)}t} e^{\lambda[\exp(t/\sqrt{\lambda})-1]}.
\end{aligned}
$$

$$
\begin{aligned}
\log_e M_Z(t) &= -\sqrt{(\lambda)}t + \lambda(e^{t/\sqrt{\lambda}} - 1) \\
&= -\sqrt{(\lambda)}t + \lambda\left(\frac{t}{\sqrt{\lambda}} + \frac{t^2}{2\lambda} + \text{terms in } 1/\sqrt{\lambda} \text{ of higher degree}\right) \\
&= -\sqrt{(\lambda)}t + \sqrt{(\lambda)}t + \frac{t^2}{2} + \frac{R}{\sqrt{\lambda}} \text{ where } R \text{ is bounded} \\
&= \frac{t^2}{2} + \frac{R}{\sqrt{\lambda}} \to \frac{t^2}{2} \text{ as } \lambda \to \infty.
\end{aligned}
$$

Hence

$$\lim_{\lambda \to \infty} M_Z(t) = e^{t^2/2}.$$

But this is the m.g.f. of a random variable which is distributed $N(0, 1)$.

Example 13

In the case of the Poisson distribution, the use of the normal approximation for moderate values of λ is apt to be disappointing. Thus for a Poisson distribution with $\lambda = 9$,

$$\Pr(7 \leqslant X \leqslant 9) = 0.1171 + 0.1318 + 0.1318 = 0.38.$$

If we take $\mu = 9$, $\sigma = \sqrt{\lambda} = 3$ and employ the continuity correction,

$$\Pr(7 \leqslant X \leqslant 9) \approx \Pr\left(\frac{6.5 - 9}{3} \leqslant \frac{X - 9}{3} \leqslant \frac{9.5 - 9}{3}\right)$$

$$= \Pr(-0.833 \leqslant Z \leqslant 0.167)$$

where Z has approximately the $N(0, 1)$ distribution. From tables this is ≈ 0.36.

Problem 16. The probability that a calculator will develop a fault within one year is $1/10$. A manufacturer sells 5000 calculators and guarantees to repair the first fault that occurs within one year.

Give an exact expression for the probability that he will have to repair more than 550. Describe the methods which are often used to approximate to the binomial distribution, giving an indication of their limitations. Using that which you consider to be more appropriate, evaluate approximately the probability in the manufacturer's problem. How many more will he need to sell before there is a 97.5% chance that he will have to repair 1000?

Queen Elizabeth College, 1982.

Problem 17. It is required to estimate to within 0.02 the proportion θ of the adult population of Utopia who intend to vote for the Notáx party in the 1980 election. There are r adults in a random sample of N of adults who say they will do so.

(i)   Assuming that these responses are independent prove that the mean and variance of r/N are θ and $\theta(1 - \theta)/N$ respectively (you may quote, without proof, results about the binomial distribution).

(ii)   Use the result of (i), and the normal approximation to the binomial, to show that when N is large, the probability that r/N differs from θ by less than 0.02 is greater than 0.95 whenever

$n > 9604\ \theta(1 - \theta)$.

(iii)   Deduce the minimum value of N that ensures whatever the value of θ, that r/N is within 0.02 of θ with a probability greater than 0.95.

Oxford & Cambridge (M.E.1.), 1979. ∎

We have seen that if a fair coin is tossed 10 times then the probability of between three and six heads (inclusive) is ≈ 0.77. That is, in 100 separate sets of ten, we expect about 77 of them to yield between three and six heads. We may say that $[3, 6]$ is a probability interval for the number of heads. It is not the only such interval with probability 0.77 since this is also the case for the interval $[4, 7]$. For prediction purposes it is convenient to find such a fixed interval which meets the following requirements.

(a)   The probability that some random variable falls in the interval is at least some stipulated level, usually 99%, 95%, or 90%.

(b)   The probabilities that this same random variable falls above or below the end points of the interval are equal.

A normal approximation is frequently used to construct such an interval. Thus if a fair coin is tossed 400 times, then the number of heads has expectation 200 and variance 100. Thus $(X - 200)/10$ has approximately the $N(0, 1)$ distribution. Since the upper and lower 2½% points of this distribution are ± 1.96,

$$\Pr\left(-1.96 \leqslant \frac{X - 200}{10} \leqslant +1.96\right) = 0.95$$

or

$$\Pr(200 - 19.6 \leqslant X \leqslant 200 + 19.6) = 0.95.$$

Thus to the nearest whole number, the number of heads will be between 180 and 220 inclusive, with probability 0.95.

Problem 18. In each of the following situations give two limits within which you would expect about 95% of the values to lie.

(a) R is the number of bacteria in successive samples of 1 cc from a well-mixed suspension containing on the average 50 bacteria per cc.

(b) R is the number of times a cricket captain wins the toss, in successive series of 100 tosses.

(c) R is the percentage of times a cricket captain wins the toss in successive series of 200 tosses.

London, B.Sc. General, 1965, Pt. I.

Problem 19. Two gamblers A and B agree to play a series of games in which each has two cards, a king and a queen. Each gambler first chooses one of his cards and places it face down on the table. Both cards are then turned over. If both are kings A *gives* B £1, while if both the queens A *gives* B £3, but if the cards on the table are a king and a queen A *receives* £2 from B. They then pick up their cards ready for the next game.

Both gamblers play their cards in a random sequence, A choosing his king with probability p and B choosing his king with probability q. Find the expected value of S, the money gained by A in a single game, in terms of p and q, and show that, if $p = 5/8$, the expected value of S is independent of q.

Further, if both p and q equal 5/8, find

(a)   the probability that A loses money in any one game.

(b)   the expected value of S^2 and hence, or otherwise, the variance of S,

(c) approximate limits (to the nearest £1), symmetrical about the mean, between which lies 95% of the distribution of A's total gain in 100 games.

Oxford & Cambridge, A.L., 1978.

BRIEF SOLUTIONS AND COMMENTS ON THE PROBLEMS

Problem 1. The required convariance can easily be written as

$$E\left[\sum a_i \left\{X_i - E(X_i)\right\}\right]\left[\sum b_j\left\{X_j - E(X_j)\right\}\right]$$

$$= E\left[\sum\sum a_i b_j \left\{X_i - E(X_i)\right\}\left\{X_j - E(X_j)\right\}\right]$$

$$= \sum\sum a_i b_j E\left[\left\{X_i - E(X_i)\right\}\left\{X_j - E(X_j)\right\}\right]$$

$$= \sum\sum a_i b_j C(X_i, X_j).$$

Problem 2.

$$V\left(\sum a_i X_i\right) = E\left[\sum a_i X_i - \sum a_i E(X_i)\right]^2$$

$$= E\left[\sum a_i \left\{X_i - E(X_i)\right\}\right]^2$$

$$= E\left[\sum a_i^2\left\{X_i - E(X_i)\right\}^2 + \right.$$

$$\left. + \sum_{i\neq j}\sum a_i a_j \left\{X_i - E(X_i)\right\}\left\{X_j - E(X_j)\right\}\right]$$

$$= \sum a_i^2 V(X_i) + \sum_{i\neq j}\sum a_i a_j C(X_i, X_j).$$

Problem 3. From the previous problem,

$$V(\bar{X}) = V\left(\sum X_i/n\right) = \sum\frac{1}{n^2} V(X_i) + \sum_{i\neq j}\sum\frac{1}{n^2} C(X_i, X_j)$$

$$= \frac{\sigma^2}{n} + \frac{n-1}{n}\rho\sigma^2,$$

since there are $n(n-1)$ terms in the double sum.

Problem 4.
(a) $(1.e^{-1}.e^{-1}.1.e^{-1}.e^{-2})/(5^2 e^{-5}/2!) = 2/25.$
(b) Required conditional probability is also

$$\Pr(X_1 + X_2 = 1)\Pr(X_3 + X_4 = 1)/\Pr(X_1 + X_2 + X_3 + X_4 = 2).$$

$$= \frac{2e^{-2}}{1!} \cdot \frac{3e^{-3}}{1!} \bigg/ \frac{5^2 e^{-5}}{2!} = \frac{12}{25}.$$

Problem 5. From section 13.2, equation (13.14), if $Y = X_1 + X_2$,

$$\phi(y) = \sum_{x_1=0}^{y} \binom{n_1}{x_1} p^{x_1} (1-p)^{n_1-x_1} \binom{n_2}{y-x_1} p^{y-x_1} (1-p)^{n_2-y+x_1}$$

$$= \sum_{x_1=0}^{y} \binom{n_1}{x_1} \binom{n_2}{y-x_1} p^y (1-p)^{n_1+n_2-y}$$

$$= \binom{n_1+n_2}{y} p^y (1-p)^{n_1+n_2-y},$$

which is the p.d.f. of a binomial distribution with parameters $n_1 + n_2, p$. The Bernoulli distribution is binomial with parameters $1, p$ hence the sum is binomial with parameters n, p (see sections 4.4, 4.5, Chapter 4).

Problem 6. The distribution of ΣX_i is Poisson with parameter $\Sigma \lambda_i$.

$$\Pr\left[X_i = x_i; \, i = 1, 2, \ldots, n \,|\, \Sigma X_i = m\right]$$

$$= \prod \Pr(X_i = x_i) / \Pr\left(\Sigma X_i = m\right) \text{ where } \Sigma x_i = m.$$

$$= \frac{\prod \lambda_i^{x_i} e^{-\lambda_i}/x_i!}{\left(\Sigma \lambda_i\right)^m e^{-\Sigma \lambda_i}/m!}$$

$$= \frac{m!}{\prod x_i!} \prod \left(\frac{\lambda_i}{\Sigma \lambda_i}\right)^{x_i}, \quad \text{where } \Sigma x_i = m,$$

which is of the required form.

Problem 7.

$$x_1 + x_2 = y \Rightarrow x_2 = y - x_1,$$

but

$$0 \leqslant x_2 \leqslant l \Rightarrow 0 \leqslant y - x_1 \leqslant l \Rightarrow y - l \leqslant x_1 \leqslant y.$$

Also $0 \leqslant x_1 \leqslant l$ hence if $y < l$, we have $0 \leqslant x_1 \leqslant y$, but if $y > l$, then

$$y - l \leqslant x_1 \leqslant l.$$

Integrating $1/l^2$ with respect to x_1 provides the required answer.

Problem 8. Altogether more tiresome! We are to evaluate

$$\frac{1}{2\pi\sigma_1\sigma_2} \int_{-\infty}^{+\infty} \exp\left\{-\frac{1}{2}\left[\left(\frac{y-x_1-\mu_2}{\sigma_2}\right)^2 + \left(\frac{x_1-\mu_1}{\sigma_1}\right)^2\right]\right\} dx_1.$$

Take outside the integral sign terms involving

$$\exp\left\{-\frac{1}{2}\left[\left(\frac{y-\mu_2}{\sigma_2}\right)^2 + \frac{\mu_1^2}{\sigma_1^2}\right]\right\}.$$

The integrand is now of the form

$$\exp\left\{-\frac{1}{2}\left[\frac{x_1^2}{\sigma_1^2} + \frac{x_1^2}{\sigma_2^2} - 2x_1\left(\frac{y-\mu_2}{\sigma_2^2} + \frac{\mu_1}{\sigma_1^2}\right)\right]\right\}.$$

Complete the squares on x_1 in the exponent of the last expression. The integration may now be completed and after collecting all the terms we have as the p.d.f. of $y = x_1 + x_2$,

$$\frac{1}{\sqrt{2\pi}\sqrt{(\sigma_1^2 + \sigma_2^2)}}\exp\left[-\frac{1}{2}(y-\mu_1-\mu_2)^2/(\sigma_1^2 + \sigma_2^2)\right].$$

Problem 9. $Y = X_1 + X_2 + X_3$ has the distribution $N(6, 169)$.

$$\Pr(Y < 19) = \Pr[(Y-6)/13 < (19-6)/13]$$
$$= \Pr(Z < 1) = 0.84,$$

since Z has the distribution $N(0, 1)$.

Problem 10. There, for example, are values $X_1 < 1.64$ and $X_2 > 1.64$ such that $(X_1 + X_2)/2 > 1.64$.

$$\Pr[(X_1 + X_2)/2 > 1.64] = \Pr(X_1 + X_2 > 3.28).$$

Now $X_1 + X_2$ is distributed $N(0, 2)$, hence required probability is also

$$\Pr(Z > 3.28/\sqrt{2}) = \Pr(Z > 2.33) = 0.01.$$

Also $6X_1$ is distributed $N(0, 6^2)$ and $8X_2$ is distributed $N(0, 8^2)$ and hence $6X_1 + 8X_2$ is distributed $N(0, 10^2)$. Hence

$$\Pr[6X_1 + 8X_2 < 7] = \Pr[Z < 7/10] = 0.758,$$

where Z is distributed $N(0, 1)$.

Problem 11

$$Y = \sum_1^9 X_i$$

has the distribution $N(9\mu, 9\sigma^2)$.

$$\Pr(Y \geqslant c_1) = \Pr[(Y-9\mu)/3\sigma \geqslant (c_1-9\mu)/3\sigma] = 0.05.$$

But $(Y - 9\mu)/3\sigma$ has the distribution $N(0, 1)$ and 1.64 is the upper 5% point of this distribution. Hence, $(c_1 - 9\mu)/3\sigma = 1.64$ or $c_1 = (4.92)\sigma + 9\mu$.

(b) From symmetry, $(c_2 - 9\mu)/3\sigma$ must be the upper 2½% point of the $N(0, 1)$ distribution which is 1.96. Hence $c_2 = (5.88)\sigma + 9\mu$.

Problem 12

$$Y = \sum_1^n X_i$$

is distributed $N(n\mu, 9n)$. Since the upper and lower 5% points of the $N(0, 1)$ distribution are ± 1.64, we have; $(c - n)/3\sqrt{n} \geq 1.64$ and $(c - 2n)/3\sqrt{n} \leq -1.64$ so that $n \geq 96.8$. The least positive integer is 97.

Problem 13

$$M_{a_1 X_1 + a_2 X_2}(t) = E(e^{a_1 X_1 t + a_2 X_2 t})$$
$$= E(e^{X_1 a_1 t}) E(e^{X_2 a_2 t}),$$

X_1, X_2 independent,

$$= M_{X_1}(a_1 t) M_{X_2}(a_2 t)$$
$$= e^{\mu_1 (a_1 t) + (1/2)\sigma_1^2 (a_1 t)^2} e^{\mu_2 (a_2 t) + (1/2)\sigma_2^2 (a_2 t)^2}$$
$$= e^{(a_1 \mu_1 + a_2 \mu_2)t + (1/2)(a_1^2 \sigma_1^2 + a_2^2 \sigma_2^2)t^2}.$$

But this is the m.g.f. of the distribution

$$N[a_1\mu_1 + a_2\mu_2, a_1^2\sigma_1^2 + a_2^2\sigma_2^2].$$

Problem 14. l, n satisfy the equations

$$l \Big/ \sqrt{\frac{0.06}{n}} = 2.0537, \quad (l - 0.2) \Big/ \sqrt{\frac{0.06}{n}} = -1.2816.$$

Bear in mind,

$$V(\bar{z}) = \frac{V(x)}{n} + \frac{4V(y)}{n}.$$

Problem 15. This rectangular distribution has mean $1/2$ and variance $1/12$. If $Z = \sqrt{48}(\bar{X} - 0.5)\sqrt{12}$, the equivalent probability is for Z between -0.48 and $+0.48$, where Z is distributed $N(0, 1)$. From tables this is about 0.37.

Problem 16. For the binomial distribution with $n = 5000, p = 1/10, np = 500$, $npq = 450$. Compare $(550 - 500)/\sqrt{450} = 2.357$ with the corresponding point on the unit normal distribution. If m is the total number sold, set

$$\frac{1000 - m/10}{3\sqrt{m/10}} = 1.96,$$

and solve for m.

Problem 17
(ii)   Consider that $0.2/\sqrt{[\theta(1-\theta)/N]}$ must be the upper 2½% point of the unit normal distribution.

(iii)   2401. For which value of θ is $\theta(1-\theta)$ a maximum?

Problem 18
(a) $50 \pm 1.96\sqrt{50} \approx 50 \pm 14, (36,64)$.

(b) $50 \pm 1.96\sqrt{25} \approx 50 \pm 10, (40,60)$.

(c) $100 \pm 1.96\sqrt{50} \approx 100 \pm 14, (86,114)$ or $(43\%, 57\%)$.

Problem 19. Observe that in this question $q \neq (1-p)$.

$$E(S) = -pq - 3(1-p)(1-q) + 2p(1-q) + 2(1-p)q.$$

If $p = q = 5/8$,

(a)   34/64   (b) $E(S) = 1/8, E(S^2) = 226/64, V(S) = 218/64$.

(c)   Total of A's gain in 100 games has expectation 100/8 and variance 100(218/64). Use normal approximation.

15

Unbiased Estimators

15.1 SAMPLE AND DISTRIBUTION

Data is usually obtained by measuring or recording. Any theoretical model of the state of nature under examination should stipulate the type of distribution which may have generated the data. In some simple situations, particular distributions will have a powerful intuitive appeal. Thus, whenever a record of events can be viewed as the outcome of a finite number of independent trials on each of which there is a constant probability of an event occurring, then the binomial distribution will be employed. Again in repeated measurements of the same fixed quantity the differences in the values obtained are usually ascribed to instrumental errors. Since any reasonable instrument should be as likely to overestimate as to underestimate and make a large error rather infrequently, the distribution generating the errors ought to be both symmetric and unimodal. The normal distribution fills the bill and efforts have been made to justify this choice on grounds other than that of convenience. Typically the error is taken to be made up of many small negative and positive contributions and the central limit theorem is invoked. If the distribution $N(0, \sigma^2)$ is indeed thought to be appropriate, then σ remains to be determined.

Any distribution incorporated in the model may display one or more unknown parameters. A numerical determination of any characteristic of a distribution is said to be its **estimate**. The mean and variance of a distribution are sufficiently important characteristics to warrant detailed consideration of their estimation.

Definition. The random variables X_1, $X_2, \ldots, X_n$ are said to be a random sample of n from the distribution with p.d.f. $f(x)$ if the X_i are mutually independent random variables each with p.d.f. $f(x)$. Hence the joint p.d.f. of X_1, $X_2, \ldots, X_n$ is $f(x_1)f(x_2)\ldots f(x_n)$ whether the random variables are discrete or continuous.

A function of the sample values which does not contain any unknown parameters is called a **statistic**. The distribution of a statistic may of course depend on unknown parameters. If θ is any unknown numerical characteristic of the joint distribution of the sample values and a statistic is used to make inferences about θ, then this statistic is also known as an **estimator** of θ and this is stressed by denoting the estimator by $\hat{\Theta}$. Any particular value of $\hat{\Theta}$ is then called an estimate of θ.

15.2 CRITERIA FOR ESTIMATORS

Definition. $\hat{\Theta}$ is said to be an *unbiased* estimator of θ if $E(\hat{\Theta}) = \theta$, whatever the value of θ. If $\hat{\Theta}$ is not unbiased, then $E(\hat{\Theta}) - \theta$ is called the bias.

Example 1

If X_1, $X_2, \ldots, X_n$ is a random sample from a distribution with mean μ, then

$$\bar{X} = \sum_1^n X_i/n$$

is an unbiased estimator of μ. For

$$E(\bar{X}) = E\left(\sum X_i/n\right) = \sum_1^n E(X_i)/n = \sum_1^n \mu/n = \mu.$$

Note that the result still holds if the X_i are not independent but have the same expectation. Except when the sample distribution has a symmetrical distribution, the median of the sample values is a biased estimator of μ.

Example 2

If X_1, $X_2, \ldots, X_n$ is a random sample from a distribution with mean μ and variance σ^2, then

$$S^2 = \sum_1^n (X_i - \bar{X})^2/(n-1)$$

is an unbiased estimator of σ^2. It is easily verified that

$$\sum_1^n (X_i - \bar{X})^2 = \sum_1^n (X_i - \mu)^2 - n(\bar{X} - \mu)^2,$$

hence

$$E\left[\sum_1^n (X_i - \bar{X})^2\right] = \sum_1^n E(X_i - \mu)^2 - nE(\bar{X} - \mu)^2$$

$$= \sum_1^n V(X_i) - nV(\bar{X})$$

$$= \sum_1^n \sigma^2 - n\frac{\sigma^2}{n}$$

$$= n\sigma^2 - \sigma^2 = (n-1)\sigma^2.$$

Hence the result holds *whether or not μ is known*.

$$\sum_1^n (X_i - \mu)^2/n$$

is another unbiased estimator of σ^2 but is not a statistic unless μ is known.

Problem 1. X has the binomial distribution with parameters n, p. Show that $X(n - X)/(n - 1)$ is an unbiased estimator of $V(X)$.

Problem 2. $X_1, X_2, \ldots, X_n$ are independently distributed $N(0, \sigma^2)$. Show that

$$\sqrt{(\pi/2)} \sum_1^n |X_i|/n$$

is an unbiased estimator of σ.

Problem 3. The discrete random variable R has the truncated Poisson distribution, that is it has p.d.f.

$$f(r) = \frac{\lambda^r}{r!(e^\lambda - 1)}, \quad r = 1, 2, \ldots.$$

A random sample of n is drawn from the distribution and it is found that n_r members of the sample have value r. Show that

$$\sum_2^\infty \frac{rn_r}{n}$$

is an unbiased estimator of λ.

15.3 COMPARING ESTIMATORS

It is always the case that any one random value from a distribution is an unbiased estimator of the mean when this exists. We must of course state which member of the sample is to be taken as the estimator *before* we draw the sample. More generally (not to mention more interestingly) we may have two unbiased estimators $\hat{\Theta}_1$, $\hat{\Theta}_2$, of the same unknown θ. If we are forced to choose between two unbiased estimators $\hat{\Theta}_1$ and $\hat{\Theta}_2$, one criterion with a plausible ring is always to choose the estimator with the smaller variance. This is in accord with the result that the variance is a measure of the spread of the probability about the mean. (See 10.1.) When $\hat{\Theta}_1$ and $\hat{\Theta}_2$ are independent, it might strike us as wasteful to throw one away and causes us to search for some best combination of the two. We restrict our attention to linear combinations and our criterion of 'best' will be that unbiased combination which has minimum variance.

Best linear combination of unbiased estimators

Suppose $\hat{\Theta}_1, \hat{\Theta}_2, \ldots, \hat{\Theta}_k$ are independent unbiased estimators of θ and

$$V(\hat{\Theta}_i) = \sigma_i^2 \ (i = 1, 2, \ldots, k).$$

Then if

$$\sum_1^k \lambda_i \hat{\Theta}_i$$

is an unbiased estimator of θ,

$$E\left(\sum_1^k \lambda_i \hat{\Theta}_i\right) = \theta.$$

But

$$E\left(\sum_1^k \lambda_i \hat{\Theta}_i\right) = \sum_1^k \lambda_i E(\hat{\Theta}_i) = \left(\sum_1^k \lambda_i\right)\theta = \theta.$$

Hence

$$\sum_1^k \lambda_i = 1.$$

The variance of

$$T = \sum_1^k \lambda_i \hat{\Theta}_i$$

is

$$\sum_1^k \lambda_i^2 V(\hat{\Theta}_i) = \sum_1^k \lambda_i^2 \sigma_i^2.$$

We are to minimize $V(T)$ subject to

$$\sum_{1}^{k} \lambda_i = 1.$$

We could use the calculus, but this and many similar problems can be solved using Cauchy's inequality. Namely: if $(a_i, b_i), i = 1, 2, \ldots, k$, are k pairs of real numbers then

$$\left(\sum_{i=1}^{k} a_i b_i\right)^2 \leqslant \left(\sum_{1}^{k} a_i^2\right) \left(\sum_{1}^{k} b_i^2\right).$$

Readers unfamiliar with this inequality should study Problem 4.

Problem 4. By arranging

$$\sum_{i=1}^{k} (a_i x - b_i)^2$$

as a quadratic form in x and exploiting the fact that it is non-negative when all the quantities are real, show that

$$\left(\sum_{1}^{k} a_i b_i\right)^2 \leqslant \left(\sum_{1}^{k} a_i^2\right) \left(\sum_{1}^{k} b_i^2\right)$$

with equality if and only if there exists c such that $a_i = cb_i, i = 1, 2, \ldots, k$. ∎

If we choose $a_i = \lambda_i \sigma_i$, $b_i = 1/\sigma_i$ we obtain

$$\sum (\lambda_i \sigma_i)^2 \sum (1/\sigma_i^2) \geqslant \left[\sum (\lambda_i \sigma_i / \sigma_i)\right]^2 = \left(\sum \lambda_i\right)^2 = 1$$

or

$$\sum \lambda_i^2 \sigma_i^2 \geqslant 1 / \left[\sum (1/\sigma_i^2)\right],$$

with equality if and only if $\lambda_i \sigma_i = c/\sigma_i \Rightarrow \lambda_i = c/\sigma_i^2$. To obtain the constant c, sum $\lambda_i = c/\sigma_i^2$ over i where

$$1 = \sum \lambda_i = c \sum (1/\sigma_i^2)$$

and finally

$$\lambda_i = \frac{1/\sigma_i^2}{\sum (1/\sigma_i^2)}. \tag{15.1}$$

The weighting in (15.1) gives the best linear combination in the sense that it attains the minimum variance.

Example 3

For a random sample of n from a distribution, each individual member of the sample is an unbiased estimator of μ and has variance σ^2. Hence, applying (15.1),

$$\lambda_i = \frac{1/\sigma^2}{\displaystyle\sum_1^n (1/\sigma^2)} = \frac{1}{n}.$$

Hence

$$T = \sum \lambda_i X_i = \sum X_i/n = \bar{X}$$

and $\bar{X}$ is the best linear estimator. It has variance σ^2/n.

Example 4

Two independent samples of n_1 and n_2 observations from the same distribution have sample means $\bar{X}_1$, $\bar{X}_2$. Find the best linear estimator of the mean of the distribution knowing only $\bar{X}_1$ and $\bar{X}_2$. We have

$$E(\bar{X}_1) = E(\bar{X}_2) = \mu,$$
$$V(\bar{X}_1) = \sigma^2/n_1,$$
$$V(\bar{X}_2) = \sigma^2/n_2.$$

Hence from (15.1),

$$\lambda_1 = \frac{n_1/\sigma^2}{n_1/\sigma^2 + n_2/\sigma^2} = \frac{n_1}{n_1 + n_2}, \quad \lambda_2 = 1 - \lambda_1 = \frac{n_2}{n_1 + n_2}.$$

Thus the required estimator is

$$(n_1 \bar{X}_1 + n_2 \bar{X}_2)/(n_1 + n_2)$$

and this is also the mean of the pooled samples.

Example 5

In measuring a physical quantity, an instrument makes a random error X which has zero mean and variance σ_1^2. An operator using this instrument makes an independent random error Y with mean zero and variance σ_2^2. Supposing the operator makes k independent readings on each of n independent settings of the instrument, what is the variance of the best unbiased estimator of the quantity, based on a linear combination of the setting averages? Suppose that Z_{ij} is the jth reading on the ith setting and that the errors are additive.

$$Z_{ij} = \mu + X_i + Y_{ij}$$

where μ is the unknown value of the quantity. Hence the k readings on the ith setting are *not* independent of each other. Write

$$\bar{Z}_i. = \sum_{j=1}^{k} Z_{ij}/k, \quad \bar{Y}_i. = \sum_{j=1}^{k} Y_{ij}/k$$

then $\bar{Z}_i.$ is an unbiased estimator of μ and has variance

$$V(X_i + \bar{Y}_i.) = V(X_i) + V(\bar{Y}_i.) = \sigma_1^2 + \sigma_2^2/k,$$

which does not depend on i. The n estimators $\bar{Z}_i.$ have a best linear combination which is their pooled mean $(\bar{Z}_1. + \bar{Z}_2. + \ldots + \bar{Z}_n.)/n$ which is the grand mean of all the observations and has variance $(\sigma_1^2 + \sigma_2^2/k)/n$. The value of this variance can be compared for different values of n and k provided σ_1, σ_2 are known or have been closely estimated from previous experiments. It follows that for a given number of observations $N = nk$, the variance is least when $k = 1, n = N$, i.e. one observations on each of the N settings.

Problem 5. $X_1, X_2, \ldots, X_n$ each have expectation μ, and variance σ^2 and the correlation coefficient between any pair of the variables is ρ. Show that the linear unbiased estimator of μ which has minimum variance is $\bar{X}$. What is the least possible value for ρ?

Problem 6. A straight line has been drawn on the floor and it is required to mark off a length of 4 feet from a given point on the line. The available measuring instruments are a 1-foot ruler and a pair of dividers (for transferring any arbitrary length from one position to another). Any measurement of one foot made with the ruler is subject to a random error distributed with a mean zero and a standard deviation of 0.5 in; any measurement transferred by the dividers is subject to a random error with a mean of zero and a standard deviation of 0.01 in. Two alternative procedures are proposed:

(a)   to mark off four successive lengths, nominally of 1 foot, with the ruler;
(b)   to mark off one nominal length of 1 foot with the ruler and to transfer this by three independent operations of the dividers, each operation extending the measured line by a nominal 1 foot. All random errors may be assumed to be independently distributed. Find the standard deviation of the measured length of 4 feet for each of these two procedures.

Comment on the following argument. The dividers are more precise than the ruler; (a) involves four operations with the ruler and (b) one operation with the ruler and three with the divider; (b) is the better procedure.

U.L. B.Sc. General, pt. I, 1966. ∎

The scope of our discussion has been limited to unbiased linear estimators with minimum variance. A more general discussion of the question of estimators lies beyond the level of the present text. (See ref. [1] for an Intermediate Level treatment.) Our next example shows that two competing and equally plausible estimators can arise in a natural way.

Example 6

X_1, X_2 are independent values from the distribution $N(0, \sigma^2)$ and it is required to estimate σ^2. Since the mean of the distribution is zero, $E(X_1^2) = E(X_2^2) = \sigma^2$. The obvious estimator is thus $T = (X_1^2 + X_2^2)/2$ which has variance

$$\frac{1}{4}[V(X_1^2) + V(X_2^2)] = \frac{1}{2}V(X^2) = \frac{1}{2}[E(X^4) - E^2(X^2)]$$

$$= \frac{1}{2}[3\sigma^4 - (\sigma^2)^2] = \sigma^4.$$

For the normal distribution $N(0, \sigma^2)$ no function of the observations X_1, X_2, which has expectation σ^2, has a variance smaller than σ^4. We have seen that

$$S^2 = \sum_1^n (X_i - \bar{X})^2/(n-1)$$

is an unbiased estimator of σ^2, whatever the value of the distribution mean. For a sample of two, $\bar{X} = (X_1 + X_2)/2$ and

$$S^2 = [X_1 - (X_1 + X_2)/2]^2 + [X_2 - (X_1 + X_2)/2]^2$$

$$= \frac{(X_1 - X_2)^2}{4} + \frac{(X_2 - X_1)^2}{4} = \frac{(X_1 - X_2)^2}{2}.$$

Now since X_1, X_2 are independent $N(0, \sigma^2)$, $X_1 - X_2$ has the distribution $N(0, 2\sigma^2)$. But $V[(X_1 - X_2)^2] = 2(\sqrt{2}\sigma)^4 = 8\sigma^4$ so finally $V[(X_1 - X_2)^2/2] = 2\sigma^4$. Thus $(X_1 - X_2)^2/2$ is a poorer estimator than $(X_1^2 + X_2^2)/2$, at least in terms of variance.

Definition. The efficiency of an unbiased estimator $\hat{\Theta}_1$ relative to another unbiased estimator $\hat{\Theta}_2$ is measured by the ratio $V(\hat{\Theta}_2)/V(\hat{\Theta}_1)$.

Thus in Example 6 the relative efficiency of S^2 compared to T is $\sigma^4/2\sigma^4 = 50\%$. Again the relative efficiency of a single member of a random sample compared to the sample mean is $(\sigma^2/n)/\sigma^2 = 1/n$. An estimator may, however, be favoured on the grounds that it is easy to compute or has a particularly manageable probability density function. Comparisons of efficiency are then useful guides to possible losses in precision.

Problem 7. $X_1, X_2, \ldots, X_n$ is a random sample from the rectangular distribution over $(0, \theta)$. Let $\hat{\Theta}_1 = 2\bar{X}$, $\hat{\Theta}_2 = [(n+1)/n] \max(X_1, X_2, \ldots, X_n)$. Given that the p.d.f. of $Y = \max(X_1, X_2, \ldots, X_n)$ is $(n/\theta)(y/\theta)^{n-1}$, $0 < y < \theta$ calculate the relative efficiency of $\hat{\Theta}_1$ to $\hat{\Theta}_2$.

REFERENCE

[1]   G. P. Beaumont, *Intermediate Mathematical Statistics*, Chapman & Hall, London, 1980.

BRIEF SOLUTIONS AND COMMENTS ON THE PROBLEMS

Problem 1

$$E[X(n-X)] = nE(X) - E(X^2)$$
$$= n(np) - [V(X) + E^2(X)] = n^2 p - V(X) - (np)^2$$
$$= n^2 p(1-p) - V(X) = (n-1) V(X),$$

since

$$V(X) = npq.$$

Problem 2

$$E(|X|) = \frac{1}{\sqrt{(2\pi)}\sigma} \int_{-\infty}^{-\infty} |x| e^{-x^2/(2\sigma^2)} \, dx$$

$$= \frac{1}{\sqrt{(2\pi)}\sigma} \int_{-\infty}^{0} -x e^{-x^2/(2\sigma^2)} \, dx$$

$$+ \frac{1}{\sqrt{(2\pi)}\sigma} \int_{0}^{\infty} x e^{-x^2/(2\sigma^2)} \, dx$$

$$= \frac{2}{\sqrt{(2\pi)}\sigma} \left\{ -\sigma^2 e^{-x^2/(2\sigma^2)} \right\}_{0}^{\infty} = \sqrt{\frac{2}{\pi}} \sigma.$$

Problem 3. $E(N_r) = nf(r)$. Hence

$$\sum_{2}^{\infty} rf(r) = \sum_{2}^{\infty} \frac{\lambda^r/(r-1)!}{(e^\lambda - 1)} = \frac{\lambda(e^\lambda - 1)}{(e^\lambda - 1)} = \lambda.$$

Problem 4

$$\sum (a_i x - b_i)^2 = \sum a_i^2 x^2 - 2\sum a_i b_i x + \sum b_i^2.$$

Since non-negative this does not change sign. That is,

$$\left(2 \sum a_i b_i \right)^2 \leqslant 4 \left(\sum a_i^2 \right) \left(\sum b_i^2 \right),$$

which is the required result. Equality occurs if and only if there is c such that $a_i c - b_i = 0$ for all i.

Problem 5. Let

$$T = \sum_1^n \lambda_i X_i$$

where

$$\sum \lambda_i = 1,$$

since T is unbiased.

$$V(T) = \sum \lambda_i^2 V(X_i) + \sum \sum_{j \neq k} \lambda_j \lambda_k \, \text{cov}(X_j, X_k)$$

$$= \sigma^2 \sum \lambda_i^2 + \rho \sigma^2 \left(\sum_j \sum_k \lambda_j \lambda_k - \sum \lambda_i^2 \right)$$

$$= \sigma^2 \sum \lambda_i^2 \, (1 - \rho) + \rho \sigma^2 .$$

As before, $\sum \lambda_i^2$ is minimized when $\lambda_i = 1/n$. Since $V(T) \geqslant 0, (1 - \rho) + n\rho \geqslant 0$
or $\rho \geqslant - 1/(n - 1)$.

Note: This removes any lurking doubt about Example 5 where

$$\text{cov}(Z_{ij}, Z_{ik}) = \text{cov}(X_i, X_i) = \sigma_1^2 .$$

Hence the readings on the same setting satisfy the conditions of this problem
and it is correct to take the setting averages.

Problem 6. If R denotes error with the ruler and D with the divider.

(a) $V[R_1 + R_2 + R_3 + R_4] = 4V(R) = 4(0.0025) = 0.01$, hence standard
deviation 0.1.

(b)   Errors are $R_1, R_1 + D_1, R_1 + D_2, R_1 + D_3$.

$$V[R_1 + R_1 + D_1 + R_1 + D_2 + R_1 + D_3]$$

$$= V[4R_1 + D_1 + D_2 + D_3]$$

$$= 16 \, V(R) + 3V(D) = 16(0.0025) + 3(0.001) = 0.043.$$

Standard deviation $\sqrt{0.043} > 0.1$. Hence (a) is better and the calculations reveals
that in (b) the dividers propagate any error made with the ruler

Problem 7. $E(X) = \theta/2$, $V(X) = \theta^2/12$. Hence $E(\bar{X}) = \theta/2$, $V(\bar{X}) = \theta^2/12n$.

$$E(Y) = \int_0^\theta \frac{ny^n}{\theta^n} \, dy = \frac{n\theta}{n + 1}.$$

$$E(Y^2) = \int_0^\theta \frac{ny^{n+1}}{\theta^n} \, dy = \frac{n}{n + 2} \, \theta^2$$

and

$$V(Y) = \frac{n\theta^2}{(n+2)(n+1)^2}.$$

Hence

$$E\left(\frac{n+1}{n}Y\right) = \theta, \quad V\left[\frac{n+1}{n}Y\right] = \frac{\theta^2}{n(n+2)}$$

$$V(\hat{\Theta}_2)/V(\hat{\Theta}_1) = 3/(n+2),$$

since $V(2\bar{x}) = \theta^2/3n$.

The superior performance, in terms of relative efficiency of the statistic based on the maximum of the sample arises from the functional dependence of the mean of the distribution on the upper limit of the distribution.

16

Sampling Finite Populations

16.1 INTRODUCTION

Suppose a population consists of N elements. With each element is associated a value of a variable x. The list of these unknown values is $x_1, x_2, \ldots, x_N$. One element is chosen at random from the population and X is the random variable equal to the value of x on the element chosen. Since X can assume one of the N values $x_1, x_2, \ldots, x_N$, it is discrete and

$$\Pr(X = x_i) = \frac{1}{N}, \quad i = 1, 2, \ldots, N.$$

For the random variable X,

$$E(X) = \sum_1^N x_i \Pr(X = x_i)$$

$$= \sum_1^N x_i/N.$$

That is, the expected value of this random variable is merely the arithmetic mean of the population values, which we denote by μ. Similarly,

$$V(X) = E(X - \mu)^2 = \sum_1^N (x_i - \mu)^2 \Pr(X = x_i)$$

$$= \sum_{1}^{N} (x_i - \mu)^2 / N = \left[\sum_{1}^{N} x_i^2 - N\mu^2 \right] \Big/ N.$$

This is the population variance, which we also denote by σ^2.

But of course we are interested in drawing a sample of several elements from the population. There are two basic methods for so doing:

(1)　We draw a first member at random from the population, note its value and then replace it. This operation is repeated until we have obtained the desired number of sample values. Clearly the random variables associated with each draw are independent and have a common distribution. Such a sample is evidently a random sample from this common distribution.

(2)　We draw a first member at random from the population, note its value and then do *not* replace it. A second member is then drawn at random from the remaining elements of the population, its value noted and *not* replaced, and so on. This operation is repeated until we have obtained the desired number of sample values, which cannot exceed N. The random variables associated with different draws are no longer independent since, for example, the outcomes available for the second draw depend on the result of the first. Hence this kind of sampling does not apparently qualify as random sampling in the sense of section 15.1. Nevertheless we shall be reluctant to abandon this nomenclature altogether since, after all, elements are selected at random from those remaining at each draw. We propose the term 'random sample drawn without replacement'. To sharpen the distinction from (1) it is common practice to describe such sampling *with* replacement as **simple random sampling.**

Simple random sampling

Suppose $X_1, X_2, \ldots, X_n$ is a simple random sample of n drawn from the finite population of values $x_1, x_2, \ldots, x_N$ with mean μ and variance σ^2. By the very nature of the sampling scheme the random variables are independent. All the results obtained for a random sample from a distribution apply. That is

$$\bar{X} = \sum X_i / n$$

is an unbiased estimator of μ with variance σ^2/n. If μ is known,

$$\sum_{1}^{n} (X_i - \mu)^2 / n$$

is an unbiased estimator of σ^2, as is

$$S^2 = \sum_{1}^{n} (X_i - \bar{X})^2 / (n - 1),$$

whether or not μ is known. In any reasonable situation, μ and σ^2 will exist. There is however a feature of finite populations which has no analogue in a probability distribution. It does make sense to talk about the *total* of the population values. Since this total is

$$\sum_1^N x_i = N\mu,$$

it can be estimated by the statistic $N\bar{X}$ (which is *not* the sample total). The unbiased estimator $N\bar{X}$ has variance $N^2 V(\bar{X}) = N^2 \sigma^2/n$ and σ^2 can be estimated by the usual methods.

16.2 RANDOM SAMPLING WITHOUT REPLACEMENT

The probability that any particular element, from a population of N, appears first in a sample is $1/N$. Given that it has so appeared, the conditional probability that any other particular element appears as the second member of the sample is $1/(N-1)$ and so on. Hence the probability that any particular n elements appear in a specified order is

$$\frac{1}{N} \cdot \frac{1}{N-1} \cdot \frac{1}{N-2} \cdot \ldots \cdot \frac{1}{N-(n-1)} = \frac{(N-n)!}{N!}.$$

Hence the probability that these same elements are drawn in one of the $n!$ possible orders is $n!(N-n)!/N! = 1/\binom{N}{n}$. It is now apparent that this probability is equivalent to giving each of the $\binom{N}{n}$ different *selections* of n elements from N an equal probability of being the sample chosen. This conclusion lends some additional force to the label 'random sampling' in this context. For after all, one of the $\binom{N}{n}$ selections is chosen at random. The reader should hasten to confirm that any particular element of the population appears in the sample with probability n/N and that any particular pair of elements appears with probability $n(n-1)/[N(N-1)]$.

We are now in a position to derive the distribution of each member of a random sample $X_1, X_2, \ldots, X_n$ drawn without replacement. The random variable X_j assumes one of the values $x_1, x_2, \ldots, x_N$. The probability that the ith element appears in the sample is n/N. The conditional probability that x_i appears in the jth place is thus

$$\frac{n}{N} \times \frac{1}{n} = \frac{1}{N}.$$

Thus

$$\Pr(X_j = x_i) = \frac{1}{N}, \quad i = 1, 2, \ldots, N.$$

Hence (as in the case of sampling with replacement),

$$E(X_j) = \sum_1^N x_i/N = \mu; \quad V(X_j) = \sum_1^N (X_i - \mu)^2/N = \sigma^2. \qquad (16.1)$$

Nevertheless the sample values are not independent. For $X_j = x_r$ and $X_k = x_s$ ($j \neq k$) if and only if the values x_r, x_s both appear in the sample and then in places j, k *in that order*. This has probability

$$\frac{n(n-1)}{N(N-1)} \times \frac{1}{n(n-1)} = \frac{1}{N(N-1)}.$$

As this does not equal $\Pr(X_j = x_r) \Pr(X_k = x_s)$, the random variables X_j, X_k are not independent, though the joint distribution of all pairs of such random variables is the same. The covariance of any such pair is

$$C(X_j, X_k) = E(X_j X_k) - E(X_j) E(X_k)$$

$$= \sum_{r \neq s} \sum x_r x_s \Pr(X_j = x_r, X_k = x_s) - \mu^2$$

$$= \frac{1}{N(N-1)} \left(\sum_1^N \sum_1^N x_r x_s - \sum_1^N x_j^2 \right) - \mu^2$$

$$= \frac{1}{N(N-1)} \left[\left(\sum_1^N x_i \right)^2 - \sum_1^N x_i^2 \right] - \mu^2$$

$$= \frac{1}{N(N-1)} \left[(N\mu)^2 - \sum_1^N x_i^2 \right] - \mu^2$$

$$= -\frac{1}{N-1} \sum_1^N \frac{x_i^2}{N} + \mu^2 \left(\frac{N}{N-1} - 1 \right)$$

$$= -\frac{1}{N-1} \left[\sum_1^N \frac{x_i^2}{N} - \mu^2 \right]$$

$$= -\frac{\sigma^2}{N-1}. \qquad (16.2)$$

Thus the covariance for each pair is $-\sigma^2/(N-1)$ and does not depend on the sample size. This side result enables us to find the variance of the sample mean.

$$V(\bar{X}) = V\left(\sum_1^n X_i/n\right) = \frac{1}{n^2} V\left(\sum_1^n X_i\right)$$

$$= \frac{1}{n^2}\left[\sum_1^n V(X_i) + \sum_{\substack{j \neq k}}^n \sum^n C(X_j, X_k)\right]$$

(see Problem 2 of Chapter 14),

$$= \frac{1}{n^2}\left[n\sigma^2 - \frac{(n^2 - n)\sigma^2}{N-1}\right]$$

$$= \frac{1}{n}\left[\frac{\sigma^2(N-1-n+1)}{N-1}\right]$$

$$= \frac{(N-n)}{(N-1)} \cdot \frac{\sigma^2}{n}. \qquad (16.3)$$

This variance is somewhat less than the corresponding result when the sampling is with replacement. Notice that the so called 'finite population factor',

$$\frac{N-n}{N-1} \to 1$$

as N increases.

Example 1

In a random sample of n drawn without replacement, a proportion of P is observed to have a particular attribute. Estimate p, the proportion in the population of N having this attribute, and find the variance of the estimator.

Suppose we attach a value $x_i = 1$ if the ith element in the population has the attribute and $x_i = 0$ otherwise, and similarly for the sample values. Then clearly $\mu = p$ and $\bar{X} = P$, and P is an unbiased estimator of p. Any sample value, X_i, takes the value 1 with probability p and is zero otherwise. Hence

$$E(X_i^2) = 1^2 p + 0^2(1-p) = p$$

and

$$V(X_i) = p - p^2 = p(1-p) = pq.$$

Hence

$$V(P) = V(\bar{X}) = \frac{N-n}{N-1} \cdot \frac{pq}{n}. \qquad (16.4)$$

Problem 1. For four values in a population, $x_1 = 1, x_2 = 3, x_3 = 5, x_4 = 7$. Write down all the ordered pairs of values which can be drawn without replacement. Verify by direct calculation that the sample means have an average of 4 and a variance of $5/3$.

Problem 2. Suppose that x_i, y_i are the numerical values of two different variables associated with the ith element in a finite population of N elements. A random sample of n elements is drawn without replacement. If $\bar{X}$, $\bar{Y}$ are the sample means for the variables, show that

$$\text{covariance } (\bar{X}, \bar{Y}) = \frac{N-n}{N} \sum_1^N \frac{(x_i - \bar{x})(y_i - \bar{y})}{n(N-1)}$$

[Hint: Consider $\bar{Z} = \bar{X} - \bar{Y}$ and apply (16.3)]

16.3 ESTIMATION OF THE POPULATION VARIANCE

In this connection it is natural to consider, for sampling without replacement,

$$S^2 = \sum_1^n (X_i - \bar{X})^2 / (n-1)$$

as a candidate. Since

$$\sum_1^n (X_i - \bar{X})^2 = \sum_1^n (X_i - \mu)^2 - n(\bar{X} - \mu)^2,$$

$$E\left[\sum_1^n (X_i - \bar{X})^2\right] = n\sigma^2 - nV(\bar{X})$$

$$= n\sigma^2 - \sigma^2 (N-n)/(N-1)$$

$$= N(n-1)\sigma^2/(N-1).$$

That is, $E(S^2) = N\sigma^2/(N-1) > \sigma^2$ and is slightly biased. However, $(N-1)S^2/N$ estimates σ^2 without bias. It is to be remarked that since

$$E(S^2) = \sum_1^N (x_i - \mu)^2 / (N-1) = s^2, \quad \text{(say)}, \tag{16.5}$$

then s^2 bears the same relationship to the population values as does S^2 to the sample values. In this notation

$$V(\bar{X}) = \frac{\sigma^2}{n} \cdot \frac{N-n}{N-1} = \frac{(N-1)s^2}{Nn} \cdot \frac{N-n}{N-1} = \frac{s^2}{n} \cdot \frac{(N-n)}{N}.$$

$$\tag{16.6}$$

Problem 3. For Example 1, find an unbiased *estimator* of the variance of the sample proportion P.

16.4 SUB-POPULATION

Drawing a random sample requires a list of the elements in the population. Suppose we are interested in the extra income earned by students who take vacation employment. These constitute a sub-population but we cannot get a list of such students to sample directly. The College office will naturally have a list of *all* students and from this we *can* take a random sample without replacement. It is only after an interview that we discover whether or not a student took vacation employment at all! Thus the elements in the sub-population of interest are not identifiable in advance. In some situations we might know the number of elements in a sub-population without being able to locate them. For example, the total number of eligible people who *actually* voted in an election becomes available but the individual names are not a matter of public record.

Suppose in a finite population of N elements, there is a sub-population of M elements whose number and location are unknown. With each element in the whole population is associated a value $x_i (i = 1, 2, \ldots, N)$ and we wish to estimate, without bias, the total of the values for those elements which belong to the sub-population. Since we cannot sample the sub-population directly, a random sample of n is drawn from the entire population of N, without replacement. When we examine the sample we discover that only m of the elements belong to the sub-population (appearances notwithstanding, m is here a random variable). Without wishing to labour the point, it is almost obvious that all samples of m from the sub-population have the same chance of being drawn.

Problem 4. Show that the conditional probability that a particular sample of m is drawn, given that a random sample of n drawn without replacement is found to contain just m from the sub-population, is $1/\binom{M}{m}$.   ∎

We are fortified by the result of the last problem, which confirms the above claim, that we can discard the values of the sample which do *not* belong to the sub-population. We shall use the letter Y to denote any X which turns out to be a member of the sub-population. That is to say if X_i is the jth member of the sample identified in the sub-population then $Y_j = X_i$. For example Y_3 may be X_5. Thus

$$\sum_1^m Y_j/m$$

estimates the mean of the sub-population as usual, and the total is estimated by

$$M \left(\sum_1^m Y_j/m \right).$$

This estimator is usable if M is known.

Number of sub-population elements unknown

If M is not known the situation is not irretrievable, for we can estimate M. Indeed m/n estimates the proportion M/N or Nm/n estimates M without bias. Substituting this estimator in

$$M \sum Y_j/m$$

we obtain

$$\frac{Nm}{n} \frac{\sum Y_j}{m} = \frac{N}{n} \sum_1^m Y_j. \tag{16.7}$$

On closer inspection, *this* estimator turns out to have an alternative interpretation. The statistic is equivalent to regarding those elements in the sample which do *not* belong to the sub-population as bearing a value of zero for the variable. This is rather different from rejecting them and having a reduced sample size (of m). We have the full sample of n, those belonging to the sub-population contribute their appropriate value of x and those which do not score zero. In this light, our random sample of n can now be regarded as drawn from *another* finite population which consists of M elements having values $y_1, y_2, \ldots, y_M$ and $N - M$ elements having value zero. We are now in a position to apply the standard results to *this* population:

(1)
$$N \sum_1^n Y_j/n = N \sum_1^m Y_j/n$$

is an unbiased estimator of

$$\sum_1^N y_j = \sum_1^M y_j,$$

since $N - M$ values are zero;

(2)   the variance of the estimator in (1) is

$$N^2 \frac{(N-n)}{N-1} \frac{\sigma_1^2}{n}, \tag{16.8}$$

where

$$\sigma_1^2 = \sum_1^M \frac{y_j^2}{N} - \left(\sum_1^M \frac{y_j}{N} \right)^2$$

and σ_1^2 is the variance of the modified population.

(3) σ_1^2 is estimated by $(N-1)S_1^2/N$, where

$$S_1^2 = \left[\sum_1^n Y_j^2 - \left(\sum_1^n Y_j \right)^2 \bigg/ n \right] \bigg/ (n-1)$$

$$= \left[\sum_1^m Y_j^2 - \left(\sum_1^m Y_j \right)^2 \bigg/ n \right] \bigg/ (n-1). \qquad (16.9)$$

The reader should strengthen his conviction in the above technique by working through the details of the next problem.

Problem 5. In a group of four persons, the values of a variable are $x_1 = 1$, $x_2 = 3, x_3 = 5, x_4 = 7$. We wish to estimate the total for the males in the group, the number of which is unknown. The true state of affairs is that there are two males, with associated values $x_1 = 1$, $x_2 = 3$ and hence total 4. A random sample of two is drawn without replacement. By considering the six possible unordered pairs show directly that the estimator given by (16.7) is unbiased and that its variance is indeed given by (16.8). Check also that the estimator of this variance is unbiased.

Number of elements in sub-population known

We now return to the question of estimating the total for the values in a sub-population when the number, M, of such elements is known. We have seen that for $m > 0$, the estimator is, in the notation of the previous section,

$$M \sum_1^m Y_j/m,$$

which has conditional variance, given m,

$$M^2 \frac{(M-m)}{M-1} \cdot \frac{\sigma_2^2}{m}. \qquad (16.10)$$

Here σ_2^2 is the variance of the sub-population values and is

$$\frac{\sum_1^M y_j^2}{M} - \left(\frac{\sum_1^M y_j}{M} \right)^2. \qquad (16.11)$$

Example 2
For Problem 5, suppose it is known that $M = 2$

Sample values		1	1	1	3	3	5
		3	5	7	5	7	7
m = number elements in sub-population		2	1	1	1	1	0
Estimate $= 2 \sum y_j/m$		4	2	2	6	6	

When $m = 2$, the only estimate is 4 which is indeed the sub-population total and has zero variance. Given $m = 1$, all four samples have conditional probability 1/4. Hence the expected value of the estimator is $(2 + 2 + 6 + 6)/4 = 4$ with conditional variance $[2^2 + 2^2 + 6^2 + 6^2 - 4(4^2)]/4 = 4$. Check this against formula (16.10).

The irritating case when $m = 0$ will rarely arise in practice since n will generally be moderately large. If M is known to be relatively small then alternative sampling methods may have to be employed. For example, continue the sampling until a specified number of elements from the sub-population have been collected.

For Problem 5, when M is not known to be two, the method of scoring zeros lead to an estimator with greater variance. In Example 2, utilizing the information that $M = 2$, whether we observe $m = 2$ or $m = 1$, our conditional estimator happened to be more precise. An overall measure of performance presents difficulties since the expected value of the *random variable* $1/m$ in (16.10) cannot be evaluated. It is frequently replaced by the approximation, $1/E(m)$. Another suggestion is always to reject a sample which contains no elements from the sub-population and draw another sample. In the current example this means that $\Pr(m = 2|m > 0) = 1/5$ and $\Pr(m = 1|m > 0) = 4/5$ so that the unconditional variance of the estimator is $(0 \times 1/5) + (4 \times 4/5) = 16/5$.

Problem 6. The values 1, 3, 4 constitute a sub-population of three in the population 1, 3, 5, 7. Random samples of two are drawn without replacement and it is known that $M = 3$. Calculate the conditional variances of the estimator of the sub-population total when $m = 2$ and $m = 1$ and compare with the estimator which dispenses with the value of M. ∎

We have lingered so long over the topic of random sampling since a firm grasp of the associated computations is needed for specialist courses. There is, however, a great variety of other ways in which a sample may be drawn. We consider one of them, still in pursuit of estimating a population total value and now sampling *with* replacement.

16.5 WEIGHTED SAMPLING WITH REPLACEMENT

When all the values attached to the elements are positive, it is sometimes possible to exploit the fact that the total must be dominated by the contribution of the larger values. It ought to be possible to increase the precision (= decrease the variance) of an estimator by increasing the chance of capturing such elements in the sample. Of course we cannot be certain of doing so, for otherwise we need not sample at all! However, we may sometimes be in possession of a set of approximate values based on previous or related studies. For example, the last census count is a good guide to the current pattern of the numbers of inhabitants in towns.

In order to simplify the algebra we shall treat the case where the sample values are drawn with replacement while reminding the reader that in practice we should not wish to see the same element twice.

Suppose then that in a finite population of N elements, the associated values of a variable are $x_1, x_2, \ldots, x_N$. The ith element is selected with fixed probability p_i and then returned. We must have $0 < p_i < 1$ and

$$\sum_{1}^{N} p_i = 1$$

so that each element has a chance of appearing in the sample and that some element must be chosen. After repeating this process n times we have the sample values $X_1, X_2, \ldots, X_n$ and seek an unbiased estimator of

$$\sum_{1}^{N} x_i.$$

It is not immediately clear how the sample values are to be treated. Suppose we consider a weighted linear combination in which, if x_j appears on any draw, then it attracts weight w_j. Then our estimator is

$$\sum_{1}^{n} W_i X_i$$

and we require

$$E\left(\sum_{1}^{n} W_i X_i\right) = \sum_{1}^{n} E(W_i X_i) = \sum_{1}^{N} x_j.$$

Now $W_i X_i$ takes the value $w_j x_j$ with probability $p_j (j = 1, 2, \ldots, N)$. Hence

$$E(W_i X_i) = \sum_{1}^{N} w_j x_j p_j$$

and thus unbiasedness requires

$$n \sum_{1}^{N} w_j x_j p_j = \sum_{1}^{N} x_j.$$

If this condition is to hold for all possible values of x_j we must equalize the coefficients on both sides of the equation and obtain

$$nw_j p_j = 1$$

or $w_j = 1/(np_j)$. That is, the weight is inversely proportional to the probability that it was selected. Thus our estimator is of the form

$$\frac{1}{n} \sum_1^n X_i/P_i \tag{16.12}$$

in which X_i/P_i takes the value x_j/p_j with probability p_j. Notice that if $p_j = 1/N$, the estimator collapses to

$$\frac{N}{n} \sum_1^n X_i,$$

as previously employed in simple random sampling.

Example 3

A finite population consists of the three values $x_1 = 5$, $x_2 = 7$, $x_3 = 12$. Suppose x_1 is chosen with probability $1/4$, x_2 with probability $1/4$ and x_3 with probability $1/2$. Then, for a single value X_1, the estimator X_1/P_1 has expectation

$$E\left(\frac{X_1}{P_1}\right) = \sum_{i=1}^3 \frac{x_i}{p_i} p_i = 5 + 7 + 12 = 24.$$

Notice that

$$E(3X_1) = 3 \sum_{i=1}^3 x_i p_i = 3 \left(\frac{5}{4} + \frac{7}{4} + \frac{12}{2}\right) = 27.$$

That is, the estimator of the total of the population values which would be appropriate for simple random sampling is here biased.

If two values, X_1, X_2, are drawn with replacement, each in accord with the above prescription, then naturally, the expected value of the mean of the separate estimators,

$$E\left[\frac{1}{2}\left\{\frac{X_1}{P_1} + \frac{X_2}{P_2}\right\}\right] = \frac{1}{2}\left[E\left(\frac{X_1}{P_1}\right) + E\left(\frac{X_2}{P_2}\right)\right] = 24.$$

The reader will find it instructive to calculate this expectation directly. Thus, $X_1 = 5$ *and* $X_2 = 12$, with probability $(1/4)(1/2) = 1/8$, while $X_1 = 12$ *and* $X_2 = 5$ also with probability $1/8$. Thus

$$\frac{1}{2}\left(\frac{X_1}{P_1} + \frac{X_2}{P_2}\right)$$

assumes the value 22 with probability $1/4$, and so on.

We now compute the variance of the estimator (16.12).

$$V\left[\frac{1}{n}\sum_1^n X_i/P_i\right] = \frac{1}{n^2}\sum_1^n V(X_i/P_i),$$

since sampling is with replacement,

$$= \frac{1}{n} V(X_1/P_1),$$

since the same probabilities are used on each draw,

$$= \frac{1}{n}\sum_1^N\left[\left(\frac{x_j}{p_j} - x_\bullet\right)^2 p_j\right], \quad x_\bullet = \sum_1^N x_j$$

$$= \frac{1}{n}\left[\sum_1^N\frac{x_j^2}{p_j} - x_\bullet^2\right],$$

since

$$\sum_1^N p_j = 1.$$

This is of the form $[V(X_i/P_i)]/n$ which is not σ^2/n where σ^2 is the 'variance' of the population value $X_1, X_2, \ldots, X_N$. That is, it reflects the variance of the random variable associated with the draw. The usual estimator of $V(X_i/P_i)$ applies for a sample of n, namely

$$S^2 = \sum_1^n\left[\frac{X_i}{P_i} - \frac{1}{n}\sum\frac{X_i}{P_i}\right]^2 \Big/ (n-1). \tag{16.13}$$

Hence the estimator of the variance of

$$\frac{1}{n}\sum_1^n X_i/P_i$$

is S^2/n.

Example 4
For Example 3,

$$E\left[\left(\frac{X_1}{P_1}\right)^2\right] = \left(\frac{5}{1/4}\right)^2\left(\frac{1}{4}\right) + \left(\frac{7}{1/4}\right)^2\left(\frac{1}{4}\right) + \left(\frac{12}{1/2}\right)^2\left(\frac{1}{2}\right)$$

$$= 584.$$

Hence, $V(X_1/P_1) = 584 - 24^2 = 8$. For two dependent values X_1, X_2, the variance of the mean of the seperate estimators is

$$V\left[\frac{1}{2}\left(\frac{X_1}{P_1} + \frac{X_2}{P_2}\right)\right] = \frac{1}{4}\left[V\left(\frac{X_1}{P_1}\right) + V\left(\frac{X_2}{P_2}\right)\right] = 4.$$

In order to estimate

$$V\left[\frac{1}{2}\left(\frac{X_1}{P_1} + \frac{X_2}{P_2}\right)\right],$$

we need the values of X_1, X_2 and apply the formula in equation (16.13), with $n = 2$, to calculate the obtained values of S^2 and hence $S^2/2$. In Table 16.1 we tabulate all the possibilities.

Table 16.1

X_1	X_2	$\Pr(X_1, X_2)$	$\frac{1}{2}\left(\frac{X_1}{P_1} + \frac{X_2}{P_2}\right)$	$S^2/2$
5	7	1/16	24	16
7	5	1/16	24	16
5	12	1/8	22	4
12	5	1/8	22	4
7	12	1/8	26	4
12	7	1/8	26	4
5	5	1/16	20	0
7	7	1/16	28	0
12	12	1/4	48	0

Hence

$$E(S^2/2) = 16 \cdot \frac{1}{16} + 16 \cdot \frac{1}{16} + 4 \cdot \frac{1}{8}$$

$$+ 4 \cdot \frac{1}{8} + 4 \cdot \frac{1}{8} + 4 \cdot \frac{1}{8} = 4.$$

which agrees with the direct computation. This calculation has been carried out only to convince the reader that the usual results for sampling a distribution apply in this case. The reader should check that if a simple random sample of two values X_1, X_2, is drawn with replacement then $3(X_1 + X_2)/2$ has variance 24.

In weighted sampling, the elements are not necessarily selected with equal probabilities. In simple random sampling each element is selected with equal probability. Weighting sampling yields a more precise estimator if there is an overall positive association between the values attached to the elements and the probabilities that they are included in the sample. In Example 3 this was the case. An illustration of the effect of a poor choice of weights is to be found in the next problem.

Problem 7. Verify that

$$\frac{1}{2} \sum_{i=1}^{N} \sum_{j=1}^{N} \left[(x_i^2 - x_j^2) \left(\frac{1}{p_i} - \frac{1}{p_j} \right) \right]$$

$$= N \sum_{i=1}^{N} \frac{x_i^2}{p_i} - \left(\sum_{i=1}^{N} x_i^2 \right) \left(\sum_{i=1}^{N} \frac{1}{p_i} \right).$$

Hence, or otherwise, prove that for any choice of the p_i such that $x_i > x_j > 0$ implies $p_i < p_j$, then the estimator

$$\frac{1}{n} \sum_{i=1}^{n} (X_i/P_i)$$

is less precise than $N\bar{X}$, where $\bar{X}$ is the mean of a simple random sample drawn with replacement.

Problem 8. From the N pairs of values $(x_1, y_1), (x_2, y_2), \ldots, (x_N, y_N)$, a random sample of n pairs is drawn without replacement. Prove that

$$E\left[\sum_{i=1}^{n} (X_i - \bar{X})(Y_i - \bar{Y}) \right] \bigg/ (n-1) = \sum_{i=1}^{N} (x_i - \bar{x})(y_i - \bar{y})/(N-1).$$

Problem 9. $X_1, X_2, \ldots, X_n$ is a random sample drawn without replacement, from a finite population of N values with variance σ^2. If

$$T_1 = \sum_{i=1}^{n} a_i X_i,$$

where the a_i are constants, show that

$$V(T_1) = \frac{\sigma^2}{N-1} \left[N \sum_{1}^{n} a_i^2 - \left(\sum_{1}^{n} a_i \right)^2 \right].$$

Hence, or otherwise, show that $\bar{X}$ is the linear unbiased estimator of the population mean with minimum variance. If

$$T_2 = \sum_{i=1}^{n} b_i X_i,$$

show further that T_1, T_2 are uncorrelated if and only if

$$\sum_{1}^{n} a_i b_i = \left(\sum_{1}^{n} a_i \right) \left(\sum_{1}^{n} b_i \right) \bigg/ N.$$

Problem 10. In a population of N elements, an unknown proportion, p, has a certain characteristic. A random sample of n is drawn without replacement. Assuming the usual result for the variance of a sample mean show that the variance of NP, the estimator of the unknown number of the population having the characteristic, takes the form $N^2 pq(N - n)/[n(N - 1)]$ and that $PQN(N - n)/(n - 1)$ is an unbiased estimator of this variance, where P is the sample proportion having the characteristic and $Q = 1 - P$.

A television company wishes to estimate the number, D, of deaf viewers who saw a popular programme which incorporated features to assist them. Although the total audience N of all viewers is known, the number, M, of those who are deaf is unknown. In a random sample of n, drawn without replacement, of all viewers let the number of deaf be m and the number of these who saw the programme be d. State an unbiased estimator of D and calculate its variance. Show further that when M is known, Md/m is also an unbiased estimator of D. State the variance of this estimator given $m = m^*$.

BRIEF SOLUTIONS AND COMMENTS ON THE PROBLEMS

[The first comment covers the confirmation demanded in section 16.2.]
    The number of samples which include a particular element is

$$\binom{N-1}{n-1} \quad \text{and} \quad \binom{N-1}{n-1} \bigg/ \binom{N}{n} = \frac{n}{N}.$$

The corresponding probability for two elements is

$$\binom{N-2}{n-2} \bigg/ \binom{N}{n} = \frac{n(n-1)}{N(N-1)}.$$

Problem 1. There are $4 \times 3 = 12$ such ordered pairs.

Sample	Sample means
1, 3	2
3, 1	2
1, 5	3
5, 1	3
1, 7	4
7, 1	4
3, 5	4
5, 3	4
3, 7	5
7, 3	5
5, 7	5
7, 5	6

The average of the sample means is

$$[2(2) + 2(3) + 4(4) + 2(5) + 2(6)]/12 = 4.$$

The variance of the means is

$$[2(2-4)^2 + 2(3-4)^2 + 4(4-4)^2 + 2(5-4)2$$
$$+ 2(6-4)^2]/12 = 5/3.$$

Since

$$\sigma^2 = [(1-4)^2 + (3-4)^2 + (5-4)^2 + (7-4)^2]/4 = 20/4 = 5,$$

then for $N = 4, n = 2$,

$$V(\bar{X}) = \left(\frac{N-n}{N-1}\right)\frac{\sigma^2}{n} = \left(\frac{4-2}{4-1}\right)\frac{5}{2} = \frac{5}{3}.$$

Problem 2. $V(\bar{Z}) = V(\bar{X} - \bar{Y}) = V(\bar{X}) - 2\,C(\bar{X},\,\bar{Y}) + V(\bar{Y})$. Also

$$V(\bar{Z}) = \frac{N-n}{N}\sum_1^N \frac{(z_i - \bar{z})^2}{(N-1)n}$$

$$= \frac{N-n}{N}\sum_1^N \frac{[x_i - y_i - (\bar{x} - \bar{y})]^2}{(N-1)n}.$$

But

$$\sum[(x_i - \bar{x}) - (y_i - \bar{y})]^2 = \sum[(x_i - \bar{x})^2 - 2(x_i - \bar{x})(y_i - \bar{y})$$
$$+ (y_i - \bar{y})^2].$$

Now apply (16.6) to the first and last terms in this last expression.

Problem 3.

$$\sum_1^n (X_i - \bar{X})^2 = \sum X_i^2 - n\bar{X}^2 = \sum X_i^2 - nP^2.$$

But $X_i = 1$ or 0 so that $X_i^2 = X_i$, hence

$$\sum X_i^2 = \sum X_i = nP \quad \text{and} \quad \sum X_i^2 - nP^2 = nP - nP^2 = nPQ.$$

Hence

$$\frac{N-1}{N} S^2 = \frac{N-1}{N} \cdot \frac{nPQ}{n-1}$$

estimates pq without bias.

$$V(P) = \frac{pq}{n} \frac{N-n}{N-1},$$

is estimated by $[PQ/(n-1)] \cdot [(N-n)/N]$.

Problem 4. Pr(particular m elements | m from sub-population)

$$= \frac{\text{Pr(particular } m \text{ elements)}}{\text{Pr}(m \text{ from sub-population})}$$

$$= \binom{N-M}{n-m} \bigg/ \left[\binom{N-M}{n-m} \binom{M}{m} \right] = 1 \bigg/ \binom{M}{m}.$$

Problem 5. Sample values $N = 4, n = 2$.

1	1	1	3	3	5	
3	5	7	5	7	7	
4	1	1	3	3	0	observed male total
8	2	2	6	6	0	estimate of total.

Remember that a pair such as $(3, 5)$ is scored as $(3, 0)$.

The expected value of the estimator is $(8 + 2 + 2 + 6 + 6 + 0)/6 = 4$ and its variance is $[(8 - 4)^2 + 2(2 - 4)^2 + 2(6 - 4)^2 + (0 - 4)^2]/6 = 8$. For the population $1, 3, 0, 0$, the mean is 1 and the variance, σ_1^2, is $3/2$.

$$N^2 \frac{(N-n)}{N-1} \frac{\sigma_1^2}{n} = 8.$$

The estimates of σ_1^2 are $3/2$, $3/8$, $3/8$, $27/8$, $27/8$, 0, with average $3/2$. Since $N^2(N - n)/[n(N - 1)] = 16/3$, variance of the estimator is estimated by $8, 2, 2, 18, 18, 0$ with average 8.

Problem 6. $1, 3, 5$, have mean 3 and variance $\sigma_2^2 = 8/3$.

$$V\left[M \sum Y_j/m\right] = M^2 \frac{\sigma_2^2}{m} \cdot \frac{(M-m)}{M-1} = \frac{24}{m} \frac{(3-m)}{2}.$$

When $m = 2$, variance is 6, when $m = 1$ variance is 24. The case $m = 0$ cannot arise.

$1, 3, 5, 0$ have mean $9/4$, and variance $\sigma_1^2 = 59/16$.

$$V\left[N \sum Y_j/n\right] = N^2 \frac{\sigma_1^2}{n} \frac{(N-n)}{N-1} = 16 \cdot \frac{59}{32} \frac{(4-2)}{4-1} = 19\frac{2}{3}.$$

Since $m = 2, m = 1$ are equi-probable, the average conditional variance is

$$(24 + 6)/2 < 19\frac{2}{3}.$$

Problem 7. The identity is straightforward. If $x_i > x_j \Rightarrow p_i < p_j$ then

$$(x_i^2 - x_j^2) \left(\frac{1}{p_i} - \frac{1}{p_j} \right) > 0,$$

since both x_i and x_j are positive. Hence

$$N \sum_1^N \frac{x_i^2}{p_i} - \left(\sum_1^N x_i^2 \right) \left(\sum_1^N \frac{1}{p_i} \right) > 0,$$

or

$$N \sum_1^N \frac{x_i^2}{p_i} > \left(\sum_1^N x_i^2 \right) \left(\sum_1^N \frac{1}{p_i} \right).$$

Now

$$V \left[\frac{1}{n} \sum_{i=1}^n (X_i/P_i) \right] = \frac{1}{n} \sum_{j=1}^N \left(\frac{x_j^2}{p_j} - x_\cdot^2 \right),$$

and is positive for all real x_j. In particular, if $x_j = 1, j = 1, 2, \ldots, N$, we have

$$\sum_{j=1}^N \left(\frac{1}{p_j} \right) \geqslant N^2.$$

Thus,

$$\sum_1^N \frac{x_i^2}{p_i} > N \sum_1^N x_j^2.$$

Now for simple random sampling, $p_i = 1/N$ $(i = 1, 2, \ldots, N)$.

$$V(N\bar{X}) = N^2 \frac{\sigma^2}{n} = N^2 \frac{\sum_1^N (x_j - \bar{x}_\cdot)^2}{nN} = \frac{N}{n} \left[\sum_1^N x_j^2 - N\bar{x}_\cdot^2 \right]$$

$$= \frac{1}{n} \left[N \sum_1^N x_j^2 - x_\cdot^2 \right],$$

and the result follows.

Problem 8.

$$E \left[\sum_1^n (X_i - \bar{X})^2 \right] \Big/ (n-1) = \sum_1^N (x_i - \bar{x})^2 / (N-1).$$

$$E\left[\sum_{1}^{n}(Y_i - \bar{Y})^2\right]\bigg/(n-1) = \sum_{1}^{N}(y_i - \bar{y})^2/(N-1).$$

$$E\left[\sum_{1}^{n}\{(X_i - \bar{X}) - (Y_i - \bar{Y})\}^2/(n-1)\right]$$

$$= \sum_{1}^{N}[(x_i - \bar{x}) - (y_i - \bar{y})]^2/(N-1).$$

Square out the l.h.s. of the third equation and apply the result of the first two equations. (Note that $E(\bar{X} - \bar{Y}) = \bar{x} - \bar{y}$.)

Problem 9.

$$V\left[\sum_{1}^{n}a_i X_i\right] = \sum_{1}^{n}a_i^2 V(X_i) + \sum\sum_{i \neq j}a_i a_j C(X_i, X_j)$$

$$= \sigma^2 \sum a_i^2 - \sigma^2 \sum\sum_{i \neq j}a_i a_j/(N-1)$$

from (16.2),

$$= N\sigma^2 \sum a_i^2/(N-1) - \sigma^2\left(\sum a_i\right)^2/(N-1).$$

If T_1 is unbiased,

$$\sum a_i = 1, \quad \text{and} \quad \sum a_i^2$$

is minimized when $a_i = 1/n$.

$$C\left[\sum a_i X_i, \sum b_j X_j\right] = \sum\sum a_i b_j C(X_i, X_j)$$

$$= \sum a_i b_i V(X_i) + \sum\sum_{i \neq j} a_i b_j C(X_i, X_j)$$

$$= \sigma^2 \sum a_i b_i - \sigma^2 \sum\sum_{i \neq j} a_i b_j/(N-1)$$

$$= \sigma^2\left[N\sum a_i b_j - \sum a_i \sum b_j\right]/(N-1) = 0,$$

and the result follows.

Problem 10. From Example 1, $E(NP) = Np$, and

$$V(NP) = N^2 V(P) = N^2 pq(N-n)/[n(N-1)].$$

From Problem 3, $N^2 V(P)$ is estimated by $N^2 PQ(N-1)/[N(n-1)]$. Let the characteristic be deaf *and* saw the programme. d/n is an unbiased estimator of D/N, hence Nd/n and use variance supplied by first part with $p = D/N$. Conditional on $m = m^*$, d/m^* estimates D/M. Hence Md/m^* estimates D with variance $M^2 p^* q^* (M - m^*)/[m^*(M-1)]$, where $p^* = D/M$.

17

Generating Random Variables

17.1 INTRODUCTION

For much of the time we have studied the properties of random variables without troubling ourselves unduly about the nature of their source. It is generally possible to envisage a mechanism which would generate values from a particular distribution. For example, in sampling with replacement, for a fixed number of times, from an urn containing only black and white balls, the number of white balls seen has a binomial distribution. The value of such a mechanism lies solely in its mimicry of important applications such as the control of the quality of industrial output by checking the performance of randomly selected units.

The statistician is mainly concerned with examining data which has been derived from a 'process' which already exists and has been operating. However, there is an important activity in which the model is given priority over the process. This may be because the design of the process is by no means finalized and the cost and time required to try out a variety of possible implementations is prohibitive. For instance, suppose an airline has to decide the number of checking-in desks that should be established in a new terminal. This is an example of a queueing process. The key features of such a problem are the patterns for passenger terminals and service time.

Even making simple assumptions, the answer to pertinent questions such as 'what is the average number of passengers waiting to check-in?' requires quite complicated mathematics. (see ref [1]). Realistic assumptions would have to take into account such factors as that:

(1)   the distribution of the service time should be truncated, since there is a minimum time necessary to handle any passenger;

(2)   there are surges in passenger arrival rates related to scheduled boarding times.

Such complications are an added inducement to simulate the progress of such a system on a computer. The state of the system is described by the values of a specified set of key indicators. An initial set of assumptions is then made both about the deterministic and stochastic behaviour of the system. Any statistical parameters are ordinarily set to values suggested by previous studies of comparable systems.

To initiate the stochastic features, the distributions of specified random variables must be sampled and the obtained values submitted to the computer. The input drives a program which changes the values of the states of the model. Most standard distribution can be sampled via a program provided for the purpose. (Consult the NAG library for an extensive range of Fortran facilities.) To cope with a situation in which the distribution is not standard, we shall study some basic procedures for generating values which appear to have been drawn from specified distributions.

An essential aid in this activity is the computer's facility for choosing a number between 0 and 1. More formally, it supplies random values from a rectangular distribution over $(0, 1)$: in this chapter, we shall adopt the alternative term 'uniform distribution', since this is the name commonly used in specialist texts on simulation. How the computer does this is outside the scope of the present text; the reader may however have realized that in view of the predictable nature of a machine, only pseudo-random variables can be generated. (see ref. [2]).

In Chapter 6, we have already studied one method of sampling a *continuous* distribution. We here repeat the result. If X has a continuous distribution, with cumulative distribution function $F(.)$, then $F(X)$ has a uniform distribution over $(0, 1)$. Hence, if U is a random value from the uniform distribution over $(0, 1)$ and the realized value, satisfies $u = F(x)$, then x is the realized value taken at random from the distribution of X. This is known as the inversion technique. Thus if $f(x) = 2x$ $(0 < x < 1)$ then $F(x) = x^2$ and $x = \sqrt{u}$. Unfortunately the inversion of $F(.)$ is generally rather difficult and will require the assistance of some methods used in numerical analysis (see the appendix on approximating roots). The methods we are going to discuss seek to avoid this kind of difficulty. For example, if U_1, U_2 are independent and uniformly distributed, then $\Pr[\max(U_1, U_2) < u] = \Pr(U_1 < u$ and $U_2 < u) = u^2$. Thus $\max(U_1, U_2)$ has the same distribution function as has a random variable with p.d.f. $2x$ $(0 < x < 1)$ and the square root can be avoided.

If we have access to random numbers between nought and one, then we have a supply of random digits from 0 to 9 inclusive. For example, the individual digits of such a random number would qualify. Alternatively, we may multiply any such random number by ten and round down. So then, once furnished with a source of random numbers, we are ready to consider how a distribution might be sampled. We can immediately tackle a discrete random variable which takes the values $0, 1, 2, \ldots, 9$ each with probability $1/10$. We merely select random digits until we have a sufficiently large sample. There are no rounding errors and nothing is wasted. What then if we wish to perform the same service for the discrete random variable which takes the values $0, 1, 2, \ldots, 5$, each with probability $1/6$? Now, when the (single) random digits are selected, the numbers 6, 7, 8, 9 cannot be directly used. They can of course be merely discarded. Thus the probability that any particular selection turns out to be usable is $6/10$ and we may 'expect' to draw $10n/6$ digits to achieve a sample size of n. The resultant wastage can be reduced by drawing *pairs* of random digits and using $00, 01, \ldots,$ 95. We assign $05, 11, 17, \ldots, 95$ the sample value 5. These are all the pairs which are obtained by adding 6 to the obvious candidate 05. In a similar way, we assign sixteen pairs of values to each of the other desired sample values and only 96, 97, 98, 99 are wasted. There is an alternative procedure which lends itself more readily to generalization. This is to take any random number and scale it so that it can be readily assigned to one of the integers $0, 1, 2, 3, 4, 5$. Thus if X is one of the digits $0, 1, 2, \ldots, 9$ chosen at random then we calculate $6X/10$ and *round down* to the nearest integer. We arrive at the following table of assignments:

Digit drawn	0	1	2	3	4	5	6	7	8	9
Value assigned	0	0	1	1	2	3	3	4	4	5

It is of course true that, although no digit is wasted, we have not achieved the desired probability of $1/6$ for each of $0, 1, 2, 3, 4, 5$. In fact the probabilities of these values, from the table, are $2/10, 2/10, 1/10, 2/10, 2/10, 1/10$. Notice that the error is never worse than $2/30$. Matters are much improved if we draw pairs of digits at random, scale the two digit numbers by $6/100$ and again round down. We arrive at the following table.

Number falls in interval	0–16	17–33	34–49	50–66	67–83	84–99
Value assigned	0	1	2	3	4	5
Probability	17/100	17/100	16/100	17/100	17/100	16/100

The correct probabilities should be $16\frac{2}{3}/100$ but the error is less than $1/100$. More generally it can be shown that if an integer is chosen at random from the set $0, 1, 2, \ldots, m-1$ and kX/m is rounded down to the nearest integer, then the probability that this nearest integer is r is $(1/k) \pm \epsilon$ where $\epsilon < 1/m (0 \leqslant r < k < m)$. Since $0 \leqslant X/m < 1$, an equivalent procedure is to draw a random number between zero and one. If this were really possible then there would be no error and the probability of the rounded integer assuming the value r would be $1/k$. In the following sections we shall assume that it *is* possible to obtain a random number, U, from the interval $(0, 1)$.

17.2 SAMPLING A DISCRETE DISTRIBUTION

We can now see how to generalize to other discrete distributions. Let X be a discrete random variable such that $\Pr(X = x) = f(x)$ and $\Pr(X \leqslant x) = F(x)$ for $x = 1, 2, \ldots$. Then $f(x) = F(x) - F(x-1)$ and $0 \leqslant F(x) \leqslant 1$. Hence, we choose a random number, U, between 0 and 1, and if $F(x-1) \leqslant u < F(x)$, then the selected value u is taken to correspond to a sample value of x. Since U is uniformly distributed, $\Pr[F(x-1) \leqslant U < F(x)] = f(x)$ and x is a value from the distribution of X. Note that when $0 \leqslant u < F(0) = f(0)$ then the sample value is taken to be zero.

Example 1
If X has p.d.f. $f(x) = 1/k$, $x = 0, 1, \ldots, k - 1$, then $F(x) = (x + 1)/k$ and $F(x-1) \leqslant u < F(x)$ reduces to $x/k \leqslant u < (x + 1)/k$.

Example 2
X has the Poisson distribution with parameter λ.

$$f(x) = \lambda^x e^{-\lambda}/x!, \quad F(x) = \sum_{i=0}^{x} f(i).$$

We start by calculating $f(0) = e^{-\lambda}$. If $0 \leqslant u < f(0)$ we assign $X = 0$. If not, we compute $f(1) = \lambda f(0)$ and test whether u satisfies $f(0) \leqslant u < f(0) + f(1)$ and so on. We stop as soon as $F(x-1) \leqslant u < F(x)$.
    Note. $f(x)/f(x-1) = \lambda/x$ which can be used in this routine. ∎

The last example shows that the labour in this method centres round locating u in the range of $F(.)$. Sometimes the method can be telescoped because the distribution function can be calculated explicitly.

Problem 1. X has the geometric distribution with parameter p. U is uniformly distributed over (0, 1). Explain how to obtain a random value from the distribution of X.

17.3 PARTICULAR METHODS FOR SOME STANDARD DISTRIBUTIONS

Binomial distribution

We can exploit the property that, if in a sequence of n independent trials the probability of a success on any trial is p, then the total number of successes has the binomial distribution with parameters n, p. Thus we obtain n independent random numbers $U_1, U_2, \ldots, U_n$ from the uniform distribution over (0, 1) and determine for each u_i whether or not $0 < u_i < p$. Any such u_i is counted as a success. The number of successes has the binomial distribution. Clearly the *first* integer N such that $u_N < p$ has the *geometric distribution*.

Poisson distribution

In Chapter 5 we showed that if the times between events have independent exponential distributions with common parameter λ, then the *number* of events in the interval $(0, 1)$ has a Poisson distribution with parameter λ. Thus if we select values T_i until the realized values $t_1, t_2, \ldots, t_{n+1}$ satisfy

$$\sum_1^n t_i \leqslant 1 < \sum_1^{n+1} t_i,$$

then n is the number of events in $(0, 1)$. If $t_1 > 1$ then $n = 0$. This method requires us to obtain random values from an exponential distribution with parameter λ.

Problem 2. Independent random values $U_1, U_2, \ldots$, are selected from the interval $(0, 1)$ *until n is the greatest integer such that*

$$\prod_1^n u_i \geqslant e^{-\lambda}.$$

Show that this procedure is equivalent to the selection process for the Poisson distribution given above.

17.4 REJECTION PROCEDURES FOR CONTINUOUS DISTRIBUTIONS

We observe that even for a random variable for which the p.d.f. is a simple function, inversion of the distribution function may pose analytic and/or computational difficulties.

Example 3
$f(x) = 6x(1 - x), 0 < x < 1$. Then

$$F(x) = \int_0^x f(y)\, dy = 3x^2 - 2x^3.$$

Suppose U is a random number between 0 and 1. Then we are to find x such that $3x^2 - 2x^3 = u$, which is not exactly convenient!

The search for alternatives to the inversion method has been strenuous. We first consider the case when $f(.)$ is defined on a finite interval (a, b) and is bounded by m.

In Fig. 17.1, the graph of $f(x)$ lies entirely within a rectangle with sides of lengths $b - a$ and m. Suppose we select pairs of values (x, y) 'at random' within the rectangle and accept only those x values of points which lie in the shaded area, rejecting all others. Then it is intuitively clear that the accepted values will have probability density function $f(.)$. For the proportion of accepted values less than some fixed value x_0 is precisely the area under the curve up to x_0 and this

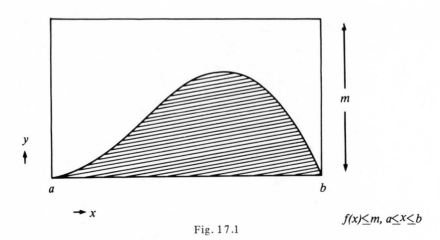

Fig. 17.1

$f(x) \leq m, \ a \leq x \leq b$

is $F(x_0)$ where $F(.)$ is the distribution function of X. We now put this argument more formally so that it can be applied regardless of any geometrical representation.

To obtain a point 'at random' we shall mean to select a pair (x, y) from the distribution of X, Y which are jointly and uniformly distributed over the rectangle. To that end, a value X is drawn from the uniform distribution over (a, b) and then an independent value Y from the distribution which is uniform over $(0, m)$. The joint p.d.f. of X, Y is thus

$$\frac{1}{m(b-a)}, \quad a \leq x \leq b, \quad 0 \leq y \leq m. \tag{17.1}$$

To say that the realized values x, y are the co-ordinates of a point inside the curve is equivalent to requiring $y \leq f(x)$. Now for accepted values of x $\Pr(X \leq x_0 | X \text{ accepted}) = \Pr(X \leq x_0 \text{ and } X \text{ accepted})/\Pr(X \text{ accepted})$. But $\Pr(X \leq x_0 \text{ and } X \text{ accepted})$

$$= \int_a^{x_0} \left\{ \int_0^{f(x)} \frac{dy}{m(b-a)} \right\} dx$$

$$= \int_a^{x_0} \frac{f(x)}{m(b-a)} \, dx.$$

Similarly

$$\Pr(X \text{ accepted}) = \int_a^b \left\{ \int_0^{f(x)} \frac{dy}{m(b-a)} \right\} dx$$

$$= \int_a^b \frac{f(x)}{m(b-a)} \, dx = \frac{1}{m(b-a)},$$

since

$$\int_a^b f(x)\,dx = 1.$$

Hence

$$\Pr(X \leqslant x_0 \,|\, X \text{ accepted}) = \int_a^{x_0} f(x)\,dx.$$

But this is the probability that a random variable with probability density function $f(.)$ is less than or equal to x_0. Since the probability of a value being accepted is $1/[m(b-a)]$, it is clearly best to choose m as the least upper bound, as shown in Fig. 17.2.

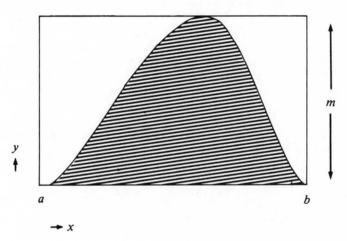

$$f(x) \leqslant m,\ a \leqslant x \leqslant b$$

Fig. 17.2

Example 4

$f(x) = 6x(1-x), 0 < x < 1$. The maximum of $f(x)$ is at $x = 1/2$ where it attains the value $3/2$. Since $a = 0, b = 1, m = 3/2$, the probability of accepting a value is $2/3$. We select X at random between 0 and 1 and, independently, Y at random between 0 and $3/2$. If $y < 6x(1-x)$, the realized value of x is accepted.

17.5 EXTENSIONS OF THE REJECTION METHOD

The method discussed in section 17.4 will fail if the probability density function $f(.)$ is bounded but is positive in an interval of infinite extent. We cannot, after all, choose a value from a distribution which is uniform over such an interval. Suppose however we can find another distribution, with probability density function $g(.)$ and a positive constant c such that

$$f(x) \leqslant cg(x) \tag{17.2}$$

for all x such that both densities are positive. A random value X is drawn from the distribution with p.d.f. $g(x)$ and an independent value U from the interval $(0, 1)$. If the realized values x, u satisfy

$$u \leqslant f(x)/[cg(x)] \tag{17.3}$$

then x is accepted. The accepted values of X have the required probability density function $f(x)$.

For any *fixed* value of x, since U is uniformly distributed over $(0, 1)$, the probability of acceptance is merely $f(x)/cg(x)$. But X has density $g(x)$. The unconditional probability of acceptance is

$$\int_{-\infty}^{+\infty} [f(x)/cg(x)]\, g(x)\, dx = \frac{1}{c} \int_{-\infty}^{+\infty} f(x)\, dx = \frac{1}{c}.$$

Hence,

$$\Pr(X \leqslant x_0 \mid X \text{ accepted}) = c\, \Pr[U \leqslant f(X)/cg(X) \text{ and } X \leqslant x_0]$$

$$= c \int_{-\infty}^{x_0} \frac{f(x)}{cg(x)} \cdot g(x)\, dx = F(x_0).$$

In passing, we observe the special case when $f(x) \leqslant m$ and $f(x)$ is positive only over the interval (a, b). We may write

$$f(x) \leqslant m(b-a) \cdot \frac{1}{(b-a)},$$

and this corresponds to $g(x) = 1/(b-a)$, $a \leqslant x \leqslant b$ and $c = m(b-a)$.

Example 5
$f(x) = 6x(1-x)$, $0 < x < 1$, $g(x) = 2x$, $0 < x < 1$. Then $6x(1-x) \leqslant 3(2x)$ with equality only when $x = 0$. (Draw a sketch.) A value X is chosen from the distribution with density $2x$ (which is easy) and an independent value U from the uniform distribution over $(0, 1)$. If $u \leqslant 6x(1-x)/6x = 1 - x$, then we retain x — otherwise we discard it and try again. The reader should check that the probability of accepting any X is $1/3$ (thus this solution is rather worse than our previous attempt).

The reader will perceive that the charm of this technique is heavily dependent on our ability to sample the distribution which has probability density function $g(.)$! This is the case in Example 5 for the distribution function $G(x) = x^2$. We have only to draw a random value between 0 and 1 and take the square root — the usual inversion technique.

Problem 3. If $f(x) = 12x^2(1-x)$, $0 < x < 1$ and $g(x) = 2x$, $0 < x < 1$, show that $f(x) \leqslant 3g(x)/2$ with equality at $x = 1/2$. Show by a direct computation that the probability of accepting a random value from the distribution with p.d.f. $g(x)$ is $2/3$.

Problem 4. Show that $f(x) = 12x^2(1 - x)$, $0 < x < 1$, has a maximum at $x = 2/3$. Apply the method of section 17.4 to sample $f(.)$.

Problem 5. If $f(x) = 12x/(1 + x)^5$, $0 < x < \infty$ and $g(x) = 2/(1 + x)^3$, $0 < x < \infty$, show that if $f(x) \leqslant cg(x)$ then $c \geqslant 3/2$. Apply the rejection technique when $c = 3/2$.

17.6 REJECTION METHOD BASED ON FACTORIZATION OF THE P.D.F.

Suppose we can factorize a p.d.f., $f(x)$, in the form

$$f(x) = cg(x) H(x), \tag{17.4}$$

where $g(x)$ is a p.d.f. and $H(x)$ is a distribution function of a random variable with p.d.f. $h(x)$. Let a random value Y be drawn from the distribution with p.d.f. $h(.)$ and an independent value X from the distribution with p.d.f. $g(.)$. When X is accepted if and only if $Y < X$, then such an X has a distribution with p.d.f. $f(.)$. For since X, Y are independent, their joint distribution has p.d.f. $g(x)h(y)$ and

$$\Pr(X \leqslant x_0 \text{ and } Y < X] = \int_{-\infty}^{x_0} g(x) \int_{-\infty}^{x} h(y) \, dy \, dx$$

$$= \int_{-\infty}^{x_0} g(x) H(x) \, dx$$

$$= \frac{1}{c} \int_{-\infty}^{x_0} f(x) \, dx.$$

Clearly $\Pr(Y < X) = 1/c$ and

$$\Pr(X \leqslant x_0 | X \text{ accepted}) = \int_{-\infty}^{x_0} f(x) \, dx,$$

or the distribution of accepted values of X has probability density function $f(x)$.

This seemingly unlikely procedure is not so far from the rejection method discussed in section 17.5. For since $0 \leqslant H(x) \leqslant 1$, if $f(x) = cg(x)H(x)$, then certainly $f(x) \leqslant cg(x)$. Furthermore to find a value of Y for the present method, we select U at random from the interval $(0, 1)$ and set $y = H^{-1}(u)$. The requirement $y < x \Rightarrow H^{-1}(u) < x \Rightarrow u < H(x) = f(x)/[cg(x)]$.

Example 6

$f(x) = 6x(1 - x)$, $0 < x < 1$. We can write $f(x) = 3.2(1 - x)x = cg(x)H(x)$, and

$$G(x) = \int_{-\infty}^{x} g(t) \, dt = \int_{0}^{x} 2(1 - t) \, dt = 1 - (1 - x)^2.$$

So we draw independent values U, V from the uniform distribution over $(0, 1)$, we set $y = u$ and find x such that $1 - (1 - x)^2 = v$. Finally we accept x if $y < x$. The (unconditional) probability that X is accepted is $1/c = 1/3$.

Problem 6. Suppose $f(x) = cg(x)[1 - H(x)]$ where $f(.), g(.)$ are probability density functions and $H(x)$ is the *distribution* function of a random variable with p.d.f. $h(.)$. A random variable Y is chosen from the distribution with density $h(.)$ and an independent value X from the distribution with density $g(.)$. Show that if x is accepted if and only if $y > x$, then the accepted X have distribution $f(.)$.

Problem 7. Suppose X has the $\Gamma(1/2, 1)$ distribution, viz.

$$f(x) = x^{(1/2)-1} e^{-x}/\Gamma(1/2).$$

If we write

$$f(x) = \sqrt{\pi} \left[\frac{1}{\pi\sqrt{x(1 + x)}} \right] (1 + x) e^{-x} = cg(x)[1 - H(x)],$$

show that $g(x)$ is a p.d.f. and $H(x)$ is the cumulative distribution function of a $\Gamma(2, 1)$ distribution. Explain how to use the result of Problem 6 to obtain random values from a $\Gamma(1/2, 1)$ distribution.

17.7 METHOD FOR COMPOUNDING DISTRIBUTIONS

Sometimes a probability density function appears to be rather complicated but can be regarded as a mixture of other distributions, which are easier to sample. sample.

Example 7
$f(x) = p\lambda e^{-\lambda x} + (1 - p)\mu e^{-\mu x}$, $x > 0, 0 < p < 1$. Here $f(x)$ is a mixture of two different exponential distributions. We draw a random value U between zero and one. If $0 < u < p$, we sample the exponential distribution with parameter λ, otherwise we sample the exponential distribution with parameter μ. The value of U can be used twice over. For if U is given to be less than p, then it is uniformly distributed over $(0, p)$ and U/p is uniformly distributed over $(0, 1)$. In that case we obtain a value of X by equating u/p to $(1 - e^{-\lambda x})$ and solving for x. On the other hand if $U > p$, then $U/(1 - p)$ can be used in conjunction with the other exponential distribution.

Problem 8. Explain how to sample the distribution with probability density function

$$f(x) = \sum_{i=0}^{\infty} \frac{\lambda^i e^{-\lambda}}{i!} \cdot (i + 1)x^i, \quad 0 < x < 1.$$

Problem 9. Show that

$$\int_0^1 \binom{n}{x} p^x (1-p)^{n-x} 2p \, dp = 2(x+1)/[(n+1)(n+2)],$$

$$x = 0, 1, 2, \ldots, n.$$

Here we have a binomial distribution compounded with a distribution which has probability density function $2p$, $(0 < p < 1)$. Explain how this result may be used to sample the discrete distribution with p.d.f.

$$2(x+1)/[(n+1)(n+2)], \quad x = 0, 1, \ldots, n.$$

17.8 THE NORMAL DISTRIBUTION

If X_1, X_2 have independent $N(0,1)$ distributions and $X_1 = R \cos \Theta$, $X_2 = R \sin \Theta$, then R has p.d.f. $f(r) = r \, e^{-r^2/2}$, $r > 0$, and Θ has p.d.f. $g(\theta) = 1/2\pi$, $0 < \theta < 2\pi$. Furthermore R, Θ are independent (Problem 9, Chapter 8). Hence if U_1, U_2 are independent random numbers from the interval $(0, 1)$ we set

$$\theta = 2\pi u_1, \quad F(r) = 1 - e^{-r^2/2} = u_2,$$

solve for θ, r and *hence* for x_1, x_2 and we have a pair of values from the distribution $N(0, 1)$.

Problem 10. The continuous random variable X has p.d.f.

$$f(x) = \sqrt{\frac{2}{\pi}} e^{-x^2/2}, \quad x > 0.$$

Show that $f(.)$ can be factorized in the form.

$$\sqrt{\frac{2e}{\pi}} [1 - H(x)] g(x)$$

where $H(x) = 1 - e^{-(x-1)^2/2}$, $g(x) = e^{-x}$ and explain how this may be exploited to provide a rejection technique for sampling the distribution with density $f(.)$.

REFERENCES

[1] G. P. Beaumont, *Introductory Applied Probability*, Ellis Horwood, Chichester, 1983.
[2] B. Conolly, *Techniques in Operational Research*, Ellis Horwood, Chichester, 1981.

BRIEF SOLUTIONS AND COMMENTS ON THE PROBLEMS

Problem 1

$$F(x) = \sum_{i=1}^{x} f(i) = \sum_{i=1}^{x} q^{i-1} p = p(1-q^x)/(1-q) = 1-q^x.$$

Hence u corresponds to a sample value x, if and only if $1-q^{x-1} \leqslant u < 1-q^x$. That is to say $q^x < 1-u \leqslant q^{x-1}$ or $x\log_e q < \log_e(1-u) \leqslant (x-1)\log_e q$. Since $\log_e q < 0$, $x > \log_e(1-u)/\log_e q \geqslant x-1$. We must take x to be the least integer greater than $\log_e(1-u)/\log_e(1-p)$.

Problem 2. If U_i is uniformly distributed over $(0, 1)$ then

$$T_i = -\frac{1}{\lambda} \log_e U_i$$

has the exponential distribution with parameter λ. Hence

$$\sum_{1}^{n} T_i \leqslant 1 < \sum_{1}^{n+1} T_i$$

is equivalent to

$$-\frac{1}{\lambda} \sum_{1}^{n} \log_e U_i \leqslant 1 < -\frac{1}{\lambda} \sum_{1}^{n+1} \log_e U_i$$

$$\sum_{1}^{n+1} \log_e U_i < -\lambda \leqslant \sum_{1}^{n} \log_e U_i$$

$$\prod_{1}^{n+1} U_i < e^{-\lambda} \leqslant \prod_{1}^{n} U_i.$$

Now

$$\prod_{1}^{n} U_i$$

is decreasing since each $U_i \leqslant 1$. Hence n is the greatest integer such that

$$\prod_{1}^{n} U_i \geqslant e^{-\lambda}.$$

If $U_1 < e^{-\lambda}$ then $n = 0$.

Problem 3. $12x^2(1-x) \leqslant 3x \leftrightarrow 4x(1-x) \leqslant 1 \leftrightarrow (2x-1)^2 \geqslant 0$, with equality when $x = 1/2$. We accept x if $u < 12x^2(1-x)/3x = 4x(1-x)$ and

$$\Pr[U < 4x(1-x)] = 4x(1-x).$$

Finally,

$$\Pr(U < X] = \int_0^1 4x(1-x)\, 2x\, dx = 2/3.$$

Problem 4. $f(2/3) = 16/9$, hence with this choice of m, the probability of a value being accepted is $9/16$. This is less than the method outlined in Problem 3 but of course we don't have to go to the trouble of sampling the distribution $g(.)$.

Problem 5. $12x/(1+x)^5 \leqslant 2c/(1+x)^3 \leftrightarrow 6x/(1+x)^2 \leqslant c$, or $cx^2 + x(2c-6) + c \geqslant 0$. Hence $(2c-6)^2 \leqslant 4c^2$ or $c \geqslant 3/2$. A random value X is drawn from the distribution with p.d.f. $2/(1+x)^3$. An independent value U is drawn from the interval $(0, 1)$. If the realized values satisfy $u < 4x/(1+x)^2$, accept x.

Problem 7. If $\tan t = \sqrt{y}$, then $(\sec^2 t)\, dt/dy = 1/(2\sqrt{y})$. Hence

$$\int_0^x \frac{dy}{\pi\sqrt{y}(1+y)} = \int_0^{\tan^{-1}\sqrt{x}} \frac{2\sec^2 t}{\pi \sec^2 t}\, dt$$

$$= \frac{2}{\pi} \tan^{-1}(\sqrt{x}) \to 1 \text{ as } x \to \infty.$$

$$\int_0^x y\, e^{-y}\, dy = -y\, e^{-y}]_0^x + \int_0^x e^{-y}\, dy$$

$$= 1 - e^{-x}(x+1),$$

which is of the required form. Select U, V from the uniform distribution over $(0, 1)$. $v = (2/\pi) \tan^{-1} \sqrt{x} \Rightarrow x = \tan^2 (\pi v/2)$, and $u = 1 - (1+y)e^{-y}$. Accept x if $y > x$.

Problem 8. $f(x)$ is of the form

$$\sum p_i g_i(x),$$

where

$$\sum_0^\infty p_i = 1 \quad \text{and} \quad \int_0^1 g_i(x)\, dx = 1$$

for each i. Sample the Poisson distribution with parameter λ, if the obtained value is i, sample the probability density function $g_i(x)$. This last step is simple since $g_i(x)$ corresponds to a distribution function x^{i+1}.

Problem 9. The integral is

$$\propto \int_0^1 p^{x+1}(1-p)^{n-x}\,dp = \Gamma(x+2)\,\Gamma(n-x+1)/\Gamma(n+3)$$

$$= (x+1)!\,(n-x)!/(n+2)!$$

and the result follows. Since

$$\int_0^p 2t\,dt = p^2,$$

select U at random from the interval $(0, 1)$. Conduct n independent trials with probability of success $\sqrt{u}$, then x is the observed number of successes. There are of course easier methods of sampling the p.d.f. $2(x+1)/[(n+2)(n+2)]$.

Problem 10. Let Y, X be independent values from the exponential distribution with parameter 1. Accept x if $y > (x-1)^2/2$. Since

$$\Pr[Y > (x-1)^2/2 \,\vert\, X = x] = e^{-(x-1)^2/2}$$

then

$$\Pr[Y > (X-1)^2/2 = \int_0^\infty e^{-(x-1)^2/2}\,e^{-x}\,dx = \sqrt{(\pi/2e)}.$$

Appendix: Approximating roots of equations

We have said that approximate methods may be used to find the value x which satisfies $F(x) = y$ where $F(.)$ is the cumulative distribution function of a continous random variable and y is a known number between 0 and 1. Interest in such methods has not entirely evaporated, in spite of the capacity computers have to provide certain function values on demand.

We know that $F(.)$ is bounded and steadily increasing, hence $G(x) = F(x) - y = 0$ has just one root. We shall not therefore be plagued by the possibility that a located root is the wrong one! A typical approximation process considers a sequence $x_1, x_2, \ldots, x_n \ldots$ of estimates of the root of the equation $G(x) = 0$. The first few members of this sequence may be good guesses but subsequent members are obtained by repeated application of a fixed rule. The process stops as soon as $|G(x_n)|$ is sufficiently small.

We begin by glancing at a simple but brutally effective technique. Suppose that x_1, x_2 are initial guesses either side of the required root. This is guaranteed if $G(x_1) < 0$ and $G(x_2) > 0$. We take $x_3 = (x_1 + x_2)/2$ and find the sign of $G(x_3)$, thereby determining on which side of the root x_3 lies. The process of bisecting intervals is continued until $|G(x_n)|$ is sufficiently small.

The reader will have remarked that the bisection method takes no notice of the relative proximity of the initial guesses to the required root. some notice of this is taken into account by the method of *false position*. Again we start with two initial values x_1, x_2 though these need *not* be on opposite sides of the root. We take as x_3, the abscissa of the point in which the line passing through $[x_1, G(x_1)]$, $[x_2, G(x_2)]$ meets the x axis (see Fig. A.1).

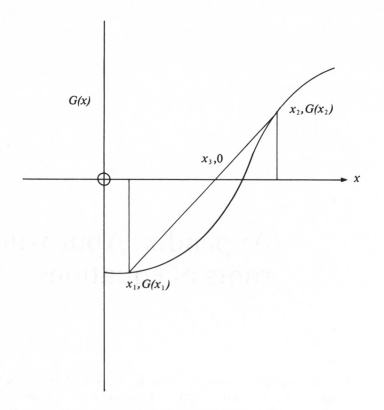

Fig. A.1

It can be readily verified that

$$x_3 = x_1 - \frac{(x_2 - x_1)}{G(x_2) - G(x_1)}\, G(x_1). \tag{A.1}$$

An important, and frequently used, extension of the method of falst position is obtained by allowing x_1 to tend towards x_2. In the limit, the line joining the two points becomes the tangent at x_2 (see Fig. A.2). Directly, or by allowing $x_1 \to x_2$ in equation (A.1), we obtain

$$x_3 = x_2 - G(x_2)/G'(x_2),$$

providing $G'(x_2)$ exists. In general,

$$x_{n+1} = x_n - G(x_n)/G'(x_n), \tag{A.2}$$

which is known as the Newton–Raphson formula.

Example 1
If $F(x) = x^2$, then $x^2 = c \Rightarrow G(x) = x^2 - c = 0$, and $G'(x) = 2x$. Hence by equation (A.2)

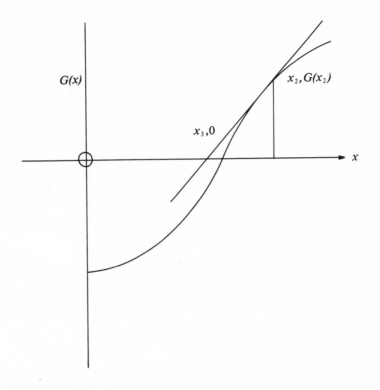

Fig. A.2

$$x_{n+1} = x_n - (x_n^2 - c)/2x_n$$

$$= \frac{1}{2}\left(x_n - \frac{c}{x_n}\right).$$

which was a popular formula for evaluating square roots on hand-calculators.

There are few difficulties in using this formula, except possibly when $G'(x) = 0$ near the required root.

A simple iteration procedure

Suppose we wish to solve the equation $H(x) = x$, where $H(x)$ is not necessarily a cumulative distribution function. Let x_n be the current estimate and $x_{n+1} = H(x_n)$, then if x_n tends to a finite limit, that limit is certainly one of the roots of $H(x) = x$. This method allows an attractive geometrical representation. Suppose we draw the graphs of $y = H(x)$ and of the line $y = x$, then the x co-ordinates of the points of intersection are the roots of $H(x) = x$. In Fig. A.3, we advance vertically from the point $(x_1, 0)$ until we meet the curve at $(x_1, H(x_1))$. We then move horizontally until we meet the line $y = x$ at the point $(H(x_1), H(x_1))$. We then bounce alternately off the curve and line until we

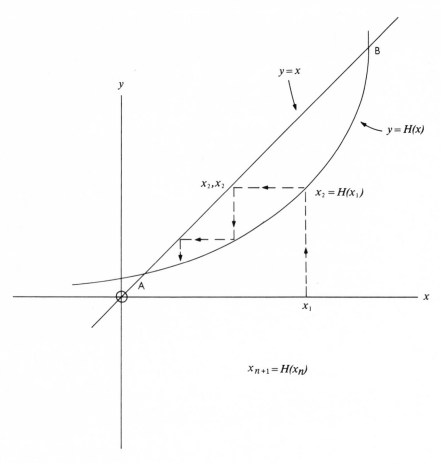

Fig. A.3

converge to the point **A**. The reader should experiment with other starting points and check that, in this case, we never end up at the other point of intersection, **B**. Thus this process is dangerous if **B** is the root we need! For a fuller discussion of the points involved see ref. [1].

We here restrict ourselves to stating a condition which guarantees convergence to the correct root. If I is an interval containing x^* such that $H(x^*) = x^*$ and

(1)   the initial value x_1 is contained in I
(2) $|H'(x)| \leqslant k < 1$ for every x in the interval I,

then $x_{n+1} = H(x_n)$ converges to x^*.

In Fig. A3 it is apparent that these conditions are not obeyed in any neighbourhood of **B**.

The second condition determines the direction in which we must move after hitting the curve and hence whether we move closer or not to the desired root.

How can we apply this technique to the matter in hand? We are to solve $F(x) = y$, where $F(.)$ is a cumulative distribution function. But we can write this equation as $x = x + y - F(x) = H(x)$. The condition $|H'(x)| \leqslant k$ now becomes $|1 - F'(x)| = |1 - f(x)| \leqslant k$, where $f(.)$ is the probability density function, or $1 - k \leqslant f(x) \leqslant 1 + k$.

Example 2
$f(x) = 2x, 0 < x < 1, F(x) = x^2$. Then

$$x_{n+1} = H(x_n) = x_n + y - F(x_n) = x_n - x_n^2 + y.$$

Since $0 < f(x) < 2$ for the whole interval $(0, 1)$, we certainly can find a sub-interval satisfying the conditions. Thus, the sequence converges to the positive root of $x^2 = y$.

REFERENCE

[1]   B. Noble, *Numerical Methods*: 1, Oliver & Boyd, 1964.

Index

Gasson, P.C.	Geometry of Spatial Forms
Goodbody, A.M.	Cartesian Tensors
Goult, R.J.	Applied Linear Algebra
Graham, A.	Kronecker Products and Matrix Calculus: with Applications
Graham, A.	Matrix Theory and Applications for Engineers and Mathematicians
Griffel, D.H.	Applied Functional Analysis
Griffel, D.H.	Linear Algebra*
Hanyga, A.	Mathematical Theory of Non-linear Elasticity
Harris, D.J.	Mathematics for Business, Management and Economics
Hoksins, R.F.	Generalised Functions
Hoskins, R.F.	Standard and Non-standard Analysis*
Hunter, S.C.	Mechanics of Continuous Media, 2nd (Revised) Edition
Huntley, I. & Johnson, R.M.	Linear and Nonlinear Differential Equations
Jaswon, M.A. & Rose, M.A.	Crystal Symmetry: The Theory of Colour Crystallography
Johnson, R.M.	Theory and Applications of Linear Differential and Difference Equations
Kim, K.H. & Roush, F.W.	Applied Abstract Algebra
Kosinski, W.	Field Singularities and Wave Analysis in Continuum Mechanics
Lindfield, G. & Penny, J.E.T.	Microcomputers in Numerical Analysis
Lord, E.A. & Wilson, C.B.	The Mathematical Description of Shape and Form
Marichev, O.I.	Integral Transforms of Higher Transcendental Functions
Massey, B.S.	Measures in Science and Engineering
Meek, B.L. & Fairthorne, S.	Using Computers
Mikolas, M.	Real Function and Orchogonal Series
Moore, R.	Computational Functional Analysis
Müller-Pfeiffer, E.	Spectral Theory of Ordinary Differential Operators
Murphy, J.A. & McShane, B.	Compution in Numerical Analysis*
Nonweiller,T.R.F	Computational Mathematics: An Introduction to Numerical Approximation
Ogden, R.W.	Non-linear Elastic Deformations
Oldknow, A. & Smith, D.	Learning Mathematics with Micros
O'Neill, M.E. & Chorlton, F.	Ideal and Incompressible Fluid Dynamics
O'Neill, M.E. & Chorlton, F.	Viscous and Compressible Fluid Dynamics*
Rankin, R.A.	Modular Forms
Ratschek, H. & Rokne, J.	Computer Methods for the Range of Functions
Scorer, R.S.	Environmental Aerodynamics
Smith, D.K.	Network Optimisation Practice: A Computational Guide
Srivastava, H.M. & Karlsson, P.W.	Multiple Gaussian Hypergeometric Series
Srivastava, H.M. & Manocha, H.L.	A Treatise on Generating Functions
Shivamoggi, B.K.	Stability of Parallel Gas Flows*
Stirling, D.S.G.	Mathematical Analysis*
Sweet, M.V.	Algebra, Geometry and Trigonometry in Science, Engineering and Mathematics
Temperley, H.N.V. & Trevena, D.H.	Liquids and Their Properties
Temperley, H.N.V.	Graph Theory and Applications
Thom, R.	Mathematical Models of Morphogenesis
Toth, G.	Harmonic and Minimal Maps
Townend, M. S.	Mathematics in Sport
Twizell, E.H.	Computational Methods for Partial Differential Equations
Wheeler, R.F.	Rethinking Mathematical Concepts
Willmore, T.J.	Total Curvature in Riemannian Geometry
Willmore, T.J. & Hitchin, N.	Global Riemannian Geometry
Wojtynski, W.	Lie Groups and Lie Algebras*

In preparation